D1345565

Statistical Process Adjustment
for Quality Control

Statistical Process Adjustment for Quality Control

ENRIQUE DEL CASTILLO

Department of Industrial and
Manufacturing Engineering
Pennsylvania State University

A Wiley-Interscience Publication
JOHN WILEY & SONS, INC.

This book is printed on acid-free paper. ∞

Library of Congress Cataloging-in-Publication Data:

Del Castillo, Enrique.
    Statistical process adjustment for quality control/Enrique del Castillo.
        p. cm. -- (Wiley series in probability and statistics)
    Includes bibliographical references and index.
    ISBN 0-471-43574-0 (cloth : alk. paper)
    1. Quality control--Statistical methods. I. Title. II. Series.

    TS156.D388 2002   2001046955

Printed in the United States of America

10 9 8 7 6 5 4 3 2 1

Para Enrique, Fernando, y Mónica

Contents

Preface

Traditionally, two different technical groups have been involved on developing process controls in industry. Control engineers focus on controlling process variables such as temperatures and flow rates using Engineering process control (EPC) techniques. Quality engineers and applied statisticians focus on monitoring product attributes (quality characteristics) through control charts and use other statistical tools for process improvement purposes. SPC ideas received renewed interest due to W. Edwards Deming influential work (particularly at Ford Motor Co.) and more recently due to the six sigma management philosophy. The situation in academia is similar, with departments of (applied) statistics and industrial engineering focusing on quality engineering, and electrical and chemical engineering departments focusing on control engineering. The two apparently opposing viewpoints need to be reconciled, and considerable work on the integration of SPC and EPC techniques has taken place in the 1990s.

OBJECTIVE AND INTENDED AUDIENCE

The purpose of this book is to present process adjustment techniques based on EPC methods and to discuss them from the point of view of controlling the quality of a *product*, emphasizing the relation with traditional SPC methods. Over the last seven years, notes based on this book have been used to support a one-semester first-year course at the graduate level in the industrial engineering departments at Penn State University and the University of Texas–Arlington. The book can be used also in senior-level/first-year-graduate courses in statistics and operations research. Portions of these notes have been used to support training courses for industry. Quality engineers and applied statisticians should benefit from reading it for self-study. Process control engineers may find the book useful because of the statistical aspects not typically discussed in more detail in more control-theoretic texts.

The two different audiences that this book may have (one academic, the other industrial) require treatment of the topics at different levels. This is always risky when writing a textbook. I have tried to accomplish these dual goals by including most of the advanced material and explanations in appendixes at the end of each chapter. I have also made frequent use of footnotes, which usually point to more advanced concepts or references. Finally, there are some sections of the book, primarily in Chapters 2 and 7 and all of Chapter 9, that are marked as more advanced, and depending on the interests of the reader, may be skipped on a first reading.

TYPES OF CONTROL PROBLEMS ADDRESSED

The aforementioned apparently opposing viewpoints of process control can be explained simply by describing, at the outset, the characteristics of the control problems we address in this book. The table below contrasts the types of systems that *process* control techniques deal with (as treated, e.g., in chemical engineering control books) compared to what we deal with in this book, namely, quality control of *products*.

	Process Control	Product Quality Control
Output(s)	Process variable(s)	Quality characteristic(s)
Input(s)	Process variable(s)	Process variable(s)
Control action	Automatic	Usually manual

As can be seen from the table, both types of control techniques adjust process variables to achieve desirable values of some output. However, in this book, the outputs of such controlled system are the quality characteristics of the product as opposed to some other process variable that may be of interest to regulate in itself, not necessarily for quality control purposes but perhaps for safety or technological reasons (e.g., reaction temperature, spindle velocity, etc.). Thus in a discrete-part manufacturing process we might be interested in adjusting a feed rate or rotation speed (process variable) to achieve a desired value of the dimension obtained after performing a machining operation. In a continuous-flow production process, we might be interested in manipulating a temperature (process variable) to get adequate values in the viscosity of some chemical product.

There is one additional distinction between traditional process control and the product quality control problems we address in this book. This has to do with the way in which we implement the control actions. In the process control literature, the adjustment (control action) is usually implemented automatically by means of sensors and actuators. In a quality control setting, adjustments are usually implemented by a process operator or a quality engineer manually, perhaps by simply dialing the new settings into a machine, as happens in modern CNC (computer numerically controlled) machines or in many semiconductor manufacturing processes.

We hasten to point out that mathematically speaking, the process adjustment methods discussed in this book are the same as those in the process control literature. The main differences, besides the application to product quality control and the emphasis on monitoring-adjustment integration, lie in how process models and transfer functions are obtained. In this book it is not assumed that a precise differential equation description of the process under control exists; rather, we show how models can be obtained empirically from data. Evidently, that is where statistical methodology, in particular, time-series analysis and control, enters the picture. There are also some economic differences between process control methods and the product quality control methods described in this book. Each type of application encounters, in practice, different cost structures.

BOOK CONTENTS

The book starts (Chapter 1) by pointing out the difficulties that SPC charts encounter in the presence of autocorrelation. Basic control charts are quickly reviewed, and performance measures such as ARL are introduced. The concepts of autocorrelation and stationarity of a time series are introduced, and comparisons between process adjustment and process monitoring are given at an introductory level, based primarily on Deming's funnel experiment, which is analyzed in detail. We also discuss integrated SPC–EPC methods. Figure 1.23 provides a guide to all the univariate process adjustment and monitoring techniques in the book and should be useful as a road map of the first eight chapters.

Chapter 2 states the main problems addressed in the book in increasing level of complexity: time-series modeling problems, transfer function modeling problems, and process adjustment problems. Two relatively technical sections provide information on difference equation models and stability of these types of models. The appendix contains a short but I hope useful review of z-transforms for those who would like to know more about what sometimes appears to be the "mysterious" backshift operator. This chapter includes very useful background results for understanding ARIMA and transfer function models, presented in detail in Chapters 3 and 4, respectively.

In Chapters 3 and 4 we discuss ARIMA and transfer function models. Model fitting using modern software is emphasized in these two chapters. For completeness, a brief review of frequency-domain analysis concepts is given as an appendix in Chapter 3. In Appendix 4A we present state-space methods, as they are touched on again in later chapters.

In Chapters 5 to 9 we present a variety of process adjustment techniques in this, which could be said to be, the main part of the book. In Chapter 5 we review minimum mean square error (or minimum variance) controllers and discuss modifications to make them more robust and practical. In Chapter 6 we give a brief introduction to the very practical PID controllers. I hope that this discussion will be well received among applied statisticians and engineers not familiar with this type of controllers. The rational for this type of controller is emphasized and the design of PI controllers is discussed as well. In Chapter 7 we describe EWMA controllers, which are closely related to integral control and popular in the semiconductor manufacturing sector. Analysis and design of these controllers are discussed.

In Chapter 8 we provide an introduction to topics in the control literature that could fill many volumes: recursive estimation and adaptive control. The recursive least squares algorithm is derived and illustrated. The closely related Kalman filter is presented in Appendix 8A. Self-tuning controllers, adaptive controllers for processes with constant but unknown parameters, are described. These controllers provide a nice framework for tackling process adjustment problems in manufacturing.

Finally, in Chapter 9 we present modeling of multivariate ARMA and ARMAX processes. This is a relatively advanced chapter, but I think the results are among the most useful. I have confined the theoretical explanations to the minimum necessary to allow readers to fit these types of models with two well-known software packages: SAS and Matlab.

Several spreadsheet simulation and optimization models and data files (some from real-life processes) are discussed throughout the text and listed at the end of the book. These files are available at Wiley's Web site.

Some historical notes together with additional references on each topic are included in the "Bibliography and Comments" sections at the end of each chapter. These will be useful for readers wishing to obtain more information on a particular subject, but no attempt was made to provide a comprehensive or exhaustive list of references.

USE OF THE BOOK

For students with no background in control but with a basic background in statistics, Chapters 1 to 8 without the chapter appendixes is usually sufficient material for an intense one-semester course. I have done this for industrial engineering majors. For students with a stronger background in statistics, starting in Section 1.2 and covering all the material until Chapter 9, including the chapter appendixes is a more reasonable course of action for a one-semester course.

The book can also be used for self-study with a view toward applications. In such a case a first reading might include all of Chapter 1, Sections 2.1, 2.2, 3.1 to 3.3, 3.9, 3.10, 4.1, 5.1 (introduction only), 5.5 and 5.6, and all of Chapter 6. Readers interested in when to use each different technique in the book should consult Figure 1.23 frequently.

A WORD ABOUT STATISTICAL SOFTWARE

Throughout the book I have tried to emphasize the use of modern statistical software. I discuss mainly three products: Minitab (for ARIMA modeling), SAS (PROC ARIMA for time-series and transfer function modeling and PROC STATESPACE to fit multivariate ARMAX processes), and Matlab Systems Identification Toolbox (for transfer function and multivariate ARMAX modeling). The first two software packages tend to be better known among statisticians and the last one tends to be better known among engineers. This is a reflection of the subject matter at the intersection of SPC and EPC.

A WORD ABOUT THE BOX ET AL. TEXTBOOKS

As he has in many other areas of applied statistics, George E. P. Box has made many fundamental contributions in the time-series control area. Together with G. M. Jenkins, a well-known electrical engineer, they summarized much of their work in the Box–Jenkins (1970, 1976), now Box–Jenkins and Reinsel (1994) book, which stands out as a classic reference. The last few chapters of the Box–Jenkins book, on process adjustment, are now complemented by the book by Box and Alberto Luceño (1997). In my opinion, these books present the very relevant work done by their authors but omit to some extent topics developed primarily by others. This is particularly true with respect to feedback control methods. Whenever I have used these books as textbooks for a class on time-series control and process adjustment, I have had to consult and complement the lectures with many other references and papers to provide my students with a more complete perspective. What I have tried to do in this book is to bring together and synthesize material from several sources in both the time-series/statistics and control theory fields in order to present it as an introductory text book for students not necessarily strong in control theory. This material includes, of course, detailed descriptions of some of the work by Box and his collaborators.

ACKNOWLEDGMENTS

I would like to thank all the reviewers of earlier drafts of this book, who made very useful comments and suggestions. In particular I would like to thank J. Stuart Hunter for his comments and encouragement. Thanks are also due to the following people with whom I have interacted in one way or another during the past ten years on topics related to various sections of the book: Douglas C. Montgomery, Arizona State University, Arnon Hurwitz (Qualtech Inc., formerly at Sematech), James Moyne (University of Michigan), Harriet Black-Nembhard (University of Wisconsin–Madison), Alberto

Luceño (Universidad de Cantabria, España), Elart von Collani and Rainer Göb (University of Würzburg, Germany), Christina Mastrangelo (University of Virginia), Ming Xie (National University of Singapore), and Bianca Colosimo (Politecnico di Milano, Italy). Thanks to Rong Pan (Penn State University) for Figures 1.18 and 1.19 and the Deming funnel simulation spreadsheet. Mr. Pan has prepared a solutions manual for this book, available at Wiley's Web site. Finally, thanks to the many graduate students that helped me debug the book. Throughout the last six years, my research has been funded by the Operations Research and Production Systems program at the National Science Foundation. I acknowledge and thank them for their financial support (NSF Grants DMI-9623669-9996031 and DMI-9988563).

CHAPTER 1

Process Monitoring versus Process Adjustment

It can be said that the birth of modern statistical process control (SPC) took place when Walter A. Shewhart, a physicist and statistician working for Bell Laboratories, developed the concept of a control chart in the 1920s. Initially, the application of control charts was confined to the fabrication of telephone sets and their components, but eventually SPC techniques became popular in other discrete-part industries, notably in metal machining. After some time, SPC techniques were adopted in continuous-process (chemical) industries.

The main idea behind SPC is to monitor the stability of one or more quality characteristics.[1] By a *quality characteristic* we mean some measurable physical attribute of a manufactured product or part that is of importance for the producer or for the consumer. Notice that the word *monitoring* is somewhat in conflict with the word *control*, which has been used historically in this area. That is, *process monitoring* usually has a more passive connotation than the more active *process control*. In SPC, a production process is thought of running in either of two mutually exclusive states: an *in-control* state, and an *out-of-control* state. Graphical devices called *control charts* were developed by Shewhart to distinguish between these two states. As long as the chart does not signal the existence of an out-of-control state, the process is thought to be operating in *statistical control*.

Two sources of variability are contemplated in SPC, *common-cause variability*, variability that is inherent in the production system and that can only be modified by altering the existing production process, and *assignable cause variability*, observed variability that can be traced to a particular problem (e.g., human error, a problem with a raw material, or a machine failure). Assignable causes occur at unpredictable times, and the goal of the chart is

---

[1]Since the purpose of this chapter is to contrast SPC methods with engineering control methods, only univariate control charts for the mean of a normally distributed process are discussed. Other types of charts, such as charts for dispersion, for attribute data, or multivariate charts can be found in most SPC books (e.g., Montgomery, 2001).

to detect them as soon as possible. The diagnosis and corrective actions that should accompany such detection are left to the process engineer or operator in charge of the process and are not modeled. In contrast, common-cause variability is predictable within certain limits. A process is in statistical control if assignable causes have been detected and corrected, so these sources of variability will not influence the process in the future and only common-cause variability will remain present. This makes the quality characteristic predictable in the sense that the fraction of product that will fall within specifications can be calculated. By applying SPC charts continuously, assignable causes are removed from the system and the quality characteristics are continuously improved.

The emphasis of this book is not, however, on process monitoring (SPC charts) but on process adjustment using engineering process control (EPC) methods. Our approach for motivating the use of EPC techniques in a quality control application is to understand the problems of serially correlated data in traditional SPC schemes. One of the key assumptions behind the use of control charts is that successive values of the quality characteristic as they are observed through time are not correlated with each other. As we discuss in this chapter, modern manufacturing methods and sensor technologies often imply that quality data are serially correlated in time (autocorrelated), and this can have a large impact on the performance of control charts. First we briefly introduce basic SPC charts and discuss how the performance of some of these charts can be assessed under ideal circumstances (i.e., in the absence of autocorrelation). Formal definitions and statistical techniques are introduced to demonstrate the autocorrelation of a process and how it can be estimated from time-series data. Then we discuss the performance of SPC control charts under the presence of autocorrelation. The performance of SPC charts is badly affected by autocorrelation, and use of engineering process control methods is proposed as an alternative to compensate for process dynamics. SPC and EPC are not techniques in conflict with each other, and the majority of researchers in this field agree that a combined SPC–EPC strategy is necessary in most industrial processes today. In fact, one could define *control* as *combining* the arts of monitoring and adjustment. In the last section of this chapter we discuss integrated SPC–EPC approaches and provide a guide to the techniques discussed elsewhere in the book.

1.1 SPC CHARTS: BRIEF OVERVIEW

In classical SPC control charts,[2] the assumed in-control model is

$$Y_t = \mu + \varepsilon_t \quad \text{for} \quad t = 1,2,\ldots \quad (1.1)$$

---

[2]Readers familiar with SPC charts can skip to Section 1.2.

a model sometimes referred to as *Shewhart's model*. Here Y_t denotes the value of the quality characteristic that was obtained from the tth sample.[3] The parameter μ is the mean of the quality characteristic, and the random variables $\{\varepsilon_t\}_{t \geq 0}$ form what is called a *white noise sequence*, which means that

$$\varepsilon_t \sim (0, \sigma^2) \quad \text{and} \quad \text{Cov}(\varepsilon_t, \varepsilon_{t+j}) = 0 \quad \text{for all time } t \text{ and all lags} \quad j \neq 0$$

where Cov denotes the covariance between two random variables separated j periods or *lags* (ε_t and ε_{t+j}, in this case). The notation $Z \sim (\mu_Z, \sigma_Z^2)$ means that the random variable Z is distributed with constant mean μ_Z and constant variance σ_Z^2. In other words, the sequence of errors is uncorrelated,[4] has zero mean, and has a variance that is constant in time. This implies that $Y_t \sim (\mu, \sigma^2)$, and that $\text{Cov}(Y_t, Y_{t+j}) = 0$ for $j \neq 0$ (See Problem 1.10).

1.1.1 Shewhart Charts for Averages

Several different types of control charts were proposed by Shewhart. For the purposes of this chapter, we illustrate Shewhart charts with the simple case of a chart used to monitor the stability of the mean of a process, usually called an $\bar{X}$ *chart* in the SPC literature, although we use the name $\bar{Y}$ *chart* for consistency of notation.

In a Shewhart chart, samples of size n parts are taken from the production process with a certain frequency. If model (1.1) is the correct description of the quality characteristic and the errors $\{\varepsilon_t\}_{t \geq 1}$ are normally distributed (N), the averages[5] are also normally distributed; that is, $\bar{Y}_t \sim N(\mu, \sigma^2/n) \equiv N(\mu, \sigma_{\bar{Y}}^2)$. To monitor the process, the averages are plotted in time order on a Shewhart $\bar{Y}$ chart with limits at

$$\mu \pm k\sigma_{\bar{Y}}.$$

If μ and $\sigma_{\bar{Y}}$ are unknown, they are usually estimated with[6]

$$\hat{\mu} = \bar{\bar{Y}} \quad \text{and} \quad \sigma_{\bar{Y}} = \frac{\bar{R}}{d_2\sqrt{n}}$$

where $\bar{\bar{Y}}$ denotes the average of the averages of the samples and $\bar{R}$ denotes the average range of the samples. Both of these quantities are computed

---

[3]In many SPC books, X denotes the values of the quality characteristic. In this book we use X to denote a controllable factor or *input* of a process. The variable Y will denote the process *output*, or quality characteristic of interest. For this same reason we will talk about a $\bar{Y}$ chart instead of an $\bar{X}$ chart.

[4]Both covariance and correlation are measures of linear association between two random variables. In Section 1.2 we define these concepts for time series.

[5]A line over a variable denotes the average of the observed values of that variable. *Average* in this book is used to denote what some authors refer to as *sample mean*, whereas *mean* in this book refers to a population parameter.

[6]Throughout this book, a hat (^) over a variable or parameter name denotes an estimate.

based on a reference set of data that was supposedly obtained while the process was running in a state of statistical control. The constant d_2 depends only on the sample size and is a bias correction factor for the estimate of $\sigma_{\bar{Y}}$. It can be read from tables found in most SPC books. For $2 \leq n \leq 9$, a good approximation is $d_2 \approx \sqrt{n}$. Other estimates of $\sigma_{\bar{Y}}$ are more efficient (see, e.g., Montgomery, 2001), but the range estimator is still very common in practice. If no subgroups are formed ($n = 1$) and a chart on the individual measurements Y_i is desired, these are plotted on a chart with limits at $\bar{Y} \pm k\bar{R}/d_2$, where $\bar{R}$ is now the average of the *moving ranges* $R_i = |Y_i - Y_{i-1}|$. In this case, since the ranges are computed by "artificial" samples of size 2, the constant d_2 used equals 1.128.

Shewhart's model (1.1) allows us to consider the following two types of process anomalies: those that are assignable to some cause and those due to random or common causes (uncontrollable noise).

As long as the sequence of random variables $\{Y_t\}_{t \geq 1}$ behaves according to model (1.1), the only variation present in the process is random and uncontrollable under the present process conditions. At time t, this variability is modeled by the random variable ε_t. In this case the process is in statistical control. For a normally distributed process, assignable causes can make μ, σ, or the distribution of Y_t vary. Charts for averages are typically used in conjunction with charts that monitor σ. We then have that *the design parameters of a Shewhart chart are n, k, and the time between samples, h.* Typically, $n = 5$ and $k = 3$ are used. The value of h is selected directly or indirectly on economic grounds, for example, based on the cost of sampling or measuring.

Example 1.1: A Shewhart Chart for Averages. Consider the fabrication of discrete metal parts in a CNC lathe machine. A single dimension of interest is monitored with a $\bar{Y}$ chart. Figure 1.1 shows the chart plotted using the Minitab software package.[7] Fifty samples of size 5 each were collected, and these were used to obtain the estimates $\bar{Y} = 5.260$, $\bar{R} = 2.108$, and $\sigma_{\bar{Y}} = 0.4053$ (since $d_2 = 2.326$ for $n = 5$), giving upper and lower control limits (CL) of UCL = 6.476 and LCL = 4.044. All measurements are in hundreds of an inch. No average falls outside the limits; thus the process operator concludes that the process was in control while the data were collected. The control limits can then be adopted to assess statistical control based on further samples. Had any points fallen outside the limits, an investigation of the cause would need to be taken. Only when the problems associated with the occurrence of these abnormal observations can be found and corrected

---

[7]Minitab is an easy-to-use statistical software tool (version 13 was used in this book). Data can simply be entered in a column of its built-in spreadsheet. The user then selects from the top menu bar the options "Stat—Control Charts" and selects the type of chart to plot. Minitab requests information about where the data are located on the spreadsheet, the sample size, and so on.

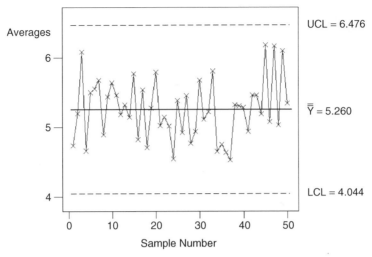

Figure 1.1 Shewhart control chart for averages.

should the abnormal points be deleted (and the control limits recomputed) after correction of the problems. □

It is recommended that the limits be recomputed periodically with recent information so that process improvements are reflected in the limits. For example, the process standard deviation will typically be reduced with time after use of an SPC chart. Recomputing the limits will promote further improvements in the process. It should be pointed out, however, that a common difficulty in practice is finding an assignable cause once an out-of-control signal is detected. Even if an assignable cause is found, management will not always solve the problem. This may lead to disenchantment with use of the chart.

Rational Subgroups
An important concept in the application of control charts relates to how samples are taken from a production process. Shewhart called his recommended sampling scheme the *rational subgroup* approach. For a chart of averages, this consists of sampling in such a way that the within-sample variability is due simply to common-cause or chance variability, whereas the variability between samples, if any, should indicate assignable or special causes of trouble in the process. A typical way of achieving these goals is to take the parts or observations that make up the sample as close together in time as possible, so that we minimize our chances of observing an assignable cause while we collect the n observations of a sample. In this way, the control limits' width is computed based on the within-sample variability, which represents common-cause variability. Further discussion of this topic can be found in Alwan (2000) and Montgomery (2001).

1.1.2   EWMA Control Charts

To monitor the mean of a process, we can instead consider using the exponentially weighted moving average (EWMA) statistic:

$$Z_t = \lambda \bar{Y}_t + (1 - \lambda)Z_{t-1} \qquad \text{for} \quad t = 1, 2, \dots \qquad (1.2)$$

where λ is a weight parameter such that $0 < \lambda < 1$. Clearly, as λ approaches 1, more weight is given to the most recent average. As λ decreases, more weight is given to older data, and in the limit when $\lambda = 0$, all the $Z_t's$ equal Z_0.

It can be shown (see Problem 1.11) that each observation $\bar{Y}_{t-j}$ receives a weight $\lambda(1 - \lambda)^j$, that is,

$$Z_t = \lambda \sum_{j=0}^{t-1} \bar{Y}_{t-j}(1 - \lambda)^j \qquad (1.3)$$

Thus the weights given to older observations decrease geometrically with age, and the geometric progression is the discrete analogy of an exponential function. For this reason, some authors refer to this chart as the geometrically weighted moving average chart, but the name EWMA is more common in practice. If $\bar{Y}_t \sim N(\mu, \sigma^2/n)$, applying the expected value and variance operators to equation (1.3), we get

$$\text{Var}[Z_t] = \frac{\sigma^2}{n} \frac{\lambda}{2 - \lambda} \left[1 - (1 - \lambda)^{2t}\right] \qquad (1.4)$$

$$E[Z_t] = \mu \left[1 - (1 - \lambda)^t\right]. \qquad (1.5)$$

Taking the limit as $t \to \infty$,

$$\text{Var}[Z_t] = \frac{\sigma^2}{n} \frac{\lambda}{2 - \lambda}$$

$$E[Z_t] = \mu$$

With these results we can set up the control limits of the chart in a manner analogous to that used for $\bar{Y}$ charts. In an EWMA control chart, the EWMA statistics Z_t are plotted ordered in time on a chart with limits at

$$\hat{\mu} \pm k \frac{\sigma}{\sqrt{n}} \sqrt{\frac{\lambda}{2 - \lambda}}$$

where $\hat{\sigma}$ can be obtained as for the $\bar{Y}$ chart. Thus it can be seen that the control limits of an EWMA can be obtained from those of an $\bar{Y}$ chart by introducing the *correction factor* $\sqrt{\lambda/(2 - \lambda)}$. There is evidence in favor of always using the limits

$$\hat{\mu} \pm k \frac{\hat{\sigma}}{\sqrt{n}} \sqrt{\frac{\lambda}{2 - \lambda} \left[1 - (1 - \lambda)^{2t}\right]}$$

Table 1.1   EWMA Chart Computations

t	Y_t	Z_t	UCL_t	LCL_t
0	—	5.2600	—	—
1	4.7416	5.1563	5.5015	5.0185
2	5.1994	5.1649	5.5693	4.9507
3	6.0937	5.3507	5.6057	4.9143
4	4.6620	5.2129	5.6272	4.8928
5	5.5053	5.2714	5.6403	4.8797
⋮	⋮	⋮	⋮	⋮
45	6.1947	5.4044	5.6625	4.8575
46	5.0761	5.3388	5.6625	4.8575
47	6.1731	5.5056	5.6625	4.8575
48	5.0338	5.4113	5.6625	4.8575
49	6.1037	5.5498	5.6625	4.8575
50	5.3477	5.5093	5.6625	4.8575

which provide a fast initial response of the chart to process upsets that occur early after process startup. Thus, as can be seen, the design parameters of an EWMA chart are n, k, λ, and h.

Example 1.2: EWMA Chart for Averages. Consider the machining process data in Example 1.1 and suppose that an EWMA chart is to be applied instead. The value $\lambda = 0.2$ is chosen. Using the initial 50 samples of size 5 each, the estimated parameters $\hat{\mu} = 5.26$ and $\hat{\sigma}_{\bar{Y}} = 0.4024$ are obtained. The EWMA recursive equation is initialized at $Z_0 = \hat{\mu} = 5.26$. Table 1.1 shows the detailed computations. As it can be seen, the control limits converge rapidly to their long-run values. Figure 1.2 displays the corresponding EWMA chart plotted using Minitab.   □

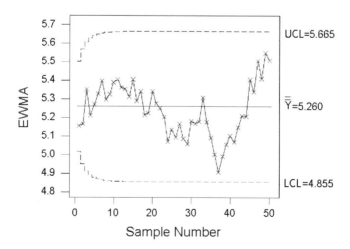

Figure 1.2   EWMA control chart for Example 1.2.

1.1.3   Cumulative Sum Control Charts

A cumulative sum (CUSUM) chart is based on the concept of sequentially probability ratio tests (SPRTs). A detailed description of the SPRT is outside the scope of this book,[8] so only an intuitive explanation of the use of the chart is presented here. In the SPC literature, V mask CUSUM charts are also discussed, but these are not easily interpretable and are difficult to design. A more modern approach, based on a *tabular CUSUM chart*, is presented here. The tabular CUSUM defines two one-sided cumulative sums, where the first one is

$$S_H(t) = \left[\overline{Y}_t - (\mu_0 + K) + S_H(t - 1) \right]^+$$

with initial value[9] $S_H(0) = 0$. The notation $(x)^+ = \max(0, x)$ means positive part and μ_0 is the assumed in-control process mean or target. This first sum accumulates deviations from μ_0 that are greater (i.e., higher) than $\mu_0 + K$. The second sum,

$$S_L(t) = \left[(\mu_0 - K) - \overline{Y}_t + S_L(t - 1) \right]^+$$

with initial value $S_L(0) = 0$, accumulates deviations from μ_0 that are smaller-lower-than $\mu_0 - K$ (see Figure 1.3). The reference value K is usually

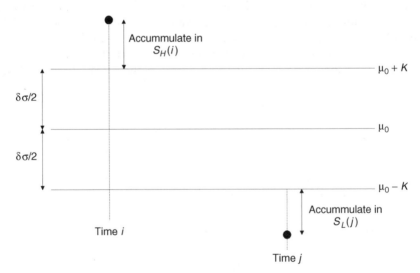

Figure 1.3   How a CUSUM chart works.

[8] For a comprehensive treatment of CUSUM charts, see the monograph by Hawkins and Olwell (1998).
[9] Other initial values can be used to provide a fast initial response to changes that occur shortly after startup (see Lucas and Crosier, 1982).

recommended to be set at half the size of the smallest shift in the mean that one wishes to detect quickly:

$$K = \frac{\delta \sigma_{\bar{Y}}}{2}$$

where δ is a multiple that allows us to measure shift sizes in terms of the standard deviation of the statistic being plotted (if individual measurements are used, we use σ instead).

If

$$S_H(i) > H \quad \text{or} \quad S_L(i) > H$$

this is evidence that the process is out of control with respect to Shewhart's model. Intuitively, if several consecutive deviations above the reference value are obtained, the process is abnormally high, and this calls for an investigation. A similar argument applies to consecutive deviations in the opposite direction. Typical values for H are four or five times the magnitude of $\sigma_{\bar{Y}}$ (or times σ if the sample size is 1). The design parameters of a tabular CUSUM chart are: n, K, H, and h.

Example 1.3: CUSUM Chart for Averages. For the same machining process data as used in Examples 1.1 and 1.2, we can plot a CUSUM chart instead. Suppose that we wish to detect changes in the mean of magnitude equal to 1 standard deviation of the averages. If the historical process standard deviation σ is 0.9 and the process target is assumed to be $\mu_0 = 5.25$, we have that $K = 0.5(0.9)\sqrt{5} = 0.2012$. The limits are set equal to $\pm H$, where we use $H = 4\sigma_{\bar{Y}} = 4(0.9)\sqrt{5} = 1.60997$. A detail of the computations involved is given in Table 1.2. The resulting chart, plotted using Minitab, is shown in

Table 1.2   CUSUM Chart Computations

t	Y_t	$S_H(t)$	$S_L(t)$
0	—	0.0000	0.0000
1	4.7416	0.0000	0.3072
2	5.1994	0.0000	0.1566
3	6.0937	0.6425	0.0000
4	4.6620	0.0000	0.3868
5	5.5053	0.0541	0.0000
⋮	⋮	⋮	⋮
45	6.1947	0.7434	0.0000
46	5.0761	0.3683	0.0000
47	6.1731	1.0901	0.0000
48	5.0338	0.6727	0.0149
49	6.1037	1.3251	0.0000
50	5.3477	1.2215	0.0000

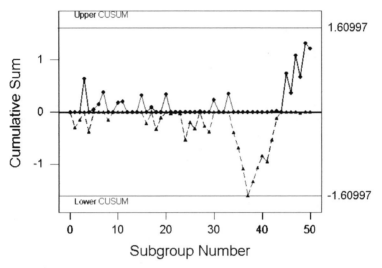

Figure 1.4   Example of a tabular CUSUM chart showing $S_H(i)$ and $-S_L(i)$.

Figure 1.4. Notice that the CUSUM chart signals an out-of-control state at sample 37. Such signals should be investigated.   □

1.1.4   Performance of Control Charts for Uncorrelated Data

A classical SPC chart can be thought of as a sequence of tests of hypothesis where the goal is to find evidence against the hypothesized Shewhart model. Therefore, it is natural to think of the probabilities associated with type I and type II errors as performance measures of how well the chart works.

Consider the simple case of a Shewhart $\bar{Y}$ chart. For this chart, the probability of a *false alarm* is

$$\alpha = P\{\text{Type I error}\} = P\{\text{one } \bar{Y}_i \text{ falls outside limits}|\text{process is in control}\}$$

Since $\bar{Y} \sim N(\mu, \sigma^2/\sqrt{n})$, assuming that the in-control mean is some known value μ_0, we have

$$\alpha = 2P\{\bar{Y} < \text{LCL}|\mu = \mu_0\} = 2P\left\{\frac{\bar{Y} - \mu_0}{\sigma/\sqrt{n}} < \frac{\text{LCL} - \mu_0}{\sigma/\sqrt{n}}\right\} = 2\Phi(-k)$$

where $\Phi(\cdot)$ denotes the standard normal distribution function. Thus the false alarm rate is determined exclusively by the width of the control limits. Figure 1.5 depicts the false alarm probability.

Example 1.4: False Alarm Probability of a Shewhart Chart for Averages. Consider a Shewhart $\bar{Y}$ chart in which 3σ control limits are used. In this case we have that $k = 3$, and therefore the false alarm probability is $\alpha = 2\Phi(-3) = 0.0027$.   □

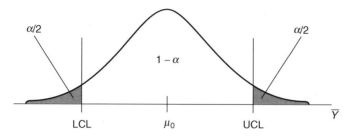

Figure 1.5   Probability of a type I error, normal distribution.

Similarly to the situation in the false alarm case, we can define the probability of not detecting a shift at a sampling time:

$$\beta = P\{\text{type II error}\} = P\{\text{one } \bar{Y} \text{ falls inside limits}|\text{process is out of control}\}$$

For computing β, we need to describe exactly what type of disturbance is not being detected. If we assume that an assignable cause shifts the process mean from μ_0 to $\mu_1 = \mu_0 + \delta\sigma$, then

$$\beta = P\{\text{LCL} \leq \bar{Y} \leq \text{UCL}|\mu = \mu_1\}$$

so

$$\beta = \Phi\left(\frac{\text{UCL} - (\mu_0 + \delta\sigma)}{\sigma/\sqrt{n}}\right) - \Phi\left(\frac{\text{LCL} - (\mu_0 + \delta\sigma)}{\sigma/\sqrt{n}}\right)$$
$$= \Phi(k - \delta\sqrt{n}) - \Phi(-k - \delta\sqrt{n})$$

where the second term is usually negligible. Note that β is a function of the sample size, the process variance, and the size of the shift in the mean. Also, if $\delta = 0$, we get $\beta = 1 - 2\Phi(-k) = 1 - \alpha$ (see Figure 1.6).

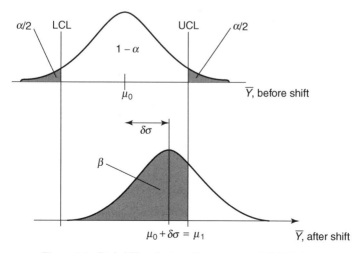

Figure 1.6   Probability of a type II error, normal distribution.

Example 1.5: Type II Error Probability for a Shewhart Chart for Averages.
Consider again a $\bar{Y}$ control chart in which samples of size 5 are taken and the width of the control limits is 3 standard deviations of the quality characteristic. Therefore, we have that $n = 5$ and $k = 3$. Suppose it is of interest to detect shifts of size equal to 2 standard deviations. Then the probability of a type II error for such a chart is $\beta = \Phi(3 - 2\sqrt{5}) - \Phi(-3 - 2\sqrt{5}) = 0.07078$.

$\square$

If we define the power of the chart as $1 - \beta \equiv$ power, it can be said that in any control chart we desire high power and low α. However, to a process engineer or operator, a discussion about probabilities of false alarms, or α's and β's, might sound esoteric. In any case, is $\alpha = 0.0027$ low enough? To avoid these issues, a much better way of describing the performance of any control chart is looking at its *run length* performance.

The run length of a chart is the *number of samples prior to observing an out-of-limits point on the chart*. Since it is assumed that the data are generated by an underlying probabilistic model, the run length is a random variable. Depending on the actual state of the system (in control or out of control), we can talk about an in-control run length (RL_{in}) and an out of control run length (RL_{out}).

In the case of a Shewhart chart for averages, it turns out that if we have enough historical process data so that we can consider μ_0 and σ to be "known," then

$$RL_{in} \sim \text{geometric}(\alpha) \quad \text{and} \quad RL_{out} \sim \text{geometric}(1 - \beta)$$

The rationale is as follows. Consider first the in-control case and compute, in sequence, the following probabilities:

$P\{\text{out of limits detected at first sample} \mid \text{process in control}\} = \alpha$

$P\{\text{out of limits detected at second sample} \mid \text{process in control}\} = (1 - \alpha)\alpha$

$$\vdots$$

$P\{\text{out of limits detected at } j\text{th sample} \mid \text{process in control}\} = (1 - \alpha)^{j-1}\alpha$

which is just the probability mass function of a geometric random variable. Thus the in-control average run length (ARL_{in}) is

$$ARL_{in} = E[RL_{in}] = \alpha \sum_{RL=1}^{\infty} RL(1 - \alpha)^{RL-1} = \frac{1}{\alpha}$$

Similarly (see Problem 1.13), for a geometrically distributed random variable with parameter α, it is known that its variance is given by $\sqrt{1 - \alpha}/\alpha$, so we can define the standard deviation of the run length (SRL_{in}) as

$$SRL_{in} = \sqrt{\text{Var}[RL_{in}]} = \sqrt{\frac{1 - \alpha}{\alpha^2}} = \frac{\sqrt{1 - \alpha}}{\alpha}$$

Therefore, the standard deviation and the average of the run length distribution will be very close if α is small.

By a similar argument, the probability mass function of the out-of-control run length distribution is $P\{RL = j\} = \beta^{j-1}(1 - \beta)$, and therefore

$$\text{ARL}_{\text{out}} = E[\text{RL}_{\text{out}}] = \frac{1}{1 - \beta}$$

and

$$\text{SRL}_{\text{out}} = \frac{\sqrt{\beta}}{1 - \beta} = \sqrt{\text{ARL}_{\text{out}}(\text{ARL}_{\text{out}} - 1)}$$

so, again, $\text{SRL}_{\text{out}} \approx \text{ARL}_{\text{out}}$ for large ARL_{out} (small power).

Example 1.6: Average Run Lengths, Shewhart Charts for Averages. Table 1.3 gives ARL values for $\bar{Y}$ charts that use different sample sizes and multipliers of different widths. Notice that the units in the table are number of samples until detection. For the sample sizes most common in practice ($n \leq 5$), a Shewhart $\bar{Y}$ chart does not detect small ($\delta \leq 1$) shifts in the mean rapidly compared to EWMA or CUSUM charts. Large shift sizes, on the order of $\delta > 2$, are detected very quickly. This is not surprising since Shewhart charts were developed at a time when large process upsets were frequent. Narrowing the control limit width results in more frequent false alarms and faster out-of-control detection. $\square$

Note that to assess the performance of a single chart, even with known parameters, the ARL is misleading, due to the thick right tail of the run length distribution. For this reason, some authors suggest looking at the median or at some percentile points of this distribution. Appendix 1C gives further details about the use of ARLs as a performance indicator in control charts.

The run length performance of EWMA and CUSUM charts is very similar if the λ weight in the EWMA is small, around 0.1 according to most recommendations. These charts detect small shifts in the mean much more

Table 1.3 Average Run Lengths, Shewhart Charts for Averages

δ	$k = 3$			$k = 2.5$		
	$n = 2$	$n = 5$	$n = 10$	$n = 2$	$n = 5$	$n = 10$
0.0	370.3	370.3	370.3	80.5	80.5	80.5
0.5	90.6	33.2	12.8	26.9	11.9	5.5
1.0	17.7	4.4	1.8	7.2	2.5	1.3
2.0	2.3	1.0	1.0	1.6	1.0	1.0
3.0	1.1	1.0	1.0	1.0	1.0	1.0

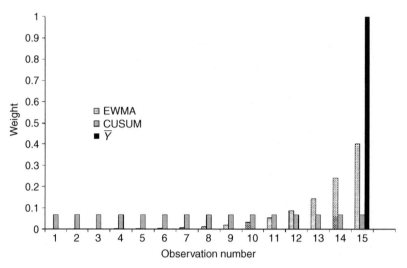

Figure 1.7 Different weights given to the last 15 observations of a process. The weights for the EWMA chart correspond to a value $\lambda = 0.4$.

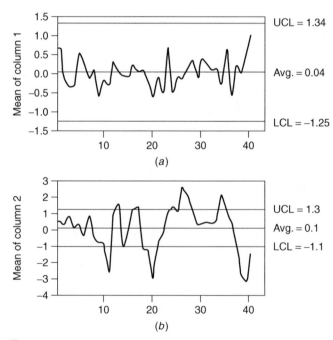

Figure 1.8 $\bar{Y}$ chart ($n = 5$, $k = 3$) applied to a process with (a) no autocorrelation and (b) positive autocorrelation.

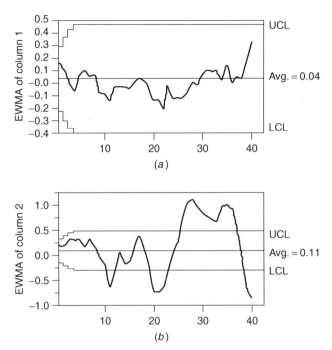

Figure 1.9 EWMA chart ($n = 5$, $k = 3, \lambda = 0.2$) applied to a process with (a) no autocorrelation and (b) positive autocorrelation.

rapidly than $\overline{Y}$ charts and should be used instead if small shift sizes in μ are suspected in a process. The derivation of the ARLs of these charts is more involved and outside the scope of this book (see Woodall, 1983, 1986; Crowder, 1987).

The reason the EWMA and the CUSUM charts can detect small shifts quickly can be traced to the way that data are weighted (Hunter, 1986). In a Shewhart $\overline{Y}$ chart (see Figure 1.7), all weight is given to the point plotted most recently (i.e., the decision whether or not to stop a process is based only on the last $\overline{Y}$ observed). In contrast, EWMA and CUSUM charts give some weight to all the previous data. This suggests that a small shift in the mean of the process will accumulate through the EWMA or CUSUM statistics and will trigger a signal sooner than relying on the last point plotted. Evidently, an $\overline{Y}$ chart is a particular case of an EWMA chart if the EWMA weight λ equals 1.0.

Figures 1.8 and 1.9 show a process where data are uncorrelated and a process where data are serially correlated. Samples of size 5 were taken and $\overline{Y}$ and EWMA charts were plotted for the data. If there is no autocorrelation, the charts perform adequately, but for high positive autocorrelation at low lags there are a considerable number of points outside limits, making interpretation difficult. (Are these true out-of-control signals, or are they expected given the autocorrelation?) We discuss these issues in detail in Section 1.3. First, some statistical definitions are needed to explain clearly what we mean by "positive autocorrelation at low lags" of a *stochastic process*.

1.2   AUTOCORRELATION OF A PROCESS

In Section 1.1 it was pointed out that classical control charts assume no correlation between successive observations of the quality characteristic. In this section we define in a more precise manner what is meant by correlation between repeatedly observed measurements of a single quality characteristic. We need to know how to estimate the serial correlation, in case it exists, and how to study its effect on classical SPC charts. To achieve these goals, the concept of a *stochastic process* is first necessary. A stochastic process $\{Y(t), t \in I\}$ is a family of indexed random variables, where I is called the *index set*. Sometimes we will refer to the mechanism generating the stochastic process simply as the process, which can be understood in the double sense of the underlying stochastic process that the quality characteristic being modeled follows, or as the production process itself, which in turn generates the stochastic process (i.e., the quality characteristic). In applications in this book, t will relate to the discrete points of time at which an observation is obtained by sampling (i.e., the index set is $I = \{\ldots, -2, -1, 0, 1, 2, \ldots\}$) and the stochastic process is said to be a *discrete-time* stochastic process. If $I = \{t : -\infty < t < +\infty\}$, the process is a continuous-time stochastic process. For discrete-time stochastic processes, it is customary to denote them as $\{Y_t\}_{t \in I}$, that is, a subscript is used for discrete-time indices. Discrete-time processes can be identified by describing the behavior of its tth element. Thus we can write, for example,

$$Y_t = \mu + \varepsilon_t$$

implying that for the given process (Shewhart's model in this case) the equation holds for all discrete points in time t. In this case, using notation introduced in Section 1.1.1, the time between observations h equals Δt.

A stochastic process can be thought of as a function of two arguments: namely, $\{Y(t, w), t \in I, w \in \mathcal{S}\}$, where $\mathcal{S}$ is the sample space of the random variables $Y(t)$. It is important to point out that for fixed $t \in I$, $Y(t, \cdot)$ is a random variable; for fixed $w \in \mathcal{S}$, $Y(\cdot, w)$ is a function of time called a *realization*, or trajectory, of the process. A *time series* is a set of observations, or realization, of a discrete-time stochastic process. In basic probability theory, this is analogous to the notion of a single observed value of a random variable (y) compared to the random variable itself (Y). The set of all possible realizations of a stochastic process is called the *ensemble* of the process, as depicted in Figure 1.10.

As in many other areas of statistics, to perform valid statistical inferences from a stochastic process, we need some notion of how repeatable the underlying random experiment is under identical conditions. For example,

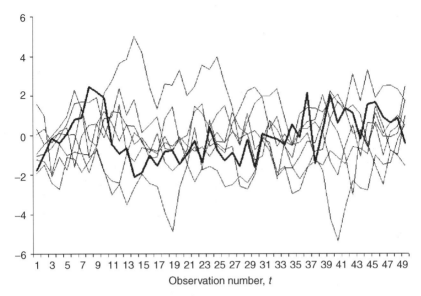

Figure 1.10 Single realization (darker line) compared to other possible realizations that make up the ensemble of a single stochastic process.

one such statistical inference could be to predict to where a quality characteristic will move in the near future. The only information available to us is a single realization of a stochastic process, the values recorded of the quality characteristic in the past. If this process is such that its random variables differ radically at different points of time, no inference could be drawn from a single realization since the probabilistic properties of the process during the period of time when the observations were taken cannot be generalized to other periods of time. For the class of models studied in this book, the notion that we need to make valid inferences is called *stationarity*.

A stochastic process is *strictly stationary* if for any integer $k \geq 1$, the joint distribution of $\{Y_{t_1}, Y_{t_2}, \ldots, Y_{t_k}\}$ is identical to the distribution of $\{Y_{t_1+\tau}, Y_{t_2+\tau}, \ldots, Y_{t_k+\tau}\}$, where $t_i \in I$, $t_i + \tau \in I$. Thus the stochastic properties of the process are unaffected by changes in the time origin. If in this definition we look at the case $k = 1$, we note that strict stationarity implies that the distribution of Y_t is the same as the distribution for $Y_{t+\tau}$. Therefore, strict stationarity implies that the distribution of the random variables occurring at different points in time is identical.

1.2.1 Mean and Variance of a Stationary Process

Since for a strictly stationary discrete-time process $\{Y_t\}_{t=-\infty}^{\infty}$ the probability density function $f(Y_t)$ is the same for all t, that is,

$$f(Y_t) = f(Y)$$

we have that

$$\mu = E[Y_t] = \int_{-\infty}^{+\infty} y f(y)\, dy$$

$$\sigma^2 = \text{Var}[Y_t] = \int_{-\infty}^{+\infty} y^2 f(y)\, dy - \mu^2.$$

Thus μ gives the long-run or asymptotic average level of the process at every time t and σ^2 gives the long-run dispersion of the process at every time t. These long-run quantities are obtained over the ensemble of the stochastic process, which is the population from which we sample. For example, $E[Y_t]$ can be thought of as the level of the process at time t averaged over an infinitely large number of possible realizations.[10] The corresponding point estimates are given by

$$\hat{\mu} = \overline{Y} = \frac{1}{N} \sum_{t=1}^{N} y_t$$

$$\hat{\sigma}^2 = \frac{1}{N} \sum_{t=1}^{N} \left(y_t - \overline{Y} \right)^2$$

where N denotes the length of the time series observed.

1.2.2   Autocovariance and Autocorrelation

The prefix *auto* means a reflexive act upon oneself, thus autocovariance is the covariance that a process has "with itself." The autocovariance function $(\gamma_{t,k})$ gives a measure of linear association (covariance) between two variables of the same process, Y_t and Y_{t+k}, that are separated k periods or lags, as a function of k. For any discrete-time process, we define

$$\gamma_{t,k} = \text{Cov}[Y_t, Y_{t+k}] = E\big[(Y_t - E[Y_t])(Y_{t+k} - E[Y_{t+k}])\big]$$
$$\text{for } k = 0, \pm 1, \pm 2, \ldots$$

For a strictly stationary process, the mean μ_t is constant for all times t, and the autocovariance reduces to

$$\gamma_{t,k} = \gamma_k = \text{Cov}[Y_j, Y_{j+k}] = E\big[(Y_t - \mu)(Y_{t+k} - \mu)\big]$$
$$\text{for } k = 0, \pm 1, \pm 2, \ldots$$

so the autocovariance depends only on the lag k. Also, note that $\gamma_0 = E[(Y_t - \mu)^2] = \text{Var}(Y_t) = \sigma^2$.

---

[10] See Appendix 1A at the end of this chapter for more technical details.

It is usually better to scale the autocovariance into a unitless quantity by dividing by the process variance. For a stationary process,

$$\rho_k = \frac{\gamma_k}{\gamma_0} \qquad \text{for } k = 0, \pm 1, \pm 2, \ldots$$

where $-1 \leq \rho_k \leq 1$. Given that $\gamma_k = \gamma_{-k}$ and $\rho_k = \rho_{-k}$ (i.e., the autocorrelation and autocovariance are *even* functions), it is common to plot these two functions only for positive lags. Both the autocovariance and the autocorrelation at lag k give a measure of the degree of linear association between two random variables of the same process that are separated k periods.

When considering relations between random variables, it is useful to recall the following implications:

1. If Y_t and Y_{t+k} are <u>independent</u>, they are <u>uncorrelated</u> (i.e., $\rho_k = 0$ for all k).
2. If Y_t and Y_{t+k} are <u>correlated</u> (i.e., $\rho_k \neq 0$ for some k), Y_t and Y_{t+k} are <u>dependent</u>.

The direction of the implications is important; uncorrelated random variables may or may not be independent. Correlation is a measure of linear association, so there might be some nonlinear association between uncorrelated variables.

It should be pointed out that the standard error of the average $(\overline{Y})$ is greatly affected by autocorrelation. Bartlett (1946) showed that

$$\sigma_{\overline{Y}} = \sqrt{\frac{\gamma_0}{N}\left[1 + 2 \sum_{k=1}^{N}\left(1 - \frac{k}{N}\right)\gamma_k\right]}$$

Thus if the stochastic process is completely uncorrelated ($\gamma_k = 0$ for all k), the standard error of the average $\overline{Y}$ is the usual $\sigma/\sqrt{N}$. As the (positive) autocorrelation increases, so does the standard error and the point estimate $\overline{Y}$ becomes less reliable.

Point estimates of the autocovariance function are given by the sample autocovariance function,[11] computed as

$$c_k = \hat{\gamma}_k = \frac{1}{N} \sum_{t=1}^{N-k} \left(y_t - \overline{Y}\right)\left(y_{t+k} - \overline{Y}\right) \qquad k = 0, 1, 2, \ldots \qquad (1.6)$$

Similarly, the sample autocorrelation function is given by

$$r_k = \hat{\rho}_k = \frac{c_k}{c_0} \qquad k = 0, 1, 2, \ldots$$

---

[11] If in the denominator we use $N - k$, rather than N we get an unbiased estimate but with larger mean square error; thus equation (1.6) is preferred. Since we defined a biased estimator of σ^2 with a denominator of N, it follows that $\hat{\gamma}_0 = \hat{\sigma}^2$.

Example 1.7: Continuous-Flow Chemical Process. Consider the data provided by Box and Jenkins (1976) representing the yields of 70 consecutive batches from a chemical process.[12] Figure 1.11 shows a time-series plot of the data and a plot of the sample autocorrelation function of the data, computed with Minitab.

Observe how the yield time-series data "jump" from batch to batch from one side of its mean (approximately located at 50%) to the other. This is an indication of negative autocorrelation at lag 1 (i.e., $r_1 < 0$ is significant). This can be seen from the sample autocorrelation plot, which indicates a significant lag 1 negative autocorrelation estimated at -0.39. Significant negative autocorrelation at low lags can indicate a process that was subject to succesive adjustments, in such a way that adjacent observations lie on opposite sides of the mean.   □

Example 1.8: A Discrete-Part, Machining Process. As a second example, consider the data in the file *Fadal.txt*, which correspond to 40 measurements of a dimension machined on aluminum parts processed on a Fadal computer numerically controlled (CNC) machine tool. The time-series data plot and the estimated autocorrelation function were obtained with Minitab and are shown in Figure 1.12.

The series is clearly nonstationary, and this reflects the autocorrelations, which are significant for many lags. A detailed description and interpretation of sample autocorrelation functions is provided in Chapter 3. In this example, the autocorrelation of the data is due to tool wear of the cutting tool. Notice that the drift in the process does not exactly follow a linear trend or ramp.[13]

□

Example 1.8 illustrates that not only in chemical processes, but also in discrete-part manufacturing, we might find considerable autocorrelation. In chemical processes and, in general, processes where material accumulates, autocorrelation in the data observed results from the dynamic or inertial elements in the production process. In a discrete-part manufacturing process, autocorrelation exists because of tool wear or because measurements were taken on dimensions generated from cutting operations that "shared" the same setup operation or some other process condition that is common to the parts being produced. Environmental temperature fluctuations can also induce autocorrelation in dimensional data of machined parts.

---

[12] The data can be found in the file <u>BJ-F.txt</u> and correspond to series F in Box et al. (1994).

[13] In fact, for this data set the random walk model, $Y_t = Y_{t-1} + \varepsilon_t$ (see Chapter 3), provides a better fit than the linear trend model, $Y_t = a + b\,t + \varepsilon_t$ (see Problem 1.1). As discussed in Chapter 3, this is an instance of a nonstationary process that can be understood as an AR(1) process with parameter $\phi = 1$.

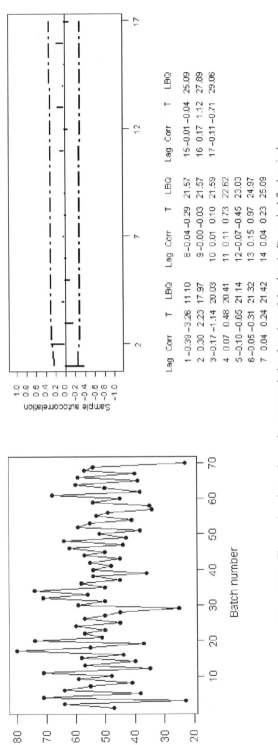

Figure 1.11 Time-series plot and sample autocorrelation function of the data in Example 1.7, chemical process data.

22

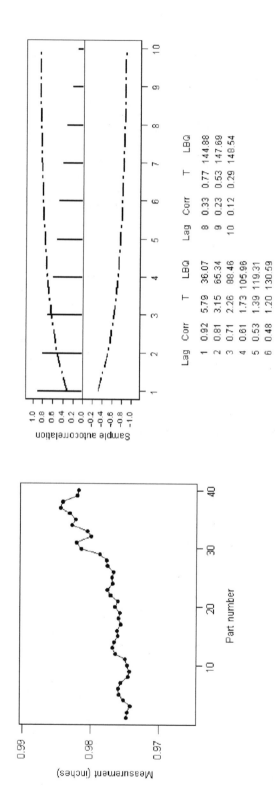

Figure 1.12 Time-series plot and sample autocorrelation function of the data in Example 1.8. machined parts data.

1.2.3   Need for Stationarity

Note that the point estimates $\overline{Y}$, $\hat{\sigma}^2$, c_k, and r_k are computed based on a single realization of the process (i.e., they are computed based on time-series data). Inferences based on a single realization are possible because the process is stationary.[14]

A different form of stationarity is called *weakly or covariance stationarity*. A process is weakly stationary if only its first two moments (mean and covariance) are finite and independent of time (in such a case the covariance depends only on the lag). If each Y_t is normally distributed and the process is weakly stationary, the process is strictly stationary. As it turns out, the type of stochastic models we study are fully characterized by their first two moments, so when we refer to stationarity in later chapters we will be referring to weak stationarity. Weak stationarity can be better understood from an engineering point of view. If the mean of the process is constant, some process engineers would say that the process is *stable*. A more exact definition of stability is given in Chapter 2 in relation to transfer function models. As will be seen, stability is a property of a *deterministic* dynamical system, whereas stationarity is a property related to a *stochastic* process.

So what should we do if the process under study *is* nonstationary? In Chapter 3 we describe different transformations that can be applied to a nonstationary process to achieve stationarity. Then μ, σ^2, or ρ_k can be estimated from the transformed series.

1.3   EFFECT OF AUTOCORRELATION ON SPC CHART PERFORMANCE

As mentioned before, SPC control charts assume that if in control, the process has a constant mean and is completely uncorrelated [process (1.1)]. An important practical question is to investigate what happens with the performance of SPC charts as the process exhibits more and more serial correlation. As one process engineer told this author, "almost every production process exhibits autocorrelation."

Positive autocorrelation at low lags is common because given the advances in sensor technologies, measurements are taken closer together in time. In discrete-part manufacturing, this sometimes implies that every part is measured. Observations that were generated close in time will tend to be similar; hence positive correlation at low lags will result. The following two examples illustrate the effect of autocorrelation on the performance of SPC charts.

---

[14] The precise condition for these time averages to be consistent estimates of the ensemble averages is called *ergodicity* (see Appendix 1A).

Example 1.9. Suppose that a process is described by

$$Z_t = \varepsilon_t + \underbrace{\lambda Z_{t-1} + \lambda(1-\lambda)Z_{t-2} + \lambda(1-\lambda)^2 Z_{t-3} + \ldots}_{\text{EWMA of past } Z's}$$

$$Y_t = \mu + Z_t \tag{1.7}$$

where $0 \le \lambda \le 1$ and Y_t is the quality characteristic. There are two extreme cases: if $\lambda = 0$, the process is just Shewhart's process [equation (1.1)], where the observations are completely uncorrelated. If $\lambda = 1$, the process is

$$Y_t = \mu + Z_t = \mu + \varepsilon_t + Z_{t-1}$$

which evidently is highly correlated[15] with

$$Y_{t-1} = \mu + Z_{t-1}.$$

Vander Weil (1996) considers this process as λ increases from 0 to 1.[16]

In the realizations of Figure 1.13*b* and *d*, a sustained shift in the mean of magnitude 5σ was induced at time 100; that is, the mean of the process changes according to

$$\mu_t = \begin{cases} \mu_0 & \text{if } t \le t_0 \\ \mu_0 + \delta\sigma & \text{if } t > t_0 \end{cases}$$

where in this example $\delta = 5$, $t_0 = 100$, and $\mu_0 = 10$. This type of shift suddenly changes the mean of the process at time t_0. As can be seen from the figures, for $\lambda = 0$, detection should be almost immediate for any SPC chart, and false alarms should be infrequent. For $\lambda \ge 0.8$, the shift becomes indistinguishable from the autocorrelation structure. Such positive autocorrelation will not necessarily worsen the detection capabilities of the charts as measured by the ARL_{out} performance criterion (Goldsmith and Whitfield, 1961; Lu and Reynolds, 1999, 2001), but the number of false alarms will certainly increase. A chart that gives frequent false alarms will soon be abandoned.   □

Example 1.10. The impact of autocorrelation on SPC charts was studied by Maragah and Woodall (1992), who consider the process

$$Y_t = \mu + \phi Y_{t-1} + \varepsilon_t \tag{1.8}$$

---

[15] In this case, the process $Z_t = Z_{t-1} + \varepsilon_t$ is called a *random walk*. The autocorrelation function of this process does not decay with the lag since the process is nonstationary.

[16] Vander Weil considered plotting on SPC charts not the original observations but the forecast errors of an IMA(1, 1) process, an approach we discuss in Chapter 3.

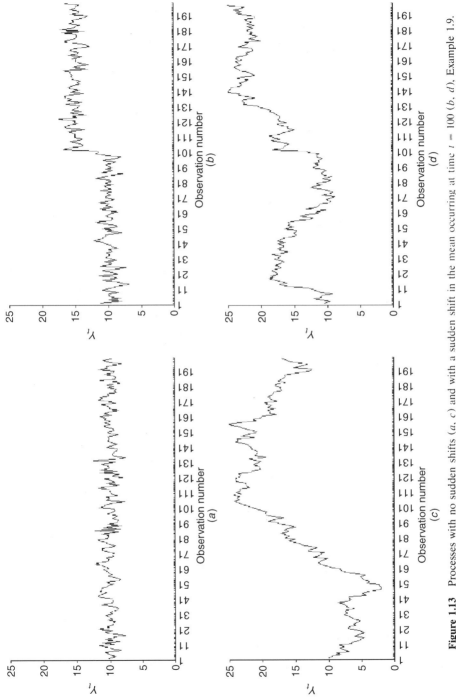

Figure 1.13 Processes with no sudden shifts (*a*, *c*) and with a sudden shift in the mean occurring at time *t* = 100 (*b*, *d*), Example 1.9. Graphs (*a*) and (*b*) display an uncorrelated series (λ = 0); graphs (*c*) and (*d*) display a highly autocorrelated (λ = 0.8) series.

Table 1.4 Number of Points Outside Limits Generated by a Shewhart Chart for Individuals Used to Monitor 25 Observations of an AR(1) Process, Obtained by Simulation

ϕ	Average	Std. Dev.
-0.9	0.001	0.035
-0.7	0.004	0.063
-0.3	0.027	0.166
-0.1	0.058	0.241
0.0	0.094	0.316
0.1	0.135	0.377
0.3	0.312	0.610
0.7	1.807	1.887
0.9	4.081	3.331

Source: Maragah and Woodall (1992).

where $-1 < \phi < 1$ is a parameter that allows us to define the autocorrelation in the $\{Y_t\}$ process. In Chapter 3 we show how this is a particular instance of the ARIMA family of stochastic processes studied by Box and Jenkins called an *AR(1) process* (Maragah and Woodall also considered other types of stochastic processes). If $\phi = 0$ in equation (1.8), we obtain Shewhart's uncorrelated process. It is shown in Chapter 3 that the autocorrelation function of this process is $\rho_k = \phi^k$, so the process is stationary as long as $|\phi| < 1$. Note that for $\phi = 1$ the process turns out to be a random walk. In the case of positive autocorrelation ($0 < \phi < 1$), the movement of the process is smoother than that of an uncorrelated process. In the case of negative autocorrelation ($-1 < \phi < 0$), the process shows a sawtooth pattern, which compared to Shewhart's process is much more crumpled.

Maragah and Woodall (1992) computed by simulation the average number of out-of-control points that a Shewhart chart for individuals will generate if 25 observations of process (1.8) with $-0.9 \leq \phi \leq 0.9$ are used to set the chart limits. A chart for individuals or a Y chart is one in which no subgroups are formed. The control limits are computed using the moving-range estimator[17] $\hat{\sigma}_Y = \overline{MR}/d_2$, where $MR_i = |Y_i - Y_{i-1}|$. Table 1.4 shows some of their results.

Note that no shift in the mean was introduced while simulating the AR(1) process. From the table it can be seen that the number of out-of-control signals increases with increasing positive autocorrelation. For $\phi = 0.9$, for example, an average of about four signals will occur in 25 samples. This is a much higher signal rate than the advertised average of one false alarm every 370 observations for Shewhart charts. Negative autocorrelation reduces the number of signals, but positive autocorrelation is much more common in

[17]This estimator was used because it is the most common estimator in control charts for individuals, but as mentioned below, it is not recommended for autocorrelated data.

practice, as mentioned before. Negative autocorrelation at low lags inflates the variance and hence the control limits width becomes too large, so detecting actual shifts becomes more difficult.　　　□

The evident problem in Example 1.10 is that the limits were computed based on a variance estimate which assumes that the process is uncorrelated. If $\phi > 0$ (positive autocorrelation), adjacent observations will tend to be similar and the moving-range estimator will underestimate the variance of the process. This will result in limits that are too narrow, producing many alarms compared to the uncorrelated case. This result is quite general: *The within-sample variance estimator will underestimate the process variance in a positively autocorrelated process.* If $\phi < 0$ (negative autocorrelation), the moving-range estimator will overestimate the true variance, resulting in limits that are too wide, so very few signals will be obtained as opposed to the uncorrelated case. Note that this will have an impact on the detection capabilities of real shifts (i.e., wide limits will make the chart take much longer to detect shifts in the process mean).

1.4 DEMING'S FUNNEL EXPERIMENT

W. Edwards Deming (1986) used to talk about an experiment aimed at showing, by analogy, the dangers of adjusting a process that is in a state of statistical control. Based on this experiment and some remarks made by Deming in his writings, some quality consultants have taken the extreme view that a process should never be adjusted and that SPC charts are always sufficient. Given the influence of Deming in the quality area, it is relevant to analyze his funnel experiment in detail. MacGregor (1990) analyzes this experiment and provides further useful information.

The experiment is conducted by mounting a funnel over a target bull's-eye placed on a flat surface. Marbles are dropped consecutively through the funnel, and their position with respect to the target is measured. The position of the funnel relative to the target can be adjusted from drop to drop (see Figure 1.14).

Deming proposed four adjustment rules that mimic, in this analogy, four scenarios sometimes found in practice:

- *Rule 1.* Leave the funnel fixed, aimed at the target, no adjustment.
- *Rule 2.* At drop k ($k = 1, 2, 3, \ldots$) the marble will come to rest at point Y_k measured from the target. Move the funnel a distance $-Y_k$ *from its last position.*
- *Rule 3.* Move the funnel a distance $-Y_k$ *from the target.*
- *Rule 4.* Set the funnel for the next drop $(k + 1)$ right over where the marble came to rest at the preceding drop.

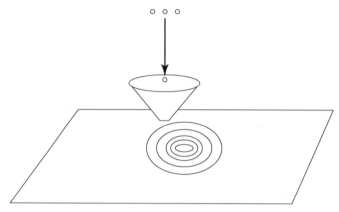

Figure 1.14 Funnel experiment setup. It is assumed that initially, the funnel is aimed at the target.

The position of the funnel is the controllable factor in a feedback mechanism, and the distance from the target is similar to the deviation of a quality characteristic from target, a variable commonly monitored in SPC. Rules 2 to 4 contain a feedback action in the sense that previous measurements are used to set the current position of the funnel. Rule 1 corresponds to SPC. Rule 4 describes a situation where a process operator is trying to make each part identical to the preceding one to achieve consistency. In that case, the process ends up "chasing" itself. Rules 2 and 3 are rather intuitive, recommending that we adjust by as much as the last observed deviation. Every person who has tried to shoot at a target probably has used either of these strategies in a sequence of trials.

Figures 1.15 to 1.17 illustrate the trajectories followed by the funnel as looked at from above the table under rules 2 to 4, respectively.

There are some important assumptions that Deming made explicit with his experiment. First, it is assumed that the process producing deviations from target is in a state of statistical control, which means that $\{Y_t\}_{t \geq 1}$ obeys Shewhart's model. Furthermore, it is assumed that the process is initially on target; that is, it is assumed that the process mean $E[Y_1]$ equals the target desired. Also, it is assumed that it is possible to aim the funnel at the target or at any other objective. Under the assumptions stated, the results are:

- *Rule 1.* This rule produces the minimum variance and a process that is on target.
- *Rule 2.* The process is on target and stable but has twice the variance as in rule 1.
- *Rule 3.* The process is nonstable and explodes in an oscillatory pattern, with oscillations being wider and wider with time.

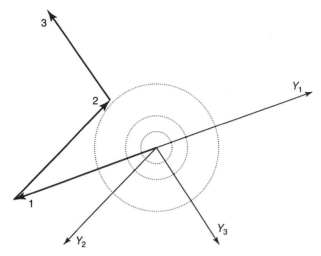

Figure 1.15 Sample trajectory of the position of the funnel under rule 2 for a process in a state of statistical control. Lighter lines, first three observed deviations from target; darker lines, funnel trajectory.

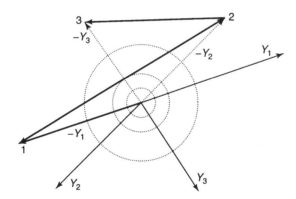

Figure 1.16 Sample trajectory of the position of the funnel under rule 3 for a process in a state of statistical control. Lighter lines, first three observed deviations from target; darker lines, funnel trajectory.

- *Rule 4.* The process explodes, with marbles moving farther away from the target as time passes (the process is actually a random walk).

The results above can be better observed from Figure 1.18, which shows computer simulations that illustrate the behavior of an in-control funnel process under the four adjustment rules. Figure 1.19 shows the corresponding analogous simulation for a time-series process that obeys Shewhart's

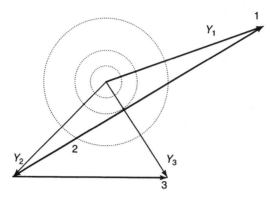

Figure 1.17 Sample trajectory of the position of the funnel under rule 4 for a process in a state of statistical control. Lighter lines, first three observed deviations from target; darker lines, funnel trajectory.

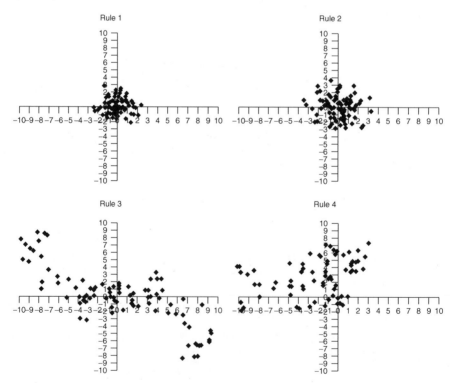

Figure 1.18 Funnel simulation of the four adjustment rules for a process that is in statistical control. The dots represent where the marbles stopped on the table.

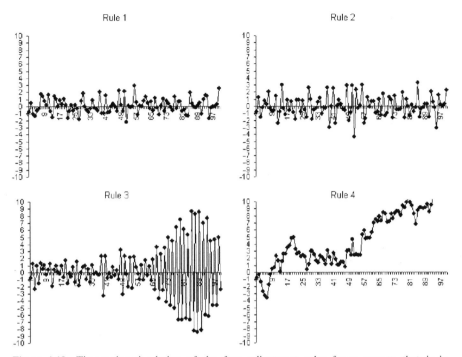

Figure 1.19 Time-series simulation of the four adjustment rules for a process that is in statistical control.

model and is adjusted by each of the rules.[18] The simulation programs used to create Figures 18 and 19 can be found in the Excel spreadsheet *Deming-funnel.xls*.

Clearly, after observing the plots it is quite evident that a process *that is on target and in a state of statistical control* should not be adjusted. Grubbs (1954) provided an adjusting mechanism for a process that is in statistical control but is initially off target. His adjusting method is described in Chapter 5.

It is simple to see how under Deming's assumptions, rule 1 produces the best results. Let the process be

$$Y_t = \varepsilon_t + X_{t-1}$$

where Y_t is the *deviation from target* observed at drop t, X_t the level of the controllable factor at run t (resetting of the funnel), and $\{\varepsilon_t\}_{t \geq 0}$ represents a

[18] The observations in the funnel experiment are an instance of multivariate (bivariate, in this case) data, given by the two coordinates of the marbles on the table, whereas the corresponding time-series simulations of Figure 1.19 are univariate since a scalar is observed at each time instance. The effect of the four rules is the same regardless of the dimension of the vector Y_t observed. Multivariate processes are discussed in Chapter 9.

white noise sequence. The adjustments will be given by the differences $X_t - X_{t-1}$. Then to get $E[Y_t] = 0$, that is, no expected deviation from the target, we simply set $X_t = 0$ for all t, which implies that no adjustments are made. With this we have $\text{Var}[Y_t] = \sigma_\varepsilon^2$, which is the minimum possible variance we can achieve for this process, since it is assumed that the variance we consider by introducing ε_t in the model is uncontrollable. If we follow any feedback control rule of the form

$$X_t = f(Y_t, Y_{t-1}, \dots)$$

it follows that regardless of the function f, this will imply that $\text{Var}[Y_t] > \sigma_\varepsilon^2$. Thus *we should not "tweak" the process if it is on-target and in statistical control*. A good question, however, is how much larger than σ_ε^2 the variance of an on-target and in-control process is under the actions of a given feedback control law. In Chapters 6 and 7 we describe how for certain control rules and processes, the inflation in variance is very moderate and always using an EPC scheme seems reasonable under some conditions. See also Problem 1.8, where a fifth adjustment rule is proposed for Deming's funnel experiment; such a rule is less drastic than rules 2 to 4 and provides a reasonable increase in variance for the assurance of protection against systematic variability of different types.

In general, if an uncontrolled process $\{Y_t\}$ exhibits autocorrelation (including the extreme case of nonstationary behavior), a feedback control rule such as rules 2 and 3 might prove beneficial. For example, suppose that we let $Y_t = \mu_t + \varepsilon_t$, where μ_t is the mean of the process. Then the process is nonstationary since the mean is "wandering" and feedback is needed, because otherwise the process, left uncontrolled, will move away from the target, depending on how μ_t moves with time. In terms of the funnel experiment, this is the case when the bull's-eye is moving; keeping the funnel fixed in such a situation is not the best alternative.

Example 1.11. Feedback adjustment might be beneficial not only for a nonstationary process. Suppose that $Y_t = \mu_o + a_t$, where $\text{Cov}(a_t, a_{t+k}) \neq 0$ for some $k \neq 0$. Then the process is stationary since the mean and autocovariance are constant in time, but the autocorrelation induces *structured* variability that we can anticipate and compensate for by feedback adjustment, a topic we discuss next.   □

1.5 INTRODUCTION TO ENGINEERING PROCESS CONTROL

By a *controller* we mean a rule, function, or algorithm that describes how the controllable factor of a process (X_t) needs be adjusted from observation to observation. In mechanical, chemical, and electrical engineering applications, controllers are usually implemented using commercially available sensors, electronic controllers, and actuators. In quality control applications, in contrast, the controller is usually implemented manually by an operator (perhaps "dialing-in" new operating conditions into a machine) once the data become available. Sometimes (see Chapter 7) the controller runs on a small computer and its suggestions are entered manually into the equipment by the process operator. Automatic closed-loop implementations are, of course, possible in quality control, but it is preferable to keep an operator "on the loop" with authority to change the settings recommended by the controller (as done, e.g., in semiconductor manufacturing; see Moyne et al., 2000). An engineering process controller (e.g., a feedback controller) takes advantage of the autocorrelation structure of the system to provide better forecasts some number of periods ahead, say $\hat{Y}_{t+k}$. The control action X_t is chosen to ensure that $\hat{Y}_{t+k}$ is close to the target T in some sense.

There are issues related to the magnitude of the adjustments that are important in practice. For example, some control schemes may recommend nonzero control actions (adjustments) even if the process is close to target. Most process operators will be reluctant to make miniscule adjustments to a process too frequently. In other processes, varying a controllable factor over a wide range of values from observation to observation may be needed if interest is only in the quality characteristic, but this is not always practical. Consider, for example, changing the temperature in a furnace. Evidently, frequent large temperature adjustments are not feasible since the furnace takes time to cool down and warm up.

If making an adjustment is expensive or frequent adjustments not feasible, it seems reasonable to wait until there is a more significant deviation from target in order to adjust. These *deadband policies* can be shown to be optimal (Box and Jenkins, 1963; Box and Kramer, 1992; Crowder, 1992; Jensen and Vardeman, 1993) if besides quadratic off-target costs, there is a fixed adjustment cost, independent of the magnitude of the adjustment, or considerable adjustment error. Some of these cost issues are considered in detail in Chapter 5.

Hunter (1994) illustrates with an example when an engineering process control (EPC) scheme should be used and when an SPC scheme should be used. Consider the following stochastic model:

$$Z_t = \phi Z_{t-1} + v_t$$
$$Y_t = Z_t + a_t \tag{1.9}$$

where $\{v_t\}$ and $\{a_t\}$ are two white noise sequences uncorrelated with each other and Y_t denotes the deviation from target of the quality characteristic. The variable Z_t is not directly observable and models the *dynamic behavior* of the deviation from the target (i.e., in this case, it models how the process mean changes with time). We observe this dynamic behavior under the presence of measurement error (given by the a_t's). The parameter ϕ determines how fast the process "moves" with time; that is, it defines the process dynamics. As shown in the following chapters, ϕ can be related to what is called a *first-order dynamical system* according to the relation[19]

$$\phi = \exp\left(\frac{-\Delta t}{T_c}\right)$$

where Δt is the time between samples. The variable T_c denotes the *time constant* of the dynamical system, about two-thirds of the steady-state or long-run level of Z_t. The shorter T_c is, the faster the process dynamics are (i.e., the faster the quality characteristic reacts to changes in the controllable factor). Clearly, if

$$\Delta t \gg T_c \quad \text{then} \quad \phi \to 0 \quad \text{and} \quad Y_t = v_t + a_t \equiv \varepsilon_t$$

which is just Shewhart's model. This means that if sampling is slow relative to the process dynamics, the process observed will be uncorrelated and a SPC chart will work adequately. Alternatively, if

$$\Delta t \ll T_c \quad \text{then} \quad \phi \to 0 \quad \text{and} \quad Z_t = Z_{t-1} + v_t$$

In this case, $Y_t = Z_t + a_t$ is the sum of a random walk and measurement noise.[20] Sampling rapidly results in an observed nonstationary process, and an SPC chart will be inappropriate. An adjustment strategy (EPC) based on predictions of Y_{t+l} is preferred in this case. However, an important time-series analysis result is that reducing the sampling rate of an nonstationary process still results in a nonstationary process (MacGregor, 1976). Reducing the sampling frequency for SPC purposes is useful only if the original process is stationary. By a similar argument, if in model (1.9) the observational noise (measurement error) $\{a_t\}$ dominates $\{Z_t\}$, an SPC chart will be adequate again over an EPC strategy. Figure 1.20 summarizes when each SPC and EPC apply better.

Some arguments have been raised against the use of EPC schemes in quality control. In their automatic form, an *automatic process controller* (APC) may mask assignable causes that need further investigation. For example, it could be difficult to see if there is a problem with thermal insulation if a temperature controller is acting on a machine. Here an assignable cause (e.g., a broken insulation seal) will not be fixed, and the operation of the process with the assignable cause will be more expensive than if the insulation problem had been detected and corrected.

In addition, large process changes may not be controllable by an EPC scheme, since the controller may become "saturated" before compensating

---

[19] Process dynamics concepts are explained in Chapters 2 and 4. As described in Chapter 4, model (1.9) is a state-space model known in the time-series literature as the *steady model*.
[20] A sum that results in a correlation pattern identical to that of an IMA(1, 1) process (see Chapter 3).

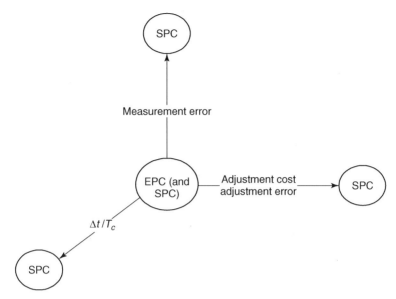

Figure 1.20 When SPC and EPC should be applied to a process. (Adapted from Hunter, 1994.)

completely for the disturbance. These issues can be resolved by an integrated EPC–SPC approach in which the SPC scheme acts as a supervisor of the EPC scheme. This is represented in the origin of the diagram in Figure 1.20. Note that the figure implies that *EPC may be needed or not, but SPC is always necessary*.

The distinction of when to use EPC and SPC tools is not as clear-cut as Figure 1.20 implies. There are cases when EPC techniques can be used under conditions when SPC is recommended by the figure. Some of these cases are discussed in Chapters 6 and 7.

1.6 COMBINED EPC–SPC APPROACHES

A considerable amount of work has appeared in the literature during the 1990s about methods that combine EPC and SPC schemes for the same process in an integrated form. The rationale is that large, abrupt changes in a process still need to be detected and corrected as soon as possible, since an EPC scheme will typically have a hard time compensating against them. With this we have the advantages of closer control to target with a smaller variance even if there is autocorrelation due to process or noise disturbances (EPC benefit). Furthermore, a continuous improvement effort is put into place through the detection and removal of assignable causes of variation that result in large, unpredictable changes (SPC benefit).

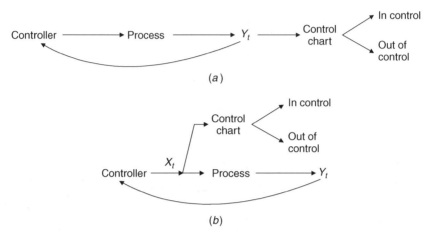

Figure 1.21 SPC–EPC integration possibilities: (*a*) SPC applied to the quality characteristic; (*b*) SPC applied to the process adjustments.

In an integrated EPC–SPC approach, it is not clear what to monitor with an SPC chart. There are at least two possibilities, illustrated in Figure 1.21.

1. *Monitor the quality characteristic.* Here the chart signals if large deviations from target are observed. The output of a feedback-controlled process may still be correlated (see, e.g., Chapter 5), and monitoring for abrupt changes in autocorrelated data becomes an important issue. [See Basseville and Nikiforov (1993) for a comprehensive treatment.]

2. *Monitor the adjustable factor.* The rationale is that a large change in the quality characteristic will result in large adjustments made, so the adjustments should have information we can use for monitoring the process. The adjustments themselves may be autocorrelated, and the comments made above about SPC and autocorrelated data also apply here. Furthermore, there will be a delay between the observed quality characteristic and the computed adjustment that should be considered.

The first of these approaches (monitoring the quality characteristic) was studied by Vander Weil et al. (1992) and Tucker et al. (1993) under the name *algorithmic statistical process control* and illustrated their approach with a batch polymerization example. Montgomery et al. (1994) also considered monitoring the quality characteristics. Application of this approach to a continuous polymerization process was reported by Capilla et al. (1999). Some of these applications are discussed in Chapter 5.

Example 1.12: Monitoring a Controlled Process. Consider the data in the file *CNC.txt*, which relates to a CNC lathe machine. Cylindrical metal pieces are produced in this operation and the quality characteristic (Y_t) is the

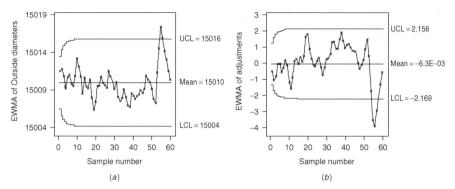

Figure 1.22 EWMA control charts used to monitor a controlled process: (*a*) SPC chart applied to the quality characteristic; (*b*) SPC chart applied to the adjustments.

outside diameter of the metal cylinders. The machine set point (aimed-at value) X_t is adjusted according to the rule $X_t - X_{t-1} = 0.3(T - Y_t)$, where T is the target diameter for the part and equals 15,010 units, where 1 unit = 0.0001 inch. Sixty parts were processed. Figure 1.22 shows EWMA control charts applied to the quality characteristics and to the adjustments. A value of $\lambda = 0.2$ was used in these charts. Observe how each chart detects an abnormal process behavior around part 55. In this example, the adjustments $\{X_t - X_{t-1}\}$ and the quality characteristic $\{Y_t\}$ are simply related (there is a strong negative correlation between these two processes, and this results in charts that work almost identically). However, there might be cases where the input–output relation is more complex, and then it will make a difference if we monitor the adjustments or the quality characteristics. Of particular importance are input–output delays, perhaps because measurements are taken in an offline laboratory after sampling. When delays are significant, an assignable cause will appear delayed in the adjustments series with respect to the quality characteristic series, and this needs to be taken into account.

□

Combinations of the two basic integrated approaches are also possible. For example, Tsung and Shi (1999) propose joint monitoring of the quality characteristic and the controllable factor of a feedback-controlled process using multivariate SPC techniques.

We can summarize by saying that the goals of having an SPC chart added to an EPC scheme include (Faltin et al., 1993) the following:

1. Verify the adequacy of the adjustment rule.

2. Help identify the causes of changes in performance.

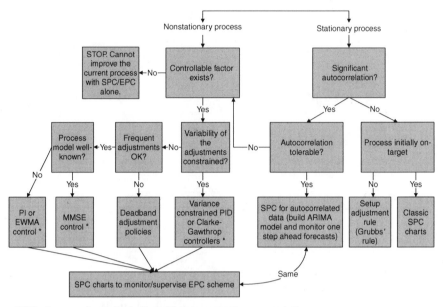

Figure 1.23   Guide to the use of EPC and SPC techniques discussed in this book.

We close this chapter with an overview of the various univariate SPC and EPC techniques described in this book and when each should be used (Figure 1.23). Additional multivariate adjustment schemes are discussed in Chapter 9.

PROBLEMS

1.1. Consider the machined parts data of Example 1.8 (see the file *Fadal.xls*). Using Minitab's regression options, fit a linear trend model $Y_t = a + bt + \varepsilon_t$ and a random walk model $Y_t = Y_{t-1} + \varepsilon_t$ to the data. Which model fits better? Try fitting the quadratic model $Y_t = a + bt + ct^2 + \varepsilon_t$ instead. Does this provide better fit than the random walk model?

1.2. Consider again the data in example 1.8 and the model $Y_t = Y_{t-1} + X_{t-1} + \varepsilon_t$, where X_t is the value aimed at (the set point). Suppose that the machine is adjusted to the target every period by setting it at $X_t = T - Y_t$ where $T = 0.975$ is the target. Draw a time-series plot for the controlled process $\{Y_t\}$.

1.3. In Problem 1.2, suppose that after observation 25, a sustained shift of magnitude 0.01 occurs in the quality characteristic. Plot the time series of the uncontrolled dimensions (as in Example 1.8) and for the controlled dimensions as in Problem 1.2.

1.4. For the step disturbance and process in problem 1.3, use a Shewhart 3σ chart applied to $\{Y_t\}$. Assume that the data are sample means with a sample size of 5.

1.5. Repeat Problem 1.4 using an EWMA chart instead with parameters $\lambda = 0.2$ and $k = 3$.

1.6. Repeat Problem 1.4, using a CUSUM chart with parameters $H = 4$, and $K = 0.005$.

1.7. Modify the spreadsheet *Deming-funnel.xls* to simulate the four rules considered by Deming under a process that obeys the model

$$Y_t = 0.5\, Y_{t-1} + X_{t-1} + \varepsilon_t$$

where $\{\varepsilon_t\}$ is a white noise sequence. (You can write your own simulation instead.) Which of the four rules seems to work best?

1.8. *Fifth rule for Deming's funnel.* Consider the second rule studied by Deming in his funnel experiment. Suppose that instead of adjusting by the full deviation from target, we adjust the funnel by only a fraction of the deviation, namely, the adjustment is $-G\, Y_t$. Simulate the performance of this adjustment rule when $G = 0.1$. Note how the adjustment rule is similar to an EWMA of the deviations.

1.9. Repeat Problem 1.8 assuming that the process is as described in Problem 1.7.

1.10. Show that if the conditions in equation (1.1) are true, then $Y_t \sim (\mu, \sigma^2)$, and $\mathrm{Cov}(Y_t, Y_{t+j}) = 0$ for $j \neq 0$.

1.11. Derive equation (1.3).

1.12. Show that the variance and expected value of the EWMA statistic are as in formulas (1.4) and (1.5).

1.13. Derive the equation for the standard deviation of the run length distribution of a Shewhart chart.

1.14. Compute the average run lengths of a $\bar{Y}$ chart for ($n = 5$, $k = 2.5$) and for ($n = 3$, $k = 3$). What effects do you notice when changing such chart design parameters?

1.15. Verify that equation (1.12) is obtained from (1.11) if $\rho_v = 0$ for $|v| > q$ (see Appendix 1B).

1.16. Show that if $\lambda = 1$, the autocorrelation function for process (1.7) is constant and equal to 1 for all lags.

1.17. Consider a stochastic process $\{Y_t\}_{t \geq 0}$ where each possible realization of the process i has a constant (in time) mean $\mu^{(i)}$ that can vary from realization to realization. Assume that the autocovariance function is constant in time and equal for all realizations. Is this process weakly stationary? Is it ergodic? Why?

BIBLIOGRAPHY AND COMMENTS

The book by Shewhart (1931) should be consulted by any person seriously interested in the main ideas behind statistical process control charts. Two modern textbooks with a wealth of information and extensive bibliographies on SPC topics are Alwan (2000), and Montgomery (2001). Properties of EWMA control charts are discussed in Lucas and Saccucci (1990). There are many books on stochastic processes, but results relating to time-series processes of the type discussed in this book are summarized in the classic books by Bartlett (1955) and Box and Jenkins (1970). The book by Siegmund (1985) provides key technical results useful for the computation of the ARLs of CUSUM charts. The book by Deming (1986) was read widely by corporate America and was very influential. This book contains an explanation of the funnel experiment and some quite radical statements *against* process adjustment. Two very important papers that anticipated much of the current discussion on adjustment methods and quality control were those of Barnard (1959) and Box and Jenkins (1963). MacGregor (1988) provides a very nice comparison between SPC and EPC methods. Further discussion of this type of integration can be found in the discussion to the paper by Box and Kramer (1992) in Box and Luceño (1997) and in del Castillo, et al. (2000).

APPENDIX 1A: STATIONARITY AND ERGODICITY

Time-based estimates $\bar{Y}$, $\hat{\sigma}^2$, c_k, and r_k converge, in the mean-square-error sense, to their ensemble expectations (μ, σ^2, γ_k, and ρ_k) provided that the process is *ergodic*. Consider, for example, the case of the process mean. For an ergodic process we have that

$$\mu_t = E[Y_t] = \int_{-\infty}^{\infty} y_t f(y_t) \, dY_t = \lim_{N \to \infty} \frac{1}{N} \sum_{t=1}^{N} y_t$$

where the quantity on the left is the ensemble mean and the quantity on the right is the long-term value of a sample function (the "sample" mean or average). In other words, if we can observe several realizations of an ergodic process and for some point in time t we average *across* the ensemble, the average so obtained will tend to μ_t as the number of realizations increases. This limit value is the ensemble mean. Thus for an ergodic process, the average computed from a single realization (sample average) tends to the ensemble mean as the length of the realization tends to infinity. A similar ensemble interpretation can be given of higher population moments.

The ergodicity conditions are quite technical.[21] If $\bar{Y}$ converges in mean square to μ, we say that the process $\{Y_t\}$ is *ergodic for the mean*, and so on for other sample functions and population parameters. If a process is weakly stationary and, in addition,

$$\lim_{N \to \infty} \frac{1}{N} \sum_{j=1}^{\infty} |\rho_j| = 0 \qquad (1.10)$$

then $\{Y_t\}$ is ergodic for the mean. The additional condition (1.10) means that the autocorrelations should eventually decay to zero for observations separated by longer and longer lags. As it turns out, for all stochastic processes studied in this book, weak stationarity implies condition (1.10), so it also implies ergodicity. Therefore, to perform valid statistical inferences based on single realizations, we require only weak stationarity.

APPENDIX 1B: STANDARD ERROR OF SAMPLE AUTOCORRELATIONS

For a weakly stationary, normally distributed process, Bartlett (1946) showed that the standard error of the sample autocorrelation is given approximately by

$$\sigma(r_k) = \sqrt{\text{Var}(r_k)} \approx \sqrt{\frac{1}{N} \sum_{v=-\infty}^{+\infty} \left(\rho_v^2 + \rho_{v+k}\,\rho_{v-k} - 4\rho_k\,\rho_v\,\rho_{v-k} + 2\rho_v^2\rho_k^2 \right)}$$

$$(1.11)$$

which implies that for increasing autocorrelation, the point estimates of the autocorrelations, $\hat{\rho}_k = r_k$, become less reliable. However, the estimates are consistent, so a large N (i.e., a long series) reduces the standard error. As a rule of thumb, N should be at least 75 to 100 observations.

---

[21] See Hamilton (1994, Chap. 3) and Fuller (1996, Chap. 6) for more details.

If in (1.11) we have $\rho_v = 0$ for $|v| > q$, all terms on the right are zero when $k > q$ except the first term, and we obtain

$$\sigma(r_k) = \sqrt{\frac{1}{N}\left(1 + 2\sum_{v=1}^{q} \rho_v^2\right)} \qquad \text{for } k > q. \qquad (1.12)$$

This result will prove useful in our discussion of moving average models (see Chapter 3). An estimate of this standard error is

$$\hat{\sigma}(r_k) = \sqrt{\frac{1}{N}\left(1 + 2\sum_{v=1}^{q} r_v^2\right)} \qquad \text{for } k > q.$$

If the process is a white noise sequence, then $q = 0$, and

$$\sigma(r_k) = \frac{1}{\sqrt{n}} \quad \text{and} \quad E[r_k] = \rho_k = 0 \qquad \text{for all } k.$$

Computing the standard error of the estimates is useful for distinguishing between significant estimated autocorrelations and nonsignificant estimated autocorrelations that are due to noise in the sampling procedure. These computations are part of the output of most time-series software packages.

The need for stationarity can also be seen from these standard error formulas. For an autocorrelated process, the point estimates will become less reliable as the autocorrelation increases. An extreme case is a nonstationary process which has an autocorrelation function that does not "decay" [see equation (1.10)], and this means that the standard error will be very large and no meaningful statistical inferences can be drawn.

APPENDIX 1C: AVERAGE RUN LENGTH AS A PERFORMANCE MEASURE

An important result in control chart design is that for geometric random variables Y_1 and Y_2 with

$$\text{ARL}_1 \leq \text{ARL}_2$$

we have that

$$P\{Y_2 \leq y\} \leq P\{Y_1 \leq y\} \qquad \text{for all } y \geq 0$$

(i.e., the statement above means that Y_1 is *stochastically larger* than Y_2). Therefore, we can use only the ARLs as a means of comparison between two or more charts with geometric-distributed run lengths. This is true only for Shewhart charts with known parameters. If parameters need be estimated, the run length distribution of an $\bar{Y}$ chart is not geometric but has a complicated form (Quesenberry, 1997; del Castillo, 1996). In such a case, higher-order moments of the run length distribution are necessary for comparing alternative chart designs.

CHAPTER 2

Modeling Discrete-Time Dynamical Processes

To adjust a process that may drift off-target if left uncontrolled, it is necessary to understand how any potential adjustments will affect the quality characteristics. It is also important to understand the drifting behavior of the process when running uncontrolled. Such understanding is based on modeling the dynamic behavior of a process or system. Therefore, some basic notions from the area of dynamical systems are needed to analyze the time-series problems we wish to consider in this book. In this chapter we present an overview of the types of problems studied in later chapters and some of the underlying mathematical tools that will be necessary to solve them. As shown in Chapters 3 and 4, the type of ARIMA and transfer function models we wish to study are in principle based on difference equations, the discrete-time counterpart of differential equations. In this chapter a comparison is made between dynamic models based on difference equations and the classical static models studied in other areas of applied statistics (e.g., regression models). Linear difference equation models are then presented as a generic tool to approximate the dynamic behavior of a variety of real-life industrial processes. Useful concepts related to the stability of a deterministic, linear dynamic system are mentioned briefly in the final section. The chapter includes appendices where more technical material related to some of the tools used in the analysis of discrete-time dynamical processes is discussed in more detail. These appendices include details about backshift operators and about z-transforms.

2.1 THREE IMPORTANT PROBLEMS

The types of problems we consider in the remainder of the book, in increasing level of complexity, are time-series analysis problems, transfer function problems, and process adjustment (control) problems. In all cases, a production process generating the quality characteristics of interest will be regarded as a *black box*. This means that only the variables that enter the process or that are generated by the process can be used for modeling purposes. Models of the internal mechanics of a process, based on first principles (chemical and physical laws) are not considered in this book. The goal will be to build statistically valid models of black box processes. The utility of such an approach cannot be underestimated. As technology advances, there is a need for rapidly characterizing new production processes. Modeling a process based on first principles is usually time-consuming and expensive. Instead, an empirical, data-based, or statistical approach to modeling is necessary. With the wide availability of sensor technologies and large databases, this approach seems more justified than ever before. We now introduce each of the problems cited in more detail.

2.1.1　Time-Series Analysis Problems

In the case of time-series analysis problems, the black-box process generates a time series $\{y_t\}$ that can be scalar- or vector-valued. With the exception of the error terms $\{\varepsilon_t\}$, we adopt the common notation of using uppercase letters to denote random variables and lowercase letters to denote actual values taken by a random variable after a random experiment is conducted.

A time-series process is supposed to be driven only by random errors or *shocks*, ε_t, that enter the system at every period. Taken as a whole, the set $\{\varepsilon_t\}_{t=-\infty}^{\infty}$ forms a white noise sequence. The output of the system is the time series, and the main problem is to characterize the properties of $\{Y_t\}$. For example, if y_t denotes the sales of a product for a given week, a company may be interested in forecasting future sales. The block diagram[1] in Figure 2.2 represents what control engineers refer to as a *filter*. This can be seen if the directions of the arrows are reversed. Then the inverse of the function H (if this exists) can be thought of as acting on the correlated process $\{Y_t\}$ to remove or filter out all autocorrelation and produce a white noise sequence $\{\varepsilon_t\}$. Time-series models are studied in Chapter 3. The goal of these models is to capture the dynamic behavior of a process. The dynamics result in autocorrelated observations. These models are useful to represent disturbances (i.e., noise) that act on a process we wish to adjust.

Figure 2.1　Time-series analysis problem.

---

[1] In this book, basic use of block diagrams is made. We introduce the basic ideas behind this type of diagram in Appendix 2A.

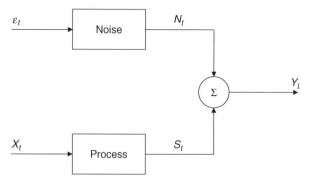

Figure 2.2 Transfer function with noise process.

2.1.2 Transfer Function Analysis Problems

In a transfer function process, there is a variable X_t that we believe has an effect on the output process $\{Y_t\}$. The controllable factor X_t can be scalar- or vector-valued, although in this book, for the most part, we focus on single-input, single-output (SISO) systems. The output is driven by a controllable factor or input, and the main problem is to determine the relation between the input and the output in the presence of noise (see Figure 2.2).

An additive model of the form

$$Y_t = S_t + N_t$$

is assumed in this book, where S_t is a signal that is a function of X_t (the transfer function proper) and N_t is a possibly correlated noise process that is driven by a white noise sequence $\{\varepsilon_t\}$. Thus the noise can be seen as a time-series subsystem with white noise at the input and correlated noise at the output. The process is specified such that if $X_t = 0$ for all t, the output equals the noise disturbance.

A typical example of a transfer function problem is in the area of econometrics, where a model is needed for some econometric variable (say, quarterly GNP) as a function of another variable (say, quarterly inflation). In a transfer function problem, the goal is not so much to control the response Y_t, but rather, to establish the input–output relation, usually to use the model for prediction of future response values. For our purposes, these input–output relations are used in conjunction with time-series noise models in order to devise adjustment policies.

2.1.3 Process Adjustment Problems

In process adjustment problems, the system is also composed of a transfer function with noise process. The main difference with the above is that it is now possible to adjust the input of the process (X_t), and the main problem is

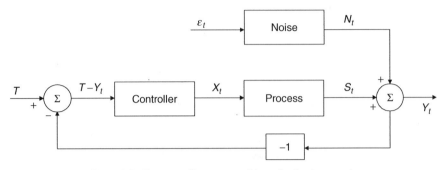

Figure 2.3 Process adjustment problem: feedback control.

to find a control rule or controller (see Section 1.5) that determines how to change the X_t's to achieve a desired performance of the output $\{Y_t\}$. In the applications described in this book, the output will be a quality characteristic and the input will be some controllable factor that can be manipulated, perhaps at some cost (see Figure 2.3). The goal is to keep the quality characteristic close to a target or set point T in some well-defined sense. The controller takes the deviations from target $T - Y_t$ as input and provides the level of the controllable factor for use as output at the next period of time. Note that the value of the quality characteristic Y_t is fed back to the controller from period to period.

To analyze a feedback process adjustment system statistically, it will be necessary to start with time-series problems and then study discrete-time transfer function models. It should be pointed out that as explained in Sections 1.5 and 1.6, SPC monitoring devices should be used in conjunction with an EPC scheme to achieve continuous improvement.

2.2   DYNAMIC VERSUS STATIC MODELS

In some areas of applied statistics (e.g., regression, design of experiments) the relation between the inputs and the outputs of a system is of the form

$$Y = f(\mathbf{X}) + \varepsilon$$

where ε is a random error and f is a function of the regressor vector $\mathbf{X}$. For example, if f is a linear function and there is only one regressor, we could be interested in a first-order polynomial model of the form

$$Y = \beta_0 + \beta_1 X + \varepsilon$$

where β_0 and β_1 are parameters. This is a *static* model in the sense that the dependent variable is a function of regressors and random variables that occur at the same point in time or independent of time, and this is why the subscript t is not shown. Time-series models are usually dynamic in that they involve noncontemporaneous relationships between variables. For example,

$$Y_t = \beta X_{t-2} + \varepsilon_t$$

which means that the effect of X is felt on Y only after two time periods, a period of time called the *system input–output delay* or *dead-time*. Alternatively, the effect of the controllable factor on the response can be felt over several periods. For example, consider the dynamic model

$$Y_t = \beta_1 X_{t-1} + \beta_2 X_{t-2} + \beta_3 X_{t-3} + \varepsilon_t.$$

In this model, β_i is the average change in the response i periods after we changed X_t by a unit impulse. That is, if at time t we set $X_t = 1$ but we set $X_j = 0$ for any other time $j \neq t$, the β's above model the response to this impulse. We discuss impulse responses in more detail when we take a closer look at transfer functions in Chapter 4.

Dynamics can also occur among the observed response or output values:

$$Y_t = \phi Y_{t-1} + \beta X_{t-1} + \varepsilon_t$$

where Y_t depends on its previous value. Finally, we can also have a dynamic relation due to the accumulation of random shocks:

$$Y_t = \beta X_{t-1} + \theta \varepsilon_{t-1} + \varepsilon_t$$

where Y_t depends on the previous and current random errors or shocks. In general, a discrete-time stochastic dynamical system model has the form

$$Y_t = f(Y_{t-1}, Y_{t-2}, \ldots; \; X_t, X_{t-1}, X_{t-2}, \ldots; \varepsilon_t, \varepsilon_{t-1}, \ldots)$$

In subsequent chapters we consider only the case where the function f is linear and time-invariant.

2.3 DIFFERENCE EQUATION DYNAMICAL MODELS[2]

Consider a deterministic dynamical system described by the difference equation

$$Y_t = a_1 Y_{t-1} + a_2 Y_{t-2} + \cdots + a_{n_a} Y_{t-n_a} + b_0 X_t + b_1 X_{t-1} + \cdots + b_{n_b} X_{t-n_b}.$$

To manipulate difference equations, it is very convenient to use some

---

[2]Sections 2.3 and 2.4 contain relatively more advanced material and may be skipped on a first reading.

operators that act on variables that are members of a given sequence. Define the *backward shift or backshift operator*[3] by

$$\mathcal{B}^k[Y_t] \equiv Y_{t-k}$$

which, for short, we write

$$\mathcal{B}^k Y_t = Y_{t-k}$$

The backshift operator will allow us to manipulate and analyze difference equations using simple algebraic operations. Details about the calculus of backshift operators are given in Appendix 2B. The difference equation can now be written as

$$\left(1 - a_1\mathcal{B} - a_2\mathcal{B}^2 - \cdots - a_{n_a}\mathcal{B}^{n_a}\right)Y_t = \left(b_0 + b_1\mathcal{B} + \cdots + b_{n_b}\mathcal{B}^{-nb}\right)X_t.$$

The terms in parentheses are polynomials in the backshift operator:

$$A(\mathcal{B}) = 1 - a_1\mathcal{B} - a_2\mathcal{B}^2 - \cdots - a_{n_a}\mathcal{B}^{n_a}$$

$$B(\mathcal{B}) = b_0 + b_1\mathcal{B} + \cdots + b_{n_b}\mathcal{B}^{-nb}.$$

Thus the difference equation can be written as

$$A(\mathcal{B})Y_t = B(\mathcal{B})X_t. \tag{2.1}$$

Similarly, we define the forward shift operator, $\mathcal{F}$, as

$$\mathcal{F}^k[Y_t] \equiv Y_{t+k}$$

which we write $\mathcal{F}^k Y_t = Y_{t+k}$. As it turns out (see Appendix 2B), the forward shift operator is, algebraically speaking, the inverse of the backshift operator; that is, $\mathcal{F}^{-1} \equiv \mathcal{B}$. Because of this, we can also write the difference equation as

$$A^*(\mathcal{F})Y_t = B^*(\mathcal{F})X_t$$

where

$$A^*(\mathcal{F}) = \mathcal{F}^{n_a}A(\mathcal{B}) = \mathcal{F}^{n_a} - a_1\mathcal{F}^{n_a-1} - \cdots - a_{n_a}$$

$$B^*(\mathcal{F}) = \mathcal{F}^{n_b}B(\mathcal{B}) = b_0\mathcal{F}^{n_b} - b_1\mathcal{F}^{n_b-1} - \cdots - b_{n_b}.$$

---

[3]Also called the *delay operator* in the control engineering literature and the *lag operator* in the econometrics literature. We use the notation and terminology of Box and Jenkins (1976).

The above defines the relation between the reciprocal polynomials A and A^* and B and B^*.

The ratios

$$\frac{B(\mathcal{B})}{A(\mathcal{B})} = H(\mathcal{B})$$

$$\frac{B^*(\mathcal{F})}{A^*(\mathcal{F})} = H^*(\mathcal{F})$$

are called the *transfer function*[4] of the system described by the difference equation. The polynomial $A(\mathcal{B})$ [or alternatively, $A^*(\mathcal{F})$] is called the *characteristic polynomial* because it determines (characterizes) the stability properties of the system. The equation $A(\mathcal{B}) = 0$ [or alternatively, $A^*(\mathcal{F}) = 0$] is called the *characteristic equation of the system*. Its roots determine the stability of the system, as we discuss next.

2.4 STABILITY OF A DIFFERENCE EQUATION MODEL

There are several definitions of stability in the control engineering literature. In the most general case, that of a nonlinear dynamical system, stability is a property of the solutions of the difference equations describing the system. For linear time-invariant systems such as those considered in this book, stability of a solution implies stability of all possible solutions, so stability is a property of the system. Two useful notions of stability are *asymptotic stability* and *bounded input–bounded output* (BIBO) *stability*. A dynamic system is said to be *asymptotically stable* if there exist some initial condition(s) such that as time tends to infinity, the solution to the difference equation converges to a finite, constant value called the *steady-state value*.

A linear time-invariant dynamic system is said to be *BIBO stable* if a bounded (finite) input signal results in a bounded output signal (Figure 2.4). Asymptotic stability implies BIBO stability, but the reverse is not always true in general. Hereafter, *stability* will mean BIBO stability unless otherwise stated.

For systems defined by a linear time-invariant difference equation, the roots of the characteristic equation (also called the *poles* of the transfer function) determine the stability of the system.[5] A system is stable if and only if the roots of the characteristic equation

$$A(\mathcal{B}) = 0$$

[4] In the discrete-time case we are considering, this is sometimes called the *pulse transfer function*, but we will use the term *transfer function* for short.
[5] The reason for this involves z-transforms. See Appendix 2C for a derivation of this result.

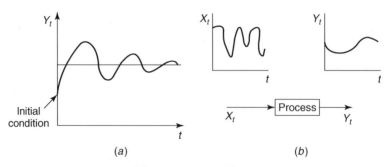

Figure 2.4 (*a*) Asymptotic and (*b*) BIBO stability.

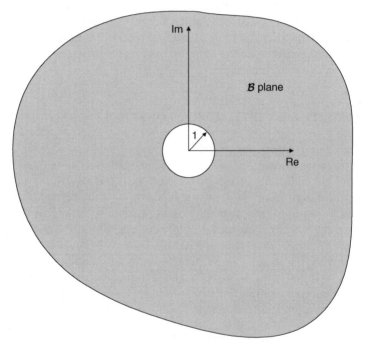

Figure 2.5 Stability region on the $\mathcal{B}$ complex plane (outside unit circle). Im, imaginary axis; Re, real axis.

are all strictly *outside* the unit circle[6] on the complex plane $\mathcal{B}$ (Figure 2.5). Equivalently, if and only if all roots of the characteristic equation

$$A^*(\mathcal{F}) = 0$$

are strictly *inside* the unit circle on the complex plane $\mathcal{F}$, the system is stable (see Figure 2.6).

---

[6] Here $\mathcal{B}$ is regarded as a complex variable instead of as an operator as we regarded it earlier. Some authors use different notations for shift operators (say, $\mathcal{B}$) and the complex variables needed to analyze stability (say, z^{-1}), but (following Box and Jenkins) we do not use that more formal convention in this book. For more details about this issue, see Appendix 2C.

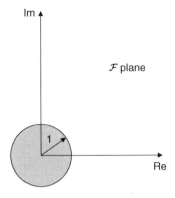

Figure 2.6 Stability region on the $\mathcal{F}$ complex plane (inside unit circle). Im, imaginary axis; Re, real axis.

Example 2.1: Stability of a Simple Difference Equation Model. Consider the difference equation model

$$Y_t = \phi Y_{t-1} + X_{t-1}$$

where $\phi = 2$. Then the characteristic equation of the system is

$$A(\mathcal{B}) = 1 - 2\mathcal{B} = 0$$

which has a single root at $\mathcal{B} = 0.5 < 1$. Since the root is inside the unit circle on the $\mathcal{B}$ plane, the system is unstable. Alternatively, $A^*(\mathcal{F}) = \mathcal{F} - 2 = 0$ has its single root at $\mathcal{F} = 2 > 1$; thus the system is unstable.

To illustrate this equation numerically, suppose we have that $Y_0 = 0$ and $X_t = 1$ for all time periods t. Then following the difference equation $Y_t = 2Y_{t-1} + X_{t-1}$, we find that

$$Y_1 = 1$$

$$Y_2 = 2(1) + 1 = 3$$

$$Y_3 = 2(3) + 1 = 7$$

$$Y_4 = 2(7) + 1 = 15$$

$$Y_5 = 2(15) + 1 = 31$$

$$\vdots$$

This shows clearly that the sequence $\{Y_t\}$ will diverge. □

Jury's Test for Stability of a Polynomial
By extension with the definition of a stable system, if a polynomial in $\mathcal{B}$ has all its roots outside (respectively, all the roots of the reciprocal polynomial in $\mathcal{F}$ are inside) the unit circle, we say that the polynomial is stable. Jury

developed a simple test[7] for assessing whether or not a polynomial has roots inside the unit circle. This will prove very useful in later chapters.

The test is as follows. Given a polynomial written in the forward shift operator,

$$A^*(\mathcal{F}) = a_0 \mathcal{F}^n - a_1 \mathcal{F}^{n-1} - \cdots - a_n$$

form the following table, where computations start on the third row of the table:

a_0	a_1	a_2	$\cdots$	a_{n-1}	a_n		
a_n	a_{n-1}	a_{n-2}	$\cdots$	a_1	a_0	let	$\alpha_n = \dfrac{a_n}{a_0}$
a_0^{n-1}	a_1^{n-1}	a_2^{n-1}	$\cdots$	a_{n-1}^{n-1}			
a_{n-1}^{n-1}	a_{n-2}^{n-1}	a_{n-3}^{n-1}	$\cdots$	a_0^{n-1}		let	$\alpha_{n-1} = \dfrac{a_{n-1}^{n-1}}{a_0^{n-1}}$
$\vdots$	$\vdots$	$\vdots$		$\vdots$			
a_0^0							

where

$$a_i^{k-1} = a_i^k - \alpha_k a_{k-i}^k$$

and

$$\alpha_k = \frac{a_k^k}{a_0^k}.$$

Thus the second row is just the first row with all elements written in reverse order, the fourth row equals the third with all elements written in reverse order, and so on. Subscripts refer to column numbers, and superscripts refer to numbers given to pairs of rows, with the number given to the first two rows (n) not shown. The computations are continued until we have a single element in the first column of a new row; every two rows the number of columns is decreased by one.

Necessary and sufficient conditions for all the roots of $A^*(\mathcal{F}) = 0$ to be inside the unit circle are that $a_0^k > 0$, for $k = 0, 1, \ldots, n - 1$. In a transfer function, this implies the stability of the system.

Example 2.2: Determining Stability Using Jury's Criterion. Let

$$Y_t = \phi_1 Y_{t-1} + \phi_2 Y_{t-2}$$

[7]For a proof of this result, see Åström (1970, pp. 118–121).

Then $A^*(\mathcal{F}) = \mathcal{F}^2 - \phi_1\mathcal{F} - \phi_2$ and $n = 2$. The corresponding Jury array is

1	$-\phi_1$	$-\phi_2$	$\alpha_2 = -\phi_2$
$-\phi_2$	$-\phi_1$	1	
$1 - \phi_2^2$	$-\phi_1(1 + \phi_2)$		$\alpha_1 = \dfrac{-\phi_1}{1 - \phi_2}$
$-\phi_1(1 + \phi_2)$	$1 - \phi_2^2$		
$1 - \phi_2^2 - \phi_1^2\dfrac{1 + \phi_2}{1 - \phi_2}$			

The following are *necessary* but not sufficient conditions for all roots to be inside the unit circle; one can check these before trying the Jury table:

$$A^*(1) > 0 \quad \text{and} \quad (-1)^n \, A^*(-1) > 0.$$

In the example, these necessary conditions result in $1 - \phi_1 - \phi_2 > 0$ (or $\phi_1 + \phi_2 < 1$) and $1 + \phi_1 - \phi_2 > 0$ (or $\phi_2 - \phi_1 < 1$), respectively. If for given ϕ_1 and ϕ_2 any of these two conditions does not hold, we can conclude that the process is unstable and stop here. Otherwise, we proceed with Jury's test.

To illustrate Jury's test, consider the computation of a_0^1, the first element (column zero) of the second pair of rows in the table above:

$$a_0^1 = 1 - (-\phi_2)(-\phi_2) = 1 - \phi_2^2$$

since $\alpha_2 = a_2/a_0 = -\phi_2$. The stability conditions indicate that the first element of every two rows should be nonnegative:

$$a_0^1 > 0 \quad \text{and} \quad a_0^0 > 0$$

In this case, from $a_0^1 = 1 - \phi_2^2 > 0$ we get

$$|\phi_2| < 1. \tag{2.2}$$

Also,

$$a_0^0 = 1 - \phi_2^2 - \phi_1^2\frac{1 + \phi_2}{1 - \phi_2} > 0$$

implies the two conditions

$$\phi_2 + \phi_1 < 1 \tag{2.3}$$

$$\phi_2 - \phi_1 < 1. \tag{2.4}$$

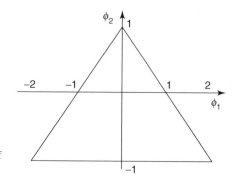

Figure 2.7 Stability region for coefficients of
a second-order polynomial.

As can be seen in Figure 2.7, conditions (2.2) to (2.4) define a triangular
region on the (ϕ_1, ϕ_2) plane. Pairs of values (ϕ_1, ϕ_2) inside the triangle of
the figure guarantee a stable dynamic behavior for this system.

We can illustrate the stability properties in this example with a couple of
numerical computations. Consider first the case when $\phi_1 = 0.1$, $\phi_2 = 0.5$
and we start the difference equation from $Y_0 = Y_1 = 1$. Then we will have,
solving the difference equation by simple substitution (these computations
are extremely easy with spreadsheet software):

$$Y_2 = 0.1(1) + 0.5(1) = 0.6$$

$$Y_3 = 0.1(0.6) + 0.5(1) = 0.56$$

$$Y_4 = 0.1(0.56) + 0.5(0.6) = 0.356$$

$$\vdots$$

and the sequence $\{Y_t\}$ will converge (to zero). Note that the parameter values
$\phi_1 = 0.1$ and $\phi_2 = 0.5$ satisfy the stability conditions (2.2) to (2.4).

As a second numerical illustration, suppose that we have $\phi_1 = 0.5$, $\phi_2 = 0.9$, and $Y_0 = Y_1 = 1$. Then, proceeding as before, we find that

$$Y_2 = 0.5(1) + 0.9(1) = 1.4$$

$$Y_3 = 0.5(1.4) + 0.9(1) = 1.6$$

$$Y_4 = 0.5(1.6) + 0.9(1.4) = 2.06$$

$$\vdots$$

and in this case the sequence $\{Y_t\}$ will diverge (to $+\infty$). Note how the
parameter values $\phi_1 = 0.5$ and $\phi_2 = 0.9$ satisfy conditions (2.2) and (2.4) but
do not satisfy condition (2.3). For the stability of this equation, all conditions
must be satisfied simultaneously.   □

PROBLEMS

2.1. Consider the difference equation $Y_t = 2 Y_{t-1} - Y_{t-2} + X_t$ and suppose that we fix the input at 10.0 for all periods t. Plot the solution (trajectory) of the process using a spreadsheet. Assume that $Y_0 = 0$.

2.2. *First-order process.* Consider a deterministic process described by the transfer function

$$Y_t = \frac{b\mathcal{B}}{1 - p\mathcal{B}} X_t$$

Using spreadsheet software, plot trajectories of this process if (a) $p > 1$; (b) $p = 1$; (c) $0 \le p \le 1$; (d) $-1 < p < 0$; (e) $p = -1$; (f) $p < -1$.[*Hint:* First write in difference equation form: $(1 - p\mathcal{B})Y_t = b\mathcal{B} X_t$.]

2.3. *Second-order process, real poles.* Consider a deterministic transfer function described by the model

$$Y_t = \frac{b_1\mathcal{B} + b_2\mathcal{B}^2}{(1 - p_1\mathcal{B})(1 - p_2\mathcal{B})} X_t$$

Using spreadsheet software, plot trajectories of this process if (a) p_1 and p_2 are real, both less than 1 in magnitude; (b) p_1 and p_2 are real, and $|p_1| = 1$; (c) p_1 and p_2 are real, and $|p_1| > 1$.

2.4. *Second-order process, complex poles.* Consider the same deterministic transfer function as in Problem 2.3. Using spreadsheet software, plot trajectories of this process if the two poles p_1 and p_2 are complex quantities and (a) $|p_1| < 1$ and $|p_2| < 1$; (b) $|p_1| = 1$; (c) $|p_1| > 1$.

2.5. Find the stability condition of $Y_t = \phi Y_{t-1} + X_{t-1}$ using Jury's test. Compare with Section 2.4.

2.6. Using computer algebra software such as Maple or Mathematica (or consulting an algebra book), find expressions for the stability conditions of the model $Y_t = \phi_1 Y_{t-1} + \phi_2 Y_{t-2} + \phi_3 Y_{t-3} + X_{t-1}$.

2.7. In Problem 2.6, if $\phi_1 = 0.8$, $\phi_2 = -0.4$, and $\phi_3 = 0.3$, is the system stable?

2.8. Solve Problem 2.6 working out Jury's test instead.

2.9. Show that if the norm of a stationary process $Y = \{Y_t\}_{t=-\infty}^{\infty}$ is defined as $\|Y\| = \max\{|Y_t|\}$, the shift operators also have unit norm.

2.10. Show that equation (2.6) in Appendix 2B follows from equation (2.5) if $|\phi| < 1$ and $t \to \infty$. What happens in these equations if $\phi = 1.0$?

2.11. Prove the final value theorem of z-transforms. [*Hint:* Use the definition of z-transform applied to $(1 - z^{-1})Y(z)$ and take limit as $z \to 1$ assuming that $Y_k = 0$ for $k < 0$.]

2.12. Obtain the solution to the difference equation in Problem 2.2 using z-transforms.

2.13. Obtain the solution to the difference equation in Problem 2.3 using z-transforms.

2.14. Find the values of ϕ_1 and ϕ_2 that will make the process $Y_t = \phi_1 Y_{t-1} + \phi_2 Y_{t-2}$ oscillate.

2.15. Can the process $Y_t = \phi Y_{t-1} + X_{t-1}$ oscillate sinusoidally?

BIBLIOGRAPHY AND COMMENTS

Many excellent books on discrete-time control discuss difference equation models and the use of the z-transform in great detail. Three of them are Ogata (1995), Åström and Wittenmark (1997), and Franklin et al. (1998). These books are tailored to control and electrical engineering applications. Books with a chemical engineering concentration that also discuss difference equations and z-transforms in detail are Seborg et al. (1989) and Ogunnaike and Ray (1994). Z-transforms were developed mainly by E. I. Jury in the United States and Y. Z. Tsypkin in the USSR. Åström and Wittenmark (1997) includes an abbreviated history of these developments.

APPENDIX 2A: ESSENTIALS OF BLOCK DIAGRAMS

In the type of modeling problems with which we are concerned, it is evident that different types of variables are transformed, added or multiplied, by a series of algebraic manipulations. *Block diagrams* are graphical representations of the information flows and transformations that affect these flows in the process under study. Block diagrams are common in many fields of engineering but are not that common in the statistics field. Here we sketch the basic ideas.

For processes observed only at discrete points in time, a block diagram shows the information available and the transformations that take place

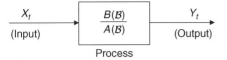

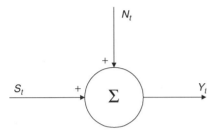

Figure 2.8 Block diagram of a deterministic transfer function.

Figure 2.9 Block diagram of a sum of variables.

sequentially at each instant of time t. Information flows (variables that are to be transformed or that result from some transformation) are represented by arrows with labels equal to the variable name. The transformations acting on these variables are represented by boxes or blocks, with labels indicating the nature (or name) of the transformation. Thus, in a simple transfer function model $Y_t = B(\mathcal{B})/A(\mathcal{B})X_t$, the corresponding block diagram is as shown in Figure 2.8.

Sometimes two or more variables need to be added or subtracted. For example, in a transfer function with noise model, the signal S_t is added to the noise N_t to give the measurement Y_t (i.e., $S_t + N_t = Y_t$). This has the block diagram illustrated in Figure 2.9, where the signs near the end of the incoming arrows indicate if each variable should be added or subtracted.

In discrete-time (digital) control textbooks, considerable emphasis is placed on sampling a continuous variable in order to discretize (digitilize) it and on reconstructing a continuous variable from sampled observations. The former operation is performed by the samplers that transform an analog measurement into a digital measurement (A/D conversion); the latter are performed by *sample and hold elements* (which perform D/A conversion). Both A/D and D/A converters have special block diagram representations. Similarly, measuring devices and actuators that implement control actions physically are also modeled as blocks since they receive or transform information within a control system.

In the types of applications in this book, A/D and D/A converters are not discussed. It will be assumed that the observations are given in discrete-time form, where the sampling is not necessarily done by a computer with A/D and D/A converters but by a quality engineer. Similarly, measurement will be assumed to take place offline, perhaps in a metrology station or lab. Finally, actuators are not modeled, as it is assumed that the process operator modifies the settings of a machine manually in order to adjust it. This process can be aided by using some type of controller designed using the methods of Chapters 5 to 9.

In this book, a very elementary use of block diagrams is followed in which we use only blocks and addition nodes, as in Figures 2.8 and 2.9. To develop a block diagram of a system, we should:

1. Identify the individual elements of the system (e.g., process, noise, controller, etc.).
2. Identify the input and output variables for each element.
3. Identify the transfer functions associated with each element.
4. Put together all elements (blocks) and input–output variables to get the final block diagram.

For more information on block diagrams, see, for example, Ogunnaike and Ray (1994, p. 464).

APPENDIX 2B: CALCULUS OF BACKSHIFT OPERATORS

Some properties of shift operators that are sometimes useful are reviewed here. For a fuller exposition at an introductory level, see Åström and Wittenmark (1997). The backshift operator is defined as

$$\mathcal{B}[Y_t] = \mathcal{B}Y_t = Y_{t-1}$$

The forward shift operator is defined as

$$\mathcal{F}[Y_t] = \mathcal{F}Y_t = Y_{t+1}$$

Since $\mathcal{B}\mathcal{F}Y_t = \mathcal{B}Y_{t+1} = Y_t$, we have that $\mathcal{B}\mathcal{F} = 1$, so it must be true that

$$\mathcal{B} \equiv \mathcal{F}^{-1}$$

To illustrate the use of the backshift operator applied to a difference equation, consider the simple difference equation

$$Y_t = Y_{t-1}$$

which we can write as

$$Y_t = \mathcal{B}Y_t.$$

Then we have, writing out the difference equation for different values of the time index t,

$$\vdots$$

$$Y_{-2} = \mathscr{B}Y_{-2} = Y_{-3}$$

$$Y_{-1} = \mathscr{B}Y_{-1} = Y_{-2}$$

$$Y_0 = \mathscr{B}Y_0 = Y_{-1}$$

$$Y_1 = \mathscr{B}Y_1 = Y_0$$

$$Y_2 = \mathscr{B}Y_2 = Y_1$$

$$\vdots$$

In other words, the operator is applied to all members of an infinite sequence. For the purposes of this book, such a sequence could be a deterministic dynamic system, a stochastic dynamic system, a stochastic process, or a time series.

If the norm of a sequence $Y = \{Y_t\}_{t=-\infty}^{\infty}$ is defined by the usual

$$\|Y\|^2 = \sum_{t=-\infty}^{\infty} Y_t^2$$

it follows that both the backshift and forward shift operators have unit norm since, clearly,

$$\|Y\|^2 = \sum_{t=-\infty}^{\infty} \mathscr{B}Y_t^2 = \sum_{t=-\infty}^{\infty} \mathscr{F}Y_t^2 = \sum_{t=-\infty}^{\infty} Y_t^2$$

because the sequence $\{Y_t\}$ is infinitely long.

The goal of operator calculus is to allow us to manipulate difference equations using regular algebra. The requirement of infinite sequences should be emphasized, because by using operators, *we are neglecting the initial conditions of the difference equations.* To see this, consider the equation

$$Y_t = \phi Y_{t-1} + X_{t-1}$$

where $|\phi| < 1$, implying that the system is stable. By repeated substitution of Y_{t-1} in the difference equation, we can solve it by going backward in time until time zero, where we assume, for illustration, that the system starts. If at time zero the process starts at Y_0, the solution to the difference equation is

$$Y_t = \phi^t Y_0 + \sum_{j=0}^{t-1} \phi^{t-j-1} X_j = \phi^t Y_0 + \sum_{i=1}^{t} \phi^{i-1} X_{t-i}. \tag{2.5}$$

In contrast, if we use backshift operators, the difference equation is

$$(1 - \phi\mathcal{B})Y_t = X_{t-1}$$

so

$$Y_t = \frac{\mathcal{B}}{1 - \phi\mathcal{B}}X_t.$$

Because $\mathcal{B}$ has unit norm, the right-hand side equals the limit of a convergent geometric series:

$$Y_t = \mathcal{B}(1 + \phi\mathcal{B} + \phi^2\mathcal{B}^2 + \phi^3\mathcal{B}^3 + \cdots)X_t = \sum_{i=1}^{\infty} \phi^{i-1}X_{t-i} \qquad (2.6)$$

Clearly, solution (2.5) equals solution (2.6) only if $t \to \infty$ (recall that $|\phi| < 1$). This implies a neglect of the initial conditions (and a neglect of the transient behavior due to the initial conditions).

In this book, operator calculus is used to solve difference equations assuming null initial conditions. When initial conditions need to be considered, other solution methods are used.

APPENDIX 2C: z-TRANSFORMS

When modeling continuous-time dynamical processes, it is customary to utilize differential equations and use Laplace transforms for their analysis and solution. Laplace transforms from the time domain to the s-domain allow us to solve differential equations by the solution of simpler algebraic equations in s. Similarly, to model discrete-time systems, we utilize difference equations and use *z-transforms* for their analysis. Analysis in the z-domain turns out to be much simpler algebraically than analysis in the time domain. As we will see, the z-transform is closely related to the backshift operator $\mathcal{B}$ used throughout this book.

Suppose that the response of interest of a process (e.g., its quality characteristic) varies through time according to some continuous-time function $Y(t)$. For process control purposes, every Δt time units a sample of product is taken and measured, resulting in the measurements $Y(k\Delta t)$ for $k = 0, 1, 2, \ldots$, which we can call Y_k following the notational convention of using subscripts for discrete-time functions. The z-transform of Y_k, denoted $\mathcal{Z}(Y_k)$, defines a new function in the z-argument:

$$\mathcal{Z}(Y_k) = Y(z) = \sum_{k=0}^{\infty} Y_k z^{-k} \qquad (2.7)$$

where it should be understood that the Y_k's on the right-hand side are given numbers (measurements), so $Y(z)$ is only a function of the variable z. Although $Y(z)$ is an infinite series, it can be written in closed form if Y_k is defined as a ratio of polynomials in $\mathcal{B}$ (this is the case of transfer functions).

Remark 2.1. For those familiar with Laplace transforms, note the analogy between z-transforms of the discrete-time function Y_k in equation (2.7) and the Laplace transform $\mathcal{L}$ of the continuous-time function $Y(t)$:

$$\mathcal{L}(Y(t)) = \int_0^\infty Y(t)\, e^{st}\, dt \tag{2.8}$$

It can be shown (see Seborg et al., 1989, p. 560; Ogunnaike and Ray, 1994, p. 850) that a z-transform can be derived by taking the Laplace transform of a discrete-time function Y_k and then making the change of variable

$$z = e^{s\Delta t}.$$

Remark 2.2. In other parts of this book it is assumed implicitly that $\Delta t = 1$; that is, the time between samples is considered the time unit. In such a case we can make $t = k\,\Delta t = k$, and we can write simply Y_t for Y_k.

Example 2.3: z-Transform of a Unit Step Function. Let us find the z-transform of the discrete unit step function:

$$Y_k = \begin{cases} 1 & \text{if } k = 0, 1, 2, \ldots \\ 0 & \text{if } k = -1, -2, \ldots \end{cases}$$

Substituting Y_k into (2.7), we get

$$Y(z) = 1 + z^{-1} + z^{-2} + z^{-3} + \cdots$$

If $|z^{-1}| < 1$, this infinite series converges to

$$Y(z) = \frac{1}{1 - z^{-1}}. \qquad \square$$

Example 2.4: z-Transform of an Exponential Function. Suppose that we wish to find the z-transform of the discrete-time function that results from sampling $Y(t) = C\, e^{t/T_c}$ every Δt time units. The function sampled is

$$Y_k = C\, e^{k\Delta t / T_c} \qquad k = 0, 1, 2, \ldots$$

where $\phi = e^{-\Delta t / T_c}$ is a constant that depends on the sampling interval. The z-transform is, from (2.7),

$$Y(z) = C \sum_{k=0}^\infty \phi^k z^{-k} = C \sum_{k=0}^\infty \left(\frac{z}{\phi}\right)^{-k}$$

which is an infinite geometric series that converges to

$$Y(z) = \frac{C}{1 - 1/\phi z^{-1}} = \frac{C}{1 - \phi z^{-1}}$$

provided that $|\phi z^{-1}| < 1$. This example explains the relation $\phi = e^{-\Delta t/T_c}$ used in Section 1.5. $\square$

Example 2.5: z-Transforms and Backshift Operators. Suppose that we wish to find the z-transform of the delayed time function $\mathcal{B}^m Y_k = Y_{k-m}$, where $Y_{k-m} = 0$ for $m > k$. We have that

$$\mathcal{L}(Y_{k-m}) = z^{-m} \sum_{j=0}^{\infty} Y_j z^{-j}$$

but from (2.7), this equals

$$\mathcal{L}(Y_{k-m}) = z^{-m} Y(z).$$

Thus we can see that the delay implied by the backshift operator $\mathcal{B}^m$ in the time domain corresponds in the z-domain to multiplying the z-transform of the original (nondelayed) function Y_k by z^{-m}. In other words, we have shown the equivalence (called the *real translation theorem*)

$$\mathcal{L}(\mathcal{B}^m Y_k) = z^{-m} Y(z) \quad \text{where} \quad Y(z) = \mathcal{L}(Y_k) \qquad \square$$

Uniqueness of z-Transforms
An important property of z-transforms is that Y_k and its z-transform $Y(z)$ constitute a unique pair. This means that from $Y(z)$ we can get Y_k applying the *inverse-z transform*, $\mathcal{L}^{-1}$, so we can shift back and forth between time-domain and z-domain. The use of z- and inverse z-transforms in the analysis of discrete-time processes is as follows:

process model in time domain $\Rightarrow \mathcal{L} \Rightarrow$ analysis in z-domain

$$\Rightarrow \mathcal{L}^{-1} \Rightarrow \text{results in time domain}$$

Final Value theorem. A useful result is that if the long-run or final value $Y_\infty = \lim_{k \to \infty} Y_k$ is finite, this value can be obtained from the z-transform:

$$Y_\infty = \lim_{k \to \infty} Y_k = \lim_{z \to 1} \left[(1 - z^{-1}) Y(z) \right]$$

See Problem 2.11.

Inversion of z-Transforms

The inverse z-transform, $\mathcal{Z}^{-1}$, takes a function in the z-domain, $Y(z)$, and returns a discrete-time-domain function:

$$\mathcal{Z}^{-1}(Y(z)) = Y_k \qquad k = 1, 2, 3, \ldots, \infty$$

Remark 2.3. Note from Example 2.5 that this implies that $\mathcal{B} = \mathcal{Z}^{-1}(z^{-1})$. Assuming that $\Delta t = 1$, we have that $\mathcal{B}Y_t = \mathcal{Z}^{-1}(z^{-1} \ Y(z))$, and for this reason some authors[8] do not use different notations for the backshift operator applied to the discrete-time-domain function Y_t, and for the z^{-1} variable, which evidently applies in the z-domain. The z variable is a complex variable because it is sometimes necessary to use complex numbers in order to analyze the convergence of the series in (2.7). With it, we can make *formal* algebraic manipulations to difference equations.

Example 2.6: Formal Manipulations. Consider the difference equation

$$Y_t = \phi Y_{t-1} + X_{t-1}$$

Using backshift operators, a convention we use in this book is to write the equation above as

$$Y_t = \frac{\mathcal{B}}{1 - \phi \mathcal{B}} X_t \tag{2.9}$$

With this convention, $\mathcal{B}$ should be regarded as a complex variable which is, in reality, the z-variable. More formally speaking, we should have derived (2.9) as follows:

$$Y(z) = \mathcal{Z}(Y_t) = \mathcal{Z}(\phi Y_{t-1} + X_{t-1}) = \phi z^{-1} Y(z) + z^{-1} X(z)$$

where the last equality follows from the linearity of the z-transform and the real translation theorem shown in Example 2.5. Then

$$Y(z) = \frac{z^{-1}}{1 - \phi z^{-1}} X(z) \tag{2.10}$$

and the transfer function is

$$H(z^{-1}) = \frac{z^{-1}}{1 - \phi z^{-1}} \qquad \square \tag{2.11}$$

---

[8] With the exception of this appendix, we follow such conventions throughout the book.

Remark 2.4. These manipulations are formal. The algebraic structure of (2.9) and (2.10) are identical provided that we consider $\mathcal{B}$ not as an operator that acts on a discrete-time sequence but as a complex number. Because of this, some authors omit use of the z variables for notational simplicity.

To get a discrete-time sequence, we should invert the z-domain function that we have. Z-transforms are usually inverted by using one of the methods described below.

Use of a z-Transform Table
For simple z-transforms, a table of z-transforms can be used. Such tables list z-domain functions $Y(z)$ and their corresponding discrete-time inverses, Y_k. For example, suppose that we wish to find the inverse of

$$Y(z) = \frac{C}{1 - \phi z^{-1}}$$

From a z-transform table we will see (see Example 2.4) that the inverse is

$$\mathcal{Z}^{-1}(Y(z)) = C\phi_k \qquad k = 0, 1, 2, \ldots$$

Inversion by Long Division
From the definition of a z-transform [equation (2.7)], any z-transform can be written as an infinite series:

$$Y(z) = \mathcal{Z}(Y_k) = Y_0 + Y_1 z^{-1} + Y_2 z^{-2} + \cdots \qquad (2.12)$$

Thus given the z-transform in this form, the discrete-time-domain function can be trivially extracted by looking at the coefficients of the right-hand side. Frequently, as we saw in this chapter, the z-transform is not given as an infinite series in z^{-1} but as a ratio of two finite polynomials in z^{-1}, as when dealing with a transfer function:

$$Y(z) = \frac{B(z^{-1})}{A(z^{-1})} = \frac{b_0 + b_1 z^{-1} + \cdots + b_{n_b} z^{-n_b}}{a_0 + a_1 z^{-1} + \cdots + a_{n_a} z^{-n_a}}$$

Thus what we need is to express this ratio in the form of the infinite series (2.12):

$$Y(z) = \frac{B(z^{-1})}{A(z^{-1})} \quad \text{or} \quad Y(z)A(z^{-1}) = B(z^{-1})$$

So we have the relation

$$\left(Y_0 + Y_1 z^{-1} + Y_2 z^{-2} + \cdots\right)\left(a_0 + a_1 z^{-1} + \cdots + a_{n_a} z^{-n_a}\right)$$
$$= b_0 + b_1 z^{-1} + \cdots + b_{n_b} z^{-n_b}$$

To obtain $\{Y_k\}_{k=0}^{\infty}$, which is the inverse z-transform of $Y(z)$, we equate coefficients of like powers of z^{-1} in each side of the previous equation:

For $(z^{-1})^0$: $\quad a_0 Y_0 = b_0 \qquad \Rightarrow \quad Y_0 = \dfrac{b_0}{a_0}$

For $(z^{-1})^1$: $\quad a_1 Y_0 + a_0 Y_1 = b_1 \quad \Rightarrow \quad Y_1 = \dfrac{b_1 - a_1 Y_0}{a_0} = \dfrac{b_1 - a_1 b_0/a_0}{a_0}$

$$\vdots$$

An expression related to this approach is the *Diophantine equation*, which gives the ratio of two polynomials taking into account the remainder. We look at Diophantine equations in Chapter 5, where they are used for optimal forecasting and control.

Partial Fractions Expansion
The method of expansion into partial fractions is a tool used when manipulating ratios of polynomials. It provides an explanation about why the roots of the characteristic equation $A(z^{-1}) = 0$ determine the stability of a process. The idea of this inversion method is to write a rational transfer function

$$Y(z) = \frac{B(z^{-1})}{A(z^{-1})}$$

as

$$
\frac{B(z^{-1})}{a_0(1 - p_1 z^{-1})(1 - p_2 z^{-1}) \cdots (1 - p_{n_a} z^{-1})}
$$
$$
= \frac{C_1}{1 - p_1 z^{-1}} + \frac{C_2}{1 - p_2 z^{-1}} + \cdots + \frac{C_{n_a}}{1 - p_{n_a} z^{-1}} \tag{2.13}
$$

where the p_i $(i = 1, \ldots, n_a)$ are the n_a roots[9] of the characteristic equation (the poles of the transfer function) and the sum of the partial fractions $C_i/(1 - p_i z^{-1})$ is called the *partial fraction expansion* of the ratio $B(z^{-1})/A(z^{-1})$. The key is to find the constants C_i, called the *residuals* associated with each pole. If we can obtain (2.13), we know that each partial fraction has inverse equal to

$$\mathscr{Z}^{-1}\left\{ \frac{C_i}{1 - p_i z^{-1}} \right\} = C_i p_i^k$$

---

[9]Evidently, there might be repeated roots. If a root is repeated l times, the right-hand side of (2.13) will contain a $C_i/(1 - p_i z^{-1})^l$ term.

and from the linearity of the inverse z-transformation, we finally obtain

$$Y_k = C_1\, p_1^k + C_2\, p_2^k + \cdots + C_{n_a}\, p_{n_a}^k \quad k = 0, 1, 2, \ldots$$

Remark 2.5. Note that Y_k will remain finite for any k if and only if *all* n_a poles p_i are less than 1 in absolute value. This implies that the roots on the z^{-1} plane are equal to $z^{-1} = 1/p_i$, which if $|p_i| < 1$, all lie *outside* the unit circle on the z^{-1} plane. This is the stability condition of a transfer function as discussed in Figure 2.5.

Remark 2.6. Some roots may be complex numbers (complex conjugates). Whenever this happens, it can be shown that the behavior of Y_k involves sinusoidal oscillations.

Remark 2.7. The unit step z-transform of the unit step function, defined as

$$U_k = \begin{cases} 1 & \text{if} \quad k = 0, 1, 2, \ldots \\ 0 & \text{if} \quad k = -1, -2, \ldots \end{cases}$$

equals

$$U(z) = \frac{1}{1 - z^{-1}}$$

(see Example 2.3). If this is entered into a discrete-time dynamical process with rational transfer function, we obtain, after taking z-transforms,

$$Y(z) = \frac{B(z^{-1})}{A(z^{-1})} U(z) = \frac{B(z^{-1})}{A(z^{-1})} \frac{1}{1 - z^{-1}}$$

Using partial fraction expansions, we get that

$$Y(z) = \frac{C_0}{1 - z^{-1}} + \frac{C_1}{1 - p_1 z^{-1}} + \cdots + \frac{C_{n_a}}{1 - p_{n_a} z^{-1}}$$

where the first term is due to the unit step input. The dynamical response of the process will be given by

$$Y_k = C_0 + C_1 p_1^k + C_2 p_2^k + \cdots + C_{n_a} p_{n_a}^k$$

If all poles are such that $|p_i| < 1$, we will have that $Y_k \to C_0$ as $k \to \infty$, so C_0 is the *steady-state value* Y_∞. Note that the steady-state value could have been found from $Y(z)$ by applying the final value theorem:

$$Y_\infty = \lim_{k \to \infty} Y_k = \lim_{z \to 1} \left[(1 - z^{-1}) Y(z) \right] = \lim_{z \to 1} [C_0] = C_0$$

Example 2.7: Inversion Using Partial Fractions Expansion. Invert the transfer function

$$Y(z) = \frac{0.5 - 0.2z^{-1}}{1 + 0.7z^{-1} - 0.08z^{-2}}$$

Since $1 + 0.7\,z^{-1} - 0.08z^{-2} = (1 - 0.1z^{-1})(1 + 0.8z^{-1})$, we want to find the right-hand side of the expression

$$Y(z) = \frac{0.5 - 0.2z^{-1}}{(1 - 0.1z^{-1})(1 + 0.8z^{-1})} = \frac{C_1}{1 - 0.1z^{-1}} + \frac{C_2}{1 + 0.8z^{-1}} \quad (2.14)$$

Multiplying each side by $(1 - 0.1\,z^{-1})$, we get:

$$(1 - 0.1z^{-1})Y(z) = C_1 + (1 - 0.1z^{-1})\frac{C_2}{1 + 0.8z^{-1}}$$

from where we see that if $z^{-1} = 1/0.1 = 10$, we would have

$$C_1 = \frac{0.5 - 0.2z^{-1}}{1 + 0.8z^{-1}}\bigg|_{z^{-1}=10} = \frac{-1.5}{9} = -0.1667$$

Similarly, to get C_2, multiply each side of (2.14) by $(1 + 0.8z^{-1})$ and get

$$(1 + 0.8z^{-1})Y(z) = (1 + 0.8z^{-1})\frac{C_1}{1 - 0.1z^{-1}} + C_2$$

If we set $z^{-1} = -1/0.8 = -1.25$, we find that

$$C_2 = \frac{0.5 - 0.2z^{-1}}{1 - 0.1z^{-1}}\bigg|_{z^{-1}=-1.25} = \frac{0.75}{1.125} = 0.6667.$$

Therefore, the inverse of $Y(z)$ is

$$Y_k = C_1 p_1^k + C_2 p_2^k = -0.1667(0.1)^k + 0.6667(-0.8)^k \qquad \square$$

Remark 2.8. In case roots are repeated, the right-hand side of (2.13) will have instead a term $C_j/(1 - p_j z^{-1})^l$ and $n_a - l$ terms $C_i/(1 - p_i z^{-1})\,(i \neq j)$. The residual C_j is obtained by multiplying both sides of (2.13) by $(1 - p_i z^{-1})^l$; the other residuals $C_i\,(i \neq j)$ are obtained in the usual way. If two roots are complex conjugates $r, r^* = a \pm bi$, the right-hand side of (2.13) will contain the corresponding two fractions with either r or r^* in the denominator. The residuals C_i are obtained in the same way as before.

CHAPTER 3

ARIMA Time-Series Models

In Chapter 2 it was pointed out how a dynamic process can be modeled with difference equations. A dynamic process, when sampled, will result in auto-correlated data, and this creates problems for SPC charts, as discussed in Chapter 1. It was mentioned that in modern manufacturing, modeling the autocorrelation structure in order to anticipate it and adjust the process is desirable. To model autocorrelated data, *stochastic difference equation models* are used. Box and Jenkins (1970) proposed a classification of stochastic difference equation models called ARIMA (autoregressive integrated moving average) which are very useful to describe a variety of real-life processes. An ARIMA model contains three components, corresponding to the "AR", "I" and "MA" parts. In this taxonomy, models are assigned a name, abbreviated in general as

$$\text{ARIMA}(\,p, d, q\,)$$

where p is the order of the AR component of the model, q the order of the MA component of the model, and d the degree of differencing[1] needed to achieve a stationary process. For the type of applications in quality control that we consider in this book, the values of p, d, and q are usually less or equal than 2. If a component is missing from the model, the name of the term and its corresponding order are taken out of the name. For example, ARIMA(0,1,1) is sometimes called an IMA(1,1) process.

In this chapter we describe ARIMA models in detail. These models will prove very useful when building transfer function models of systems we wish to control or adjust in the presence of correlated (possibly nonstationary) noise. Identification and offline estimation are illustrated with the SAS and Minitab statistical software packages. In the final section we comment briefly on SPC methods for monitoring autocorrelated processes based on ARIMA models.

Before we start a detailed description of ARIMA models, consider the following numerical illustrations. Suppose that in a process the value of the

---

[1]A transformation used to get a stationary process (see Section 3.4).

Table 3.1 Illustration of an AR(1) Process, $Y_t = 0.8Y_{t-1} + \varepsilon_t$

t	ε_t	Y_t
0	0	0
1	1.123	1.123
2	−0.171	0.727
3	−0.572	0.001
4	0.721	0.729
5	−0.252	0.331

quality characteristic observed at time t, is correlated with the value of the quality characteristic observed immediately one period before. Suppose that you take observations Y_t, Y_{t+1} and estimate their correlation to be 0.8. We could then hypothesize the model

$$Y_{t+1} = 0.8Y_t + \varepsilon_{t+1}$$

or

$$Y_t = 0.8Y_{t-1} + \varepsilon_t \qquad (3.1)$$

where $\varepsilon_t \sim (0, \sigma_\varepsilon^2)$ is a random error or shock that accounts for any other source of variability that we have not identified so far. Table 3.1 shows how a series $\{Y_t\}$ would evolve for five hypothetical shocks, assuming that $t = 1$ is our first observation. Model (3.1) is an instance of an autoregressive model of first order, or AR(1) model.

A quite different model would be one in which the error observed at time t, ε_t, affects not only the current observation but also the next observation (thus the process "remembers" the previous shock). An instance of such a model is

$$Y_t = 0.3\varepsilon_{t-1} + \varepsilon_t \qquad (3.2)$$

Table 3.2 shows how the first five observations of the process would evolve if the same errors as in Table 3.1 were realized again. Model (3.2) is an instance of a moving average (MA) time-series process of first order, since

Table 3.2 Illustration of an MA(1) Process, $Y_t = 0.3\varepsilon_{t-1} + \varepsilon_t$

t	ε_t	Y_t
0	0	0
1	1.123	1.123
2	−0.171	0.166
3	−0.572	−0.522
4	0.721	0.564
5	−0.252	−0.0827

the errors are averaged in the model to produce the observations. Combined AR and MA models are also possible. Next we describe the class of ARIMA models in more detail.

3.1 AUTOREGRESSIVE MODELS

An AR(p) model [also called an ARIMA(p,0,0) model] is given in its general form by the linear stochastic difference equation

$$Y_t = \xi + \phi_1 Y_{t-1} + \phi_2 Y_{t-2} + \cdots + \phi_p Y_{t-p} + \varepsilon_t$$

where the ϕ_i's are parameters, $\{\varepsilon_t\}$ is a white noise sequence, and therefore $\{Y_t\}$ is a stochastic process. The form of this model is the same as that found in a linear regression model, with the variable ξ playing the role of the intercept. The main difference with linear regression models is that the regressors are lagged variables of the same process, hence the name *autoregressive*. If the parameters ϕ_i are such that this process is stationary, we could write a simpler model for the deviations from the mean of the process,[2] that is:

$$\tilde{Y}_t \equiv Y_t - \mu = \phi_1 \tilde{Y}_{t-1} + \phi_2 \tilde{Y}_{t-2} + \cdots + \phi_p \tilde{Y}_{t-p} + \varepsilon_t \qquad (3.3)$$

which can be written as

$$A(\mathcal{B})\tilde{Y}_t = \varepsilon_t$$

where $A(\mathcal{B}) = 1 - \phi_1 \mathcal{B} - \phi_2 \mathcal{B}^2 - \cdots - \phi_p \mathcal{B}^p$.

The *stationarity* conditions of linear, stochastic difference equations are exactly the same as the conditions for the *stability* of a deterministic linear difference equation model. Thus an AR(p) process is stationary if and only if the $A(\mathcal{B})$ polynomial is stable, which means that it has all roots strictly outside the unit circle on the $\mathcal{B}$ complex plane.

Example 3.1: AR(1) Process. This model, sometimes called a *Markov process*, is

$$\tilde{Y}_t = \phi_1 \tilde{Y}_{t-1} + \varepsilon_t.$$

We can drop the subscript of ϕ because $p = 1$. Thus

$$(1 - \phi\mathcal{B})\tilde{Y}_t = \varepsilon_t$$

---

[2] A notational convention used by Box and Jenkins (1976). We reserve the tilde notation for processes that have a constant, finite mean.

and the characteristic equation is

$$A(\mathcal{B}) = 1 - \phi\mathcal{B} = 0$$

so the only root is $\mathcal{B} = 1/\phi$, and from $|1/\phi| > 1$ we see that

$$|\phi| < 1$$

is the stationarity condition for this process.   □

In Appendix 2C we showed the mathematical details explaining why is it that the roots of the characteristic equation determine the stationarity of a process. That explanation is based on z-transforms. We can provide some further rationale for this important fact from a more intuitive point of view. Consider solving the AR(1) process by repeated substitution of $\tilde{Y}_{t-k}$ in the recursive equation. We get

$$\tilde{Y}_t = \phi\tilde{Y}_{t-1} + \varepsilon_t = \phi\left(\phi\tilde{Y}_{t-2} + \varepsilon_{t-1}\right) + \varepsilon_t = \phi^2\tilde{Y}_{t-2} + \phi\varepsilon_{t-1} + \varepsilon_t$$

$$= \phi^2\left(\phi\tilde{Y}_{t-3} + \varepsilon_{t-2}\right) + \phi\varepsilon_{t-1} + \varepsilon_t = \phi^3\tilde{Y}_{t-3} + \phi^2\varepsilon_{t-2} + \phi\varepsilon_{t-1} + \varepsilon_t$$

$$\vdots$$

$$= \sum_{j=0}^{\infty} \phi^j \varepsilon_{t-j}$$

assuming that we can go back in time indefinitely. This means that

$$\tilde{Y}_t = \sum_{j=0}^{\infty} \phi^j \mathcal{B}^j \varepsilon_t = \frac{1}{1 - \phi\mathcal{B}} \varepsilon_t \tag{3.4}$$

where the geometric sum converges to the last term on the right if and only if $|\phi \mathcal{B}| < 1$. But since $|\mathcal{B}| = 1$ (see Appendix 2B), this condition is the same as $|\phi| < 1$. From the right-hand side of equation (3.4), we see that the stability of the $A(\mathcal{B})$ polynomial allows us to divide by $A(\mathcal{B})$ as if we were using standard algebra.[3]

Equation (3.4) also allows us to get a better understanding of what is meant by stationarity. The condition $|\phi| < 1$ implies that previous shocks ε_{t-j} affect the current random variable of interest ($\tilde{Y}_t$) less and less as the age of the shock (j) increases. This is similar to the exponential weights given by the EWMA statistic (see Section 1.1). If we have a process with $|\phi| > 1$, shocks will have an increasing effect (weight) on the current observation the

---

[3]A problem with division of polynomials having nonstationary roots on the unit circle is described in Appendix 3A.

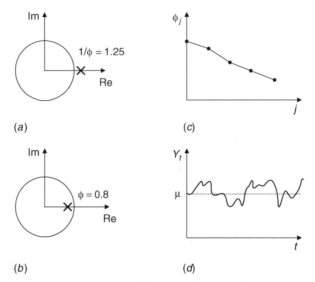

Figure 3.1 Behavior of a stationary AR(1) process ($\phi = 0.8$): (*a*) $\mathcal{B}$ plane; (*b*) $\mathcal{F}$ plane; (*c*) autocorrelation function; and (*d*) sample realization.

older they are, a less plausible situation for quality control data. If $\phi = 1$, we have a borderline case that turns out to be very useful in practice. It is analyzed in detail in Section 3.4.

More information can be obtained by looking at two instances of AR(1) processes. Suppose first that $\phi = 0.8$. Since

$$\tilde{Y}_t = \frac{1}{1 - \phi\mathcal{B}}\varepsilon_t = \frac{\mathcal{F}}{\mathcal{F} - \phi}\varepsilon_t$$

the characteristic equation has its root at $\mathcal{F} = \phi = 0.8$, which is inside the unit circle on the $\mathcal{F}$ plane, and therefore the system is stationary. Figure 3.1 illustrates the location of the roots on the $\mathcal{F}$ and $\mathcal{B}$ complex planes, the behavior of the weights $\phi^j = 0.8^j$ as the age j increases, and a sample realization.

Similarly, consider the case of an AR(1) process with $\phi = 1.1$. The root $\phi = 1.1$ is outside the unit circle on the $\mathcal{F}$ complex plane and the process is nonstationary. The root on the $\mathcal{B}$ plane is inside the unit circle (at $1/\phi = 0.9090$). Figure 3.2 illustrates this situation. In this case we get explosive behavior regardless of ξ, with the process mean tending to infinity as t tends to infinity. Finally, consider the case of an AR(1) process with $\phi = 1.0$. In this case, the process is borderline nonstationary. If $\xi = 0$, the mean of the process is zero, but if $\xi \neq 0$, the process mean moves with a slope of ξ (see Figure 3.3).

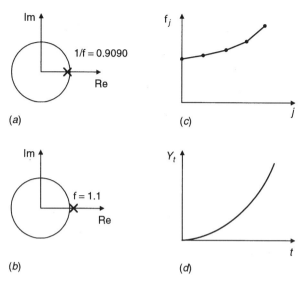

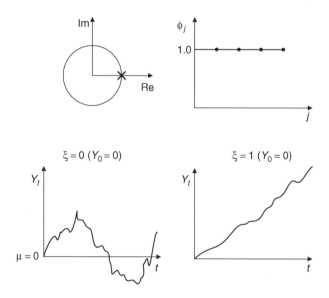

Figure 3.2 Behavior of a nonstationary AR(1) process ($\phi = 1.1$): (*a*) $\mathcal{B}$ plane; (*b*) $\mathcal{F}$ plane; (*c*) autocorrelation function; and (*d*) sample realization. The process is explosive regardless of ξ.

Figure 3.3 Behavior of a nonstationary AR(1) process with ($\phi = 1.0$). Location on $\mathcal{B}$ plane = location on $\mathcal{F}$ plane.

Example 3.2: AR(2) Process. Sometimes called a *Yule process*, an AR(2) process is described by the model

$$\tilde{Y}_t = \phi_1 \tilde{Y}_{t-1} + \phi_2 \tilde{Y}_{t-2} + \varepsilon_t.$$

Thus we can write it as

$$A(\mathcal{B}) = \left(1 - \phi_1 \mathcal{B} - \phi_2 \mathcal{B}^2\right) Y_t = \varepsilon_t$$

and applying Jury's criterion for the stability of a polynomial, we get that the stationarity conditions for an AR(2) process are

$$|\phi_2| < 1$$
$$\phi_2 + \phi_1 < 1$$
$$\phi_2 - \phi_1 < 1.$$

These inequalities define a triangular region in the (ϕ_1, ϕ_2) plane identical to the one in Example 2.2.   □

3.1.1   Moments of an AR(p) Process

Mean of an AR(p) Process
The expected value of a stationary AR(p) process[4] $\{Y_t\}$ is given by

$$E[Y_t] = \mu = E\left[\xi + \phi_1 Y_{t-1} + \phi_2 Y_{t-2} + \cdots + \phi_p Y_{t-p} + \varepsilon_t\right]$$

Since $E[Y_j] = \mu$ for all j, we have that

$$\mu = \xi + \sum_{j=1}^{p} \phi_j \mu$$

Thus

$$\mu = \frac{\xi}{1 - \sum_{j=1}^{p} \phi_j}$$

Note that for an AR(1), if $\phi_1 = 1$, the mean diverges, a consequence of nonstationarity. For $|\phi_1| > 1$, the ratio above is not interpretable.

Autocovariance and Autocorrelation of an AR(p) Process
To find the autocovariance, multiply each element of the process by $\tilde{Y}_{t-k}$ and take expectations:

$$\gamma_k = E\left[\tilde{Y}_t \tilde{Y}_{t-k}\right]$$
$$= E\left[\phi_1 \tilde{Y}_{t-1} \tilde{Y}_{t-k}\right] + E\left[\phi_2 \tilde{Y}_{t-2} \tilde{Y}_{t-k}\right] + \cdots + E\left[\phi_p \tilde{Y}_{t-p} \tilde{Y}_{t-k}\right] + E\left[\varepsilon_t \tilde{Y}_{t-k}\right]$$

[4]Clearly, $E[\tilde{Y}_t] = 0$.

The last term on the right is zero since the error term in $\tilde{Y}_{t-k}$ (which is ε_{t-k}) is uncorrelated with ε_t. From this we obtain the autocovariance function

$$\gamma_k = \phi_1 \gamma_{k-1} + \phi_2 \gamma_{k-2} + \cdots + \phi_p \gamma_{k-p} \qquad k = 1, 2, 3, \ldots \quad (3.5)$$

Dividing by $\gamma_0 = \sigma_Y^2$, we get the autocorrelation function

$$\rho_k = \frac{\gamma_k}{\gamma_0} = \phi_1 \rho_{k-1} + \phi_2 \rho_{k-2} + \cdots + \phi_p \rho_{k-p} \qquad k = 1, 2, 3, \ldots \quad (3.6)$$

Notice that both the autocovariance and the autocorrelation functions of an AR(p) process obey the same difference equation as the process. For a stationary process, this means that the autocorrelations will eventually be zero for large k. We will say that *for stationary AR(p) processes, the autocorrelation function "decays exponentially" or "tails off."* In fact, the actual decay depends on the solution to the AR(p) difference equation, and an oscillatory pattern can occur frequently. However, the decay of $|\rho_k|$ will be exponential.

Equations (3.6) for $k = 1, 2, \ldots, p$ are called the *Yule–Walker equations*. They can be used for estimating the ϕ_i's given estimates of the autocorrelations.

Variance of an AR(p) Process
The variance is obtained from equation (3.5) for $k = 0$:

$$\gamma_0 = \phi_1 \gamma_{-1} + \phi_2 \gamma_{-2} + \cdots + \phi_p \gamma_{-p} + E\left[\varepsilon_t \tilde{Y}_t\right]$$

where in this case the last term contains the expectation of the product of two random variables that have a common error variable (ε_t), so it is not zero:

$$\sigma_Y^2 = \phi_1 \gamma_1 + \phi_2 \gamma_2 + \cdots + \phi_p \gamma_p + \sigma_\varepsilon^2.$$

This follows because the autocovariance function is an even function (i.e., $\gamma_j = \gamma_{-j}$). Dividing both sides by $\gamma_0 = \sigma_Y^2$, we get, after solving,

$$\sigma_Y^2 = \gamma_0 = \frac{\sigma_\varepsilon^2}{1 - \sum_{j=1}^p \phi_j \rho_j}$$

Example 3.3: Moments of an AR(1) Process. The process is

$$\tilde{Y}_t = \phi \tilde{Y}_{t-1} + \varepsilon_t$$

Since $p = 1$, the formulas at the beginning of this section provide the following moments:

$$\mu = \frac{\xi}{1 - \phi}$$

$$\gamma_0 = \sigma_Y^2 = \frac{\sigma_\varepsilon^2}{1 - \phi\rho_1}$$

$$\gamma_1 = \phi\gamma_0 = \frac{\phi\sigma_\varepsilon^2}{1 - \phi\rho_1}$$

$$\gamma_2 = \phi\gamma_1 = \frac{\phi^2\sigma_\varepsilon^2}{1 - \phi\rho_1}$$

$$\vdots$$

$$\gamma_k = \phi\gamma_{k-1} = \frac{\phi^k\sigma_\varepsilon^2}{1 - \phi\rho_1}.$$

Similarly, the autocorrelations are

$$\rho_1 = \frac{\gamma_1}{\gamma_0} = \phi$$

$$\rho_2 = \frac{\gamma_2}{\gamma_0} = \phi^2$$

$$\vdots$$

$$\rho_k = \phi^k.$$

Therefore, the autocorrelation function of an AR(1) process decays geometrically (exponentially if time is "continuous") if it is stationary (i.e., if $|\phi| < 1$).

$\square$

Figure 3.4 illustrates some stationary AR(1) processes where ϕ is positive (so we have positive autocorrelation) and when ϕ is negative (so the autocorrelation at lag 1 is negative). High negative autocorrelation at lag 1 is usually easy to identify from the time series plot, since the process jumps back and forth around the mean from sample to sample. Positively autocorrelated processes move much more smoothly, with longer excursions on one side of the mean the larger the positive autocorrelation is. In the limit, when ϕ is 1, the process does not "stick" to its mean and we have a nonstationary process (a random walk).

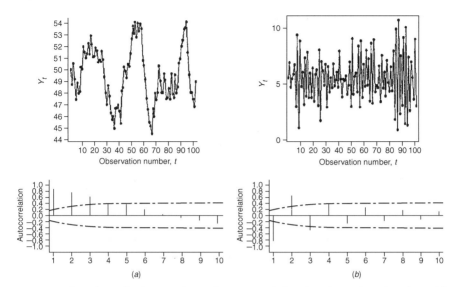

Figure 3.4 Sample realizations and sample autocorrelation functions of stationary AR(1) processes: (a) $\phi = 0.8$; (b) $\phi = -0.8$.

A better understanding of the AR(1) process can be obtained if we look at the original process:

$$Y_t = \xi + \phi Y_{t-1} + \varepsilon_t$$

Solving the difference equation by repeated substitution, we get

$$
\begin{aligned}
Y_t &= \xi + \phi(\xi + \phi Y_{t-2} + \varepsilon_{t-1}) + \varepsilon_t \\
&= \xi + \phi\xi + \phi^2 Y_{t-2} + \phi\varepsilon_{t-1} + \varepsilon_t \\
&= \xi + \phi\xi + \phi^2 (\xi + \phi Y_{t-3} + \varepsilon_{t-2}) + \phi\varepsilon_{t-1} + \varepsilon_t \\
&= \xi + \phi\xi + \phi^2\xi + \phi^3 Y_{t-3} + \phi^2 \varepsilon_{t-2} + \phi\varepsilon_{t-1} + \varepsilon_t \\
&\;\;\vdots \\
&= \xi \sum_{j=0}^{\infty} \phi^j + \sum_{j=0}^{\infty} \phi^j \mathcal{B}^j \varepsilon_t
\end{aligned}
$$

assuming that the process can go back an infinite number of periods. If $|\phi| < 1$, this equals

$$Y_t = \frac{\xi}{1 - \phi} + \frac{\varepsilon_t}{1 - \phi\mathcal{B}}.$$

Thus, if we apply the expected value and variance operators, we get, respectively,

$$\mu = E[Y_t] = \frac{\xi}{1 - \phi}$$

and

$$\sigma_Y^2 = \text{Var}[Y_t] = \text{Var}\left(\varepsilon_t + \phi\varepsilon_{t-1} + \phi^2 \varepsilon_{t-2} + \cdots \right)$$

$$= \sigma_\varepsilon^2 \left(1^2 + \phi^2 + \phi^4 + \phi^6 + \cdots \right) = \sigma_\varepsilon^2 \sum_{j=0}^{\infty} \left(\phi^2 \right)^j = \frac{\sigma_\varepsilon^2}{1 - \phi^2}$$

where the last equality follows provided that $\phi^2 < 1$. Thus if $|\phi| \geq 1$, both the mean and the variance become infinitely large as $t \to \infty$.

We can also check the nonstationary behavior of the moments of an AR(1) process when $\phi = 1$ and $\xi \neq 0$, the case usually described as a random walk *with drift*:

$$Y_t = \xi + Y_{t-1} + \varepsilon_t$$

where the constant ξ is called the *drift parameter* and determines the average slope of the process. By repeated substitution and assuming that the process starts at time zero with a value equal to $Y_0 = 0$, we get

$$Y_t = \sum_{i=0}^{t-1} \mathcal{B}^i \varepsilon_t + t\xi.$$

Thus

$$E[Y_t] = t\xi = \mu_t$$

so the process mean is not constant but is a linear function of time with slope equal to ξ and

$$\text{Var}[Y_t] = t\sigma_\varepsilon^2.$$

It can be shown (see Problem 3.8) that if $Y_0 = 0$, the autocovariance function is

$$\gamma_{k,t} = E\left[(Y_t - \mu_t)(Y_{t+k} - \mu_{t+k}) \right] = t\sigma_\varepsilon^2 \tag{3.7}$$

and $\rho_k = 1$ for all lags k. Thus the population autocorrelation function does not decay. As discussed in Section 3.6, a nondecaying or slowly decaying sample autocorrelation function gives evidence that the process is nonstationary or near nonstationarity.

Example 3.4: Moments of an AR(2) Process. The process is described by

$$Y_t = \xi + \phi_1 Y_{t-1} + \phi_2 Y_{t-2} + \varepsilon_t$$

or if $\tilde{Y}_t = Y_t - \mu$ (assuming stationarity in the mean), this can be written as

$$\tilde{Y}_t = \phi_1 \tilde{Y}_{t-1} + \phi_2 \tilde{Y}_{t-2} + \varepsilon_t.$$

From the AR(p) formulas, with $p = 2$, we get

$$\mu = \frac{\xi}{1 - \phi_1 - \phi_2}$$

$$\sigma_Y^2 = \gamma_0 = \frac{\sigma_\varepsilon^2}{1 - \phi_1 \rho_1 - \phi_2 \rho_2}.$$

The autocorrelation function obeys the difference equation

$$\rho_k = \phi_1 \rho_{k-1} + \phi_2 \rho_{k-2}$$

which for $k = 1$ and $k = 2$ equal, respectively,

$$\rho_1 = \phi_1 + \phi_2 \rho_1$$
$$\rho_2 = \phi_1 \rho_1 + \phi_2$$

since $\rho_0 = 1$. These are the Yule–Walker equations for an AR(2) process. These two equations can be solved for ϕ_1 and ϕ_2, giving

$$\phi_1 = \frac{\rho_1(1 - \rho_2)}{1 - \rho_1^2} \quad \text{and} \quad \phi_2 = \frac{\rho_2 - \rho_1^2}{1 - \rho_1^2}. \tag{3.8}$$

Thus with the estimates of the autocorrelations at lags 1 and 2, we can obtain point estimates for ϕ_1 and ϕ_2. These method-of-moments estimates can be generalized for AR(p) processes. $\square$

3.2   MOVING AVERAGE MODELS

If a process can, instead, be represented by a finite weighted sum of random shocks $\{\varepsilon_t\}$, it is called *a moving average process*. A general q-order MA process, denoted MA(q), is given by

$$Y_t = \mu - \theta_1 \varepsilon_{t-1} - \theta_2 \varepsilon_{t-2} - \cdots - \theta_q \varepsilon_{t-q} + \varepsilon_t.$$

If, as before, we define $\tilde{Y}_t = Y_t - \mu$, an MA(q) process is

$$\tilde{Y}_t = -\theta_1 \varepsilon_{t-1} - \theta_2 \varepsilon_{t-2} - \cdots - \theta_q \varepsilon_{t-q} + \varepsilon_t.$$

The term *moving average* comes from thinking of $\tilde{Y}_t$ as an average of the current and previous q shocks. However, since the weights θ_i do not necessarily add up to 1, this terminology, although standard in the time-series analysis literature, is somewhat misleading. Introducing backshift operators, we get

$$\tilde{Y}_t = \left(1 - \theta_1 \mathcal{B} - \theta_2 \mathcal{B}^2 - \cdots - \theta_q \mathcal{B}^q\right) \varepsilon_t \equiv C(\mathcal{B}) \varepsilon_t$$

Thus, comparing MA(q) and AR(p) processes, we see that MA(q) processes have an $A(\mathcal{B})$ polynomial equal to 1 and AR(p) processes have a $C(\mathcal{B})$ polynomial equal to 1. For this reason, MA processes are always stationary, since their $A(\mathcal{B})$ polynomial is inverted trivially [i.e., $1/A(\mathcal{B})$ is trivially equal to 1] and the moments of $\tilde{Y}_t$ do not depend on time, implying stationarity. Such $A(\mathcal{B})$ polynomial is evidently not a function of the backshift operator and therefore it does not have roots, including roots inside or on the unit circle on the $\mathcal{B}$ complex plane, that would make the process nonstationary.

Invertibility is a condition similar mathematically to stationarity and relates to the $C(\mathcal{B})$ polynomial of MA processes. Invertibility allows us to divide by the $C(\mathcal{B})$ polynomial, in the same way that stationarity allows us to divide by the $A(\mathcal{B})$ polynomial. A time-series process is invertible if it can be written as an AR process of the form

$$\tilde{Y}_t = \pi_1 \tilde{Y}_{t-1} + \pi_2 \tilde{Y}_{t-2} + \cdots + \varepsilon_t$$

where the sum on the right can be finite or infinite but must converge to a finite value. Invertibility means that the influence of previous observations $\tilde{Y}_{t-j}$ on $\tilde{Y}_t$ decreases with age. To see this, consider an MA(1) process:

$$(1 - \theta \mathcal{B})^{-1} \tilde{Y}_t = \varepsilon_t$$

That is,

$$\sum_{j=0}^{\infty} \theta^j \mathcal{B}^j \tilde{Y}_t = \varepsilon_t$$

Thus if $|\theta| > 1$, the weight given to observation Y_{t-j} *increases* with j, a rather unnatural situation.

Note the duality between stationarity and invertibility:

- *Stationarity* The influence of previous ε_i's on the current observation decreases with age.
- *Invertibility* The influence of previous Y_t's on the current observation decreases with age.

A process is invertible if its $C(\mathcal{B})$ polynomial is stable. Thus an MA(q) process is invertible if all roots of $C(\mathcal{B})$ are strictly outside the unit circle. In contrast, all AR(p) processes are invertible, since their $C(\mathcal{B})$ polynomial equals 1 and this is trivially inverted. Then the π_j-weights implied by the definition of invertibility simply equal the ϕ_j autoregresive parameters. Such a zero-order polynomial does not have roots, unstable or otherwise. Note the duality between AR(p) and MA(q) processes:

- *MA(q) processes.* Always stationary; invertibility depends on $C(\mathcal{B})$.
- *AR(p) processes.* Always invertible; stationarity depends on $A(\mathcal{B})$.

Moments of an MA(q) Process

Since MA(q) processes are always stationary, their mean is given by the constant term in the defining difference equation; thus

$$E[Y_t] = \mu.$$

The variance is obtained immediately from applying the variance operator:

$$\gamma_0 = \sigma_Y^2 = \text{Var}[Y_t] = \left(1 + \theta_1^2 + \theta_2^2 + \cdots + \theta_q^2\right)\sigma_\varepsilon^2.$$

The autocovariance function is obtained from

$$\gamma_k = E\left[\left(\varepsilon_t - \theta_1\varepsilon_{t-1} - \cdots - \theta_q\varepsilon_{t-q}\right)\left(\varepsilon_{t-k} - \theta_1\varepsilon_{t-k-1} - \cdots - \theta_q\varepsilon_{t-k-q}\right)\right].$$

Only the cross-products with contemporaneous ε's are nonzero, so we get

$$\gamma_k = \begin{cases} 0 & \text{if } k > q \\ \left(-\theta_k + \theta_1\theta_{k+1} + \theta_2\theta_{k+2} + \cdots + \theta_{q-k}\theta_q\right)\sigma_\varepsilon^2 & \text{if } k = 1, 2, \ldots, q \end{cases}$$

Thus, in MA(q) models, two observations separated a number of periods k greater than q do not share any shock ε_{t-k}; thus, by definition, those observations cannot be correlated. The autocorrelation function is

$$\rho_k = \begin{cases} 0 & \text{if } k > q \\ \dfrac{-\theta_k + \theta_1\theta_{k+1} + \theta_2\theta_{k+2} + \cdots + \theta_{q-k}\theta_q}{1 + \theta_1^2 + \theta_2^2 + \cdots + \theta_q^2} & \text{if } k = 1, 2, \ldots, q \end{cases}$$

As discussed in Section 3.6, an autocorrelation function that seems to have zero values after lag q indicates that the process under study is an MA(q) process. We will say that for MA(q) processes, the autocorrelation function *cuts off* after lag q.

Example 3.5: Moments of an MA(1) Process. In this case we have $q = 1$, so the model is

$$\tilde{Y}_t = (1 - \theta\mathcal{B})\varepsilon_t$$

and $E[Y_t] = \mu$. We also have that

$$\mathrm{Var}[Y_t] = \gamma_0 = (1 + \theta^2)\sigma_\varepsilon^2$$

The autocovariance function has a very simple form:

$$\gamma_k = \begin{cases} 0 & \text{if } k > 1 \\ -\theta\sigma_\varepsilon^2 & \text{if } k = 1 \end{cases}$$

and the autocorrelation function is

$$\rho_k = \begin{cases} 0 & \text{if } k > 1 \\ \dfrac{-\theta}{1 + \theta^2} & \text{if } k = 1 \end{cases}$$

Thus it can be seen that the autocorrelation function cuts off after lag 1. This function can be explained rather intuitively. Write the model at times t and $t-1$:

$$\tilde{Y}_t = -\theta\varepsilon_{t-1} + \varepsilon_t$$

$$\tilde{Y}_{t-1} = -\theta\varepsilon_{t-2} + \varepsilon_{t-1}$$

Thus $\tilde{Y}_t$ and $\tilde{Y}_{t-1}$ share the shock ε_{t-1}. Therefore, $\mathrm{Cov}(\varepsilon_{t-1}, -\theta\varepsilon_{t-1}) = -\theta\sigma_\varepsilon^2$, and we get the result indicated above. Note that the autocorrelation at lag 1 for this process cannot be greater than 0.5 (see Problem 3.11). Figure 3.5 shows a plot of ρ_1 versus values of θ. From the graph it is evident that

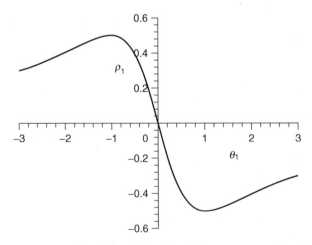

Figure 3.5 Lag 1 autocorrelation (ρ_1) as a function of the parameter θ_1 in an MA(1) process.

both an invertible ($|\theta| < 1$) and an noninvertible ($|\theta| > 1$) choice of θ would be consistent with a given estimate of the autocorrelation ρ_1. But as mentioned before, the noninvertible choice of θ reflects the strange phenomenon that older Y's have a stronger influence on the current observation than do more recent observations, so the invertible choices should be used for the applications with which we are concerned. □

Invertibility Conditions of an MA(1)
As mentioned before, since $C(\mathcal{B}) = 1 - \theta\mathcal{B} = 0$ has its root at $\mathcal{B} = 1/\theta$, we need $|\theta| < 1$ for invertibility.

Example 3.6: Moments of an MA(2) Process. The process is

$$\tilde{Y}_t = \left(1 - \theta_1\mathcal{B} - \theta_2\mathcal{B}^2\right) \varepsilon_t$$

The mean and the variance are $E[Y_t] = \mu$ and $\sigma_Y^2 = (1 + \theta_1^2 + \theta_2^2)\sigma_\varepsilon^2$. The autocovariance function is

$$\gamma_k = \begin{cases} -(\theta_1 + \theta_1\theta_2)\sigma_\varepsilon^2 & \text{if } k = 1 \\ \theta_2\sigma_\varepsilon^2 & \text{if } k = 2 \\ 0 & \text{if } k > 2 \end{cases}$$

and the autocorrelation function is

$$\rho_k = \begin{cases} \dfrac{\theta_1 + \theta_2\theta_1}{1 + \theta_1^2 + \theta_2^2} & \text{if } k = 1 \\ \dfrac{-\theta_2}{1 + \theta_1^2 + \theta_2^2} & \text{if } k = 2 \\ 0 & \text{if } k > 2 \end{cases}.$$

Thus the autocorrelation function cuts off after lag 2. Figures 3.6 and 3.7 illustrate the possible values for ρ_1 and ρ_2 for choices of θ_1 and θ_2 in the $(-2, 2)$ and $(-1, 1)$ ranges, respectively. Not all combinations of (θ_1, θ_2) on the graph are invertible (see below for invertibility conditions) Clearly, if the autocorrelations at lags 1 and 2 are relatively high, say higher than $2/3$ in absolute value, an MA(q) process (with $q = 1, 2$) can be ruled out. □

Invertibility Conditions of an MA(2)
Invertibility conditions of an MA(2) are analogous to the stationarity conditions of an AR(2) process. The polynomial

$$C(\mathcal{B}) = 1 - \theta_1\mathcal{B} - \theta_2\mathcal{B}^2$$

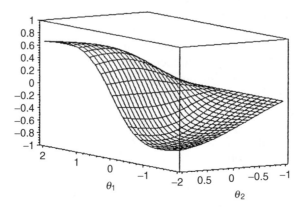

Figure 3.6 Lag 1 autocorrelation (ρ_1) as a function of the parameters θ_1 and θ_2 in an MA(2) process.

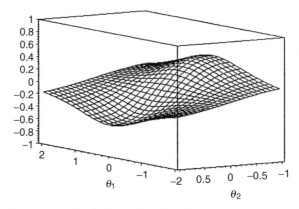

Figure 3.7 Lag 2 autocorrelation (ρ_2) as a function of the parameters θ_1 and θ_2 in a MA(2) process.

must be stable for invertibility of the process. From Jury's test, this results in the conditions

$$\theta_1 + \theta_2 < 1$$
$$\theta_2 - \theta_1 < 1$$
$$|\theta_2| < 1.$$

3.3 ARMA PROCESSES

A model can also contain both AR and MA components for reasons of *parsimony*, a modeling principle which says that models with few parameters are preferable since they are easier to interpret. Since a stationary AR(1) process is equivalent to an infinite-order MA process, instead of having an MA model with a high q, it can be better to use a model with an AR(1) term. Similarly, instead of having an AR model with a large p, it might be better to include an MA(1) term in the model. This will keep the number of parameters low. The general stationary ARMA(p, q) model is

$$\tilde{Y}_t = \phi_1 \tilde{Y}_{t-1} + \phi_2 \tilde{Y}_{t-2} + \cdots + \phi_p \tilde{Y}_{t-p} - \theta_1 \varepsilon_{t-1} - \theta_2 \varepsilon_{t-2} - \cdots - \theta_q \varepsilon_{t-q} + \varepsilon_t$$

$$(3.9)$$

which can be written as

$$\left(1 - \phi_1 \mathcal{B} - \phi_2 \mathcal{B}^2 - \cdots - \phi_p \mathcal{B}^p\right) \tilde{Y}_t = \left(1 - \theta_1 \mathcal{B} - \theta_2 \mathcal{B}^2 - \cdots - \theta_q \mathcal{B}^q\right) \varepsilon_t$$

or

$$A(\mathcal{B}) \tilde{Y}_t = C(\mathcal{B}) \varepsilon_t.$$

An ARMA(p, q) model is stationary if the AR component is stationary, that is, if the $A(\mathcal{B})$ polynomial has all roots outside the unit circle on the $\mathcal{B}$ plane. Similarly, the process is invertible if the MA part is invertible [i.e., if the $C(\mathcal{B})$ polynomial has all roots outside the unit circle].

Moments of an ARMA(p, q) Process
The mean of an ARMA process $\{Y_t\}$ is given by

$$E[Y_t] = \mu = \frac{\xi}{1 - \sum_{i=1}^{p} \phi_i}$$

The autocovariance is obtained by multiplying (3.9) by $\tilde{Y}_{t-k}$ and taking expectations:

$$\gamma_k = \phi_1 \gamma_{k-1} + \cdots + \phi_p \gamma_p + \gamma_{Y\varepsilon}(k) - \theta_1 \gamma_{Y\varepsilon}(k-1) - \cdots - \theta_q \gamma_{Y\varepsilon}(k-q)$$

where $\gamma_{Y\varepsilon}(k) = E[\tilde{Y}_{t-k}\, \varepsilon_t] = E[\tilde{Y}_t\, \varepsilon_{t+k}]$ is the (cross) covariance[5] between Y and ε, which is zero if $k > 0$. In such a case Y will lag behind ε and the two terms do not share any error term. If $k \leq 0$, we have that $\gamma_{Y\varepsilon}(k) \neq 0$ since Y is ahead of ε and given that the AR term can be expressed as a function of all previous and current errors, it will be a function of the error ε_t, so the two variables will be correlated. With this, we get

$$\gamma_k = \phi_1 \gamma_{k-1} + \cdots + \phi_p \gamma_p \qquad \text{if } k \geq q + 1$$

$$\rho_k = \phi_1 \rho_{k-1} + \cdots + \phi_p \rho_p \qquad \text{if } k \geq q + 1$$

[5]See Chapter 4 for more details on cross-covariance functions.

Thus, *apart from the first q lags, an ARMA(p,q) process has an autocorrelation function that resembles an AR(p)*. This will make identification of ARMA processes difficult based solely on the ACF. Statistical techniques have been developed specifically for ARMA process identification. In practice, ARMA processes are usually fit after trying either a pure MA or pure AR process.

Example 3.7: Moments of an ARMA(1, 1) Process. For the type of processes with which we are concerned, it is usually not recommended to use models with orders above an ARMA(1, 1), for parsimony reasons. The ARMA(1, 1) process is

$$\tilde{Y}_t = \phi \tilde{Y}_{t-1} - \theta \varepsilon_{t-1} + \varepsilon_t$$

The autocovariance function for $k = 0, 1, \ldots$ is

$$\gamma_0 = E\left[\tilde{Y}_t \tilde{Y}_t\right] = E\left[\phi \tilde{Y}_{t-1}\tilde{Y}_t - \theta \varepsilon_{t-1}\tilde{Y}_t + \varepsilon_t \tilde{Y}_t\right] = \phi \gamma_1 - \theta \gamma_{Y\varepsilon}(-1) + \sigma_\varepsilon^2$$

(3.10)

$$\gamma_1 = E\left[\tilde{Y}_t \tilde{Y}_t\right] = E\left[\phi \tilde{Y}_{t-1}\tilde{Y}_{t-1} - \theta \varepsilon_{t-1}\tilde{Y}_{t-1} + \varepsilon_t \tilde{Y}_{t-1}\right] = \phi \gamma_0 - \theta \sigma_\varepsilon^2 \quad (3.11)$$

Similarly,

$$\gamma_2 = E\left[\tilde{Y}_t \tilde{Y}_{t-2}\right] = \phi \gamma_1$$

and for $k \geq 2$, we have $\gamma_k = \phi \gamma_{k-1}$. To find $\gamma_{Y\varepsilon}(-1)$, we compute

$$\gamma_{Y\varepsilon}(-1) = E\left[\tilde{Y}_{t+1}\varepsilon_t\right] = E\left[\varepsilon_t \phi \tilde{Y}_t - \varepsilon_t \theta \varepsilon_t + \varepsilon_t \varepsilon_{t+1}\right] = \sigma_\varepsilon^2(\phi - \theta).$$

Solving (3.10) and (3.11) for γ_0 and γ_1, we get

$$\gamma_0 = \frac{1 + \theta^2 - 2\phi\theta}{1 - \theta^2}\sigma_\varepsilon^2$$

$$\gamma_1 = \frac{(1 - \phi\theta)(\phi - \theta)}{1 - \phi^2}\sigma_\varepsilon^2$$

$$\gamma_k = \phi \gamma_{k-1} \quad \text{if } k \geq 2.$$

The autocorrelation function is then

$$\rho_1 = \frac{(1 - \phi\theta)(\phi - \theta)}{1 + \theta^2 - 2\phi\theta}$$

$$\rho_k = \phi \rho_{k-1} \quad \text{if } k \geq 2$$

Therefore, the autocorrelation function decays exponentially after lag 2 provided that $|\phi| < 1$. $\qquad\square$

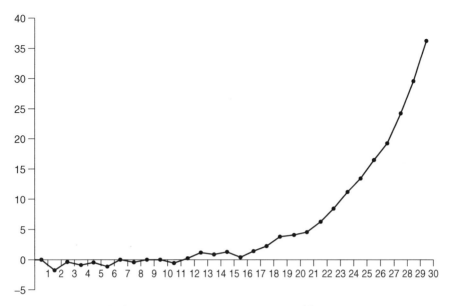

Figure 3.8 One realization of an explosive nonstationary AR(1) process with $\phi = 1.2$. The errors follow a standard normal distribution.

3.4 NONSTATIONARY PROCESSES: ARIMA

Most industrial processes are nonstationary in that if left to itself uncontrolled, the process will wander or drift off target. However, a nonstationary ARMA (ARIMA) process exhibits "explosive" behavior for roots strictly inside the unit circle. For example, an AR(1) process with $\phi = 1.2$ looks as shown in Figure 3.8.

Box and Jenkins (1976) propose the use of AR processes that although nonstationary in the sense that they do not have constant mean or variance, do not explode. Since roots inside the unit circle are ruled out, the only possibility left is to allow for roots *on* the unit circle. Box and Jenkins referred to these models as *homogeneous nonstationary models*, since apart from the wandering level of the process, portions of realizations over nonoverlapping time intervals appear to have a similar behavior. Homogeneous nonstationary processes do not have explosive behavior since their expected value or mean is zero. However, their higher-order moments can be nonconstant.

Example 3.8. Consider an AR(1) process with $\phi = 1$ and $\xi = 0$, also called a *random walk*:

$$\tilde{Y}_t = \tilde{Y}_{t-1} + \varepsilon_t$$

Compare with Section 3.1, where a random walk with drift was described. Figure 3.9 shows a realization of this process. Clearly, the random walk

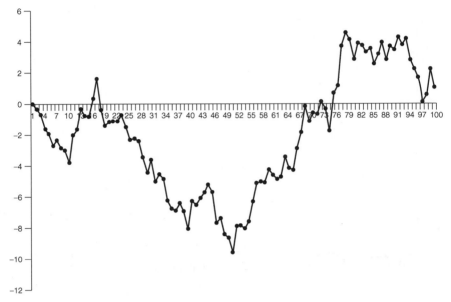

Figure 3.9 One realization of a random walk process, equivalent to an AR(1) process with $\phi = 1, \xi = 0$, standard normal errors.

process has a root on the unit circle:

$$(1 - \mathcal{B})\tilde{Y}_t = \varepsilon_t \quad \text{so } A(\mathcal{B}) = 1 - \mathcal{B}; \text{ root at } \mathcal{B} = 1$$

This is an instance of a process that is nonstationary in the variance but is stationary for the mean.   □

In general, homogeneous nonstationary AR processes have d roots on the unit circle and all other roots outside. Their general form is

$$A(\mathcal{B})Y_t = A'(\mathcal{B})(1 - \mathcal{B})^d Y_t = \varepsilon_t$$

where $A'_p(\mathcal{B})$ is a p-order polynomial will all roots *inside* the unit circle on the $\mathcal{B}$ plane. If we have that $\xi = 0$, these processes are stationary for the mean since the mean equals zero and we could use the tilde notation, as in the random walk example above. The general homogeneous nonstationary ARMA(p, q) process is

$$A(\mathcal{B})Y_t = A'_p(\mathcal{B})(1 - \mathcal{B})^d Y_t = C_q(\mathcal{B})\varepsilon_t$$

which is referred to as an ARIMA(p, d, q) process.

To analyze an ARIMA(p, d, q) process, we first transform it to a stationary process by *differencing*.[6] Define

$$Y_t - Y_{t-1} = (1 - \mathcal{B})Y_t = \nabla Y_t$$

where ∇ (nabla)[7] is called the *difference operator*. By definition, $\nabla = 1 - \mathcal{B}$. The inverse of the difference operator, ∇^{-1}, is called the *summation operator* ($\mathcal{S}$), defined as

$$\mathcal{S}Y_t \equiv \sum_{j=-\infty}^{t} Y_j = (1 + \mathcal{B} + \mathcal{B}^2 + \cdots)Y_t = \frac{Y_t}{1 - \mathcal{B}} = \frac{1}{\nabla}Y_t = \nabla^{-1}Y_t$$

assuming that the sequence $\{Y_t\}$ is infinitely long.[8] The continuous analog of a sum is an integral, and thus the reason for the phrase *integrated of order d* in the term ARIMA(p, d, q).

Common time-series modeling practice is to apply the difference operator as needed on a a time series until we get what appears to be a stationary process. Figures 3.10 to 3.12 show the effect of differencing on a deterministic series $\{Y_t\}$ defined by the difference equation $\nabla^2 Y_t = 1$. For such explosive

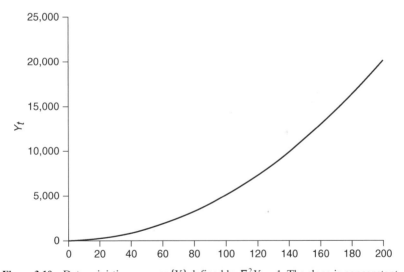

Figure 3.10   Deterministic sequence $\{Y_t\}$ defined by $\nabla^2 Y_t = 1$. The slope is nonconstant.

[6]Nonstationary variance is sometimes removed by taking a logarithmic transformation. In other instances, direct modeling of the nonconstant variance is preferred, as in financial applications (ARCH/GARCH models; see Hamilton, 1994).
[7]The symbol nabla does not come from Greece but apparently owes its form to a certain harp used in Middle Eastern Arab countries.
[8]See Appendix 3B for a discussion of finite summation operators.

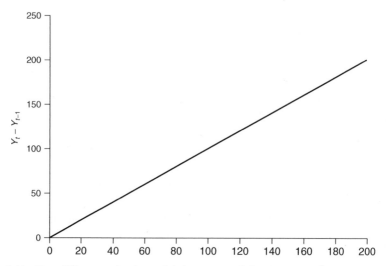

Figure 3.11   First difference of a deterministic sequence $\{Y_t\}$ defined by $\nabla^2 Y_t = 1$. A constant slope is achieved.

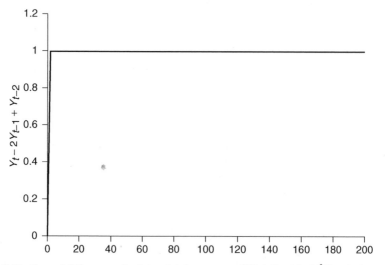

Figure 3.12   Second difference of a deterministic sequence $\{Y_t\}$ defined by $\nabla^2 Y_t = 1$. A constant (mean) value of 1.0 is achieved.

behavior, two differences are usually necessary. Note that we take differences on Y_t, not on $\nabla^2 Y$. Since we can write $Y_t = 1/\nabla^2$, a first difference of Y_t yields the model $Z_t = \nabla Y_t = 1/\nabla$, shown in Figure 3.11. One more difference yields the equation $w_t = \nabla^2 Y_t = 1$, shown in Figure 3.12.

Figures 3.13 to 3.15 show the effect of successive differencing on an ARIMA(0, 2, 0) process. Note that for $d = 2$ the process is not explosive. In fact, as long as $\xi = 0$, ARIMA(p, d, q) processes have a mean of zero. In

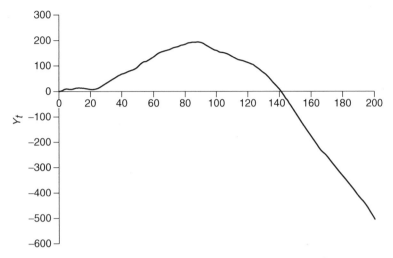

Figure 3.13 One realization of a nonstationary process $\{Y_t\}$ defined by $\nabla^2 Y_t = \varepsilon_t$. Notice the random changes in slope.

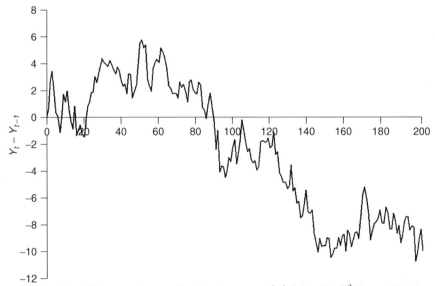

Figure 3.14 First difference of one realization of a process $\{Y_t\}$ defined by $\nabla^2 Y_t = \varepsilon_t$. The slope is constant, but the mean wanders.

what follows, by *nonstationary ARIMA models* we refer to homogeneous nonstationarity [i.e., we refer to a process with d roots on the unit circle and $\xi = 0$ (no drift)]. Appendix 3B elaborates further on processes with drift.

Contrast the deterministic sequences of Figures 3.10 to 3.12 with the effect of differencing an analogous stochastic series such as an ARIMA(0,2,0) process. ARIMA($p, 2, q$) models with $\xi = 0$ (i.e., with no drift) are sometimes useful to model random changes in slope.

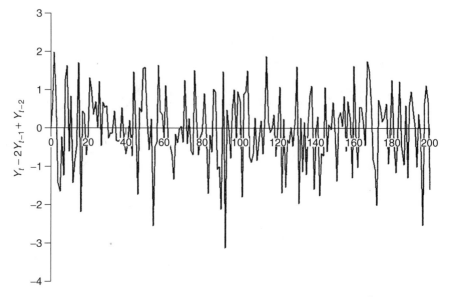

Figure 3.15 Second difference of one realization of a process $\{Y_t\}$ defined by $\nabla^2 Y_t = \varepsilon_t$. This is a stationary process equal to the white noise sequence $\{\varepsilon_t\}$.

Example 3.9. Consider a random walk process:

$$\tilde{Y}_t = \tilde{Y}_{t-1} + \varepsilon_t$$

If we subtract $\tilde{Y}_{t-1}$ from each side of the equation, we will be in effect computing the first difference of $\tilde{Y}_t$ on the left-hand side:

$$\tilde{Y}_t - \tilde{Y}_{t-1} = \nabla \tilde{Y}_t = \varepsilon_t$$

If we consider the new process $w_t = \nabla \tilde{Y}_t = \varepsilon_t$, it is evident that this equals a white noise process $\{\varepsilon_t\}$ which is stationary.   □

Example 3.10. An important nonstationary process that occurs frequently in industrial process data is the ARIMA(0,1,1) = IMA(1,1) process. This is just equal to an MA(1) process differenced once:

$$\nabla \tilde{Y}_t = -\theta \varepsilon_{t-1} + \varepsilon_t$$

As discussed next, the EWMA provides optimal forecasts for this process.   □

3.5 FORECASTING USING ARIMA TIME-SERIES MODELS

Let $\hat{Y}_{t+l|t}$ denote the forecast of the random variable Y_{t+l} computed at time t. The variable l is called the *lead time*. The most common criterion for computing forecasts is the mean square error (MSE). If a forecast is selected such that it minimizes the MSE, it is called a *minimum mean square error* (MMSE) forecast. To find MMSE forecasts, we should solve

$$\min_{\hat{Y}_{t+l|t}} \text{MSE}\left(\hat{Y}_{t+l|t}\right) = \min_{\hat{Y}_{t+1|t}} E_t\left[\left(Y_{t+l} - \hat{Y}_{t+l|t}\right)^2\right] \qquad (3.12)$$

The symbol $E_t[\cdot]$ denotes expectation taken at time t. In other words, any random variable with index $j > t$ is in the future with respect to the current time t. For any random variable with index $j \leq t$, it is assumed that we have already observed a realization of it at the end of period t.

Expanding the binomial term on the right-hand side and taking expectation, after some algebra it can be shown (see Problem 3.13) that

$$\text{MSE}\left(\hat{Y}_{t+l|t}\right) = \text{Var}(Y_{t+l}) + \left(E_t[Y_{t+l}] - \hat{Y}_{t+l|t}\right)^2 \qquad (3.13)$$

where the last term equals the square of the bias of $\hat{Y}_{t+l|t}$. We have the familiar decomposition of mean square error: variance plus squared bias. Note, however, that the variance obtained is the variance of the variable we wish to predict, and this is a random variable to be observed in the future, not under our control. All we can do to minimize the MSE is to choose $\hat{Y}_{t+l|t}$ equal to $E_t[Y_{t+l}]$, which gives us precisely the general MMSE forecasting formula

$$Y_{t+l|t} = E_t[Y_{t+l}]$$

To obtain $E_t[Y_{t+l}]$, we write the ARIMA model in difference equation form at time $t + l$ and take the expected value at time t. The latter is equivalent to the following simple rules:

1. Replace Y_{t-j} $(j \geq 0)$ by values observed.
2. Replace Y_{t+j} $(j \geq 1)$ by $\hat{Y}_{t+j|t}$.
3. Replace ε_{t-j} $(j \geq 0)$ by $\hat{\varepsilon}_t = e_t = Y_{t-j} - \hat{Y}_{t-j|t-j-1}$.
4. Replace ε_{t+j} $(j \geq 1)$ by $E_t[\varepsilon_{t+j}] = 0$.

Example 3.11: Forecasting an ARIMA(1,1,0) Process. The model is

$$\underbrace{\left(1 - \phi\mathcal{B}\right)}_{AR(1)} \underbrace{\left(1 - \mathcal{B}\right)}_{I(1)} Y_t = \varepsilon_t$$

or $Y_t = (1 + \phi)Y_{t-1} - \phi Y_{t-2} + \varepsilon_t$. Thus we have the following forecasts for various lead times.

$l = 1$:

$$Y_{t+1} = (1 + \phi)Y_t - \phi Y_{t-1} + \varepsilon_{t+1}$$

$$\hat{Y}_{t+1|t} = E_t[Y_{t+1}] = (1 + \phi)Y_t - \phi Y_{t-1}$$

$l = 2$:

$$Y_{t+2} = (1 + \phi)Y_{t+1} - \phi Y_t + \varepsilon_{t+2}$$

$$\hat{Y}_{t+2|t} = E_t[Y_{t+2}] = (1 + \phi)\hat{Y}_{t+1|t} - \phi Y_t$$

since at time t both Y_{t+1} and ε_{t+2} are unknown future variables.

$l = 3$:

$$Y_{t+3} = (1 + \phi)Y_{t+2} - \phi Y_{t+1} + \varepsilon_{t+3}$$

$$\hat{Y}_{t+3|t} = E_t[Y_{t+3}] = (1 + \phi)\hat{Y}_{t+2|t} - \phi \hat{Y}_{t+1|t}$$

$l \geq 3$:

$$\hat{Y}_{t+l|t} = E_t[Y_{t+3}] = (1 + \phi)\hat{Y}_{t+l-1|t} - \phi \hat{Y}_{t+l-2|t}$$

We can illustrate the computation of the forecasts numerically. Suppose it is known that $Y_0 = 0$, $Y_1 = 5$, and $\phi = 0.7$, and let us assume that samples are taken every hour. Then the one-hour-ahead forecast is given by

$$\hat{Y}_{2|1} = (1 + 0.7)(5) - 0.7(0) = 8.5$$

The two-hour-ahead forecast is given by

$$\hat{Y}_{3|1} = (1 + 0.7)(8.5) - 0.7(5) = 10.95$$

and the three-hour-ahead forecast is

$$\hat{Y}_{4|1} = (1 + 0.7)(10.95) - 0.7(8.5) = 12.665.$$

If one hour later it is observed that $Y_2 = 7.2$, the one-, two-, and three-hour-ahead forecasts are updated as follows:

$$\hat{Y}_{3|2} = (1 + 0.7)(7.2) - 0.7(5) = 8.74$$

$$\hat{Y}_{4|2} = (1 + 0.7)(8.74) - 0.7(7.2) = 9.81$$

$$\hat{Y}_{5|2} = (1 + 0.7)(9.81) - 0.7(8.74) = 10.56$$

and so on.   □

Example 3.12: Forecasting an IMA(1, 1) Process. The model is

$$(1 - \mathcal{B})Y_t = (1 - \theta\mathcal{B})\varepsilon_t$$

or $Y_t = Y_{t-1} - \theta\varepsilon_{t-1} + \varepsilon_t$. The forecasts are obtained as follows:

$l = 1$:

$$Y_{t+1} = Y_t - \theta\varepsilon_t + \varepsilon_{t+1}$$

$$\hat{Y}_{t+1|t} = E_t[Y_{t+1}] = Y_t - \theta\left(Y_t - \hat{Y}_{t|t-1}\right)$$

which rearranging terms becomes

$$\hat{Y}_{t+1|t} = (1 - \theta)Y_t + \theta\hat{Y}_{t|t-1}$$

Thus the MMSE one-step-ahead forecast of an IMA(1,1) is given by an EWMA of the data with weight $\lambda = 1 - \theta$. It can be shown (see Problem 3.14) that the EWMA also provides MMSE forecasts for an IMA(1,1) process for *any* lead times.   □

3.6   IDENTIFICATION OF ARIMA(p, d, q) MODELS

The purpose of identification is to find the values of p, d, and q from a given time series $\{y_t\}$. Box and Jenkins recommended an iterative approach to model building in which conjectures about the model are tested empirically based on data:

identification and estimation $\Leftrightarrow$ diagnostic checking $\Rightarrow$ use of model

Tentative model identification is based primarily on two statistical tools: the sample autocorrelation function (SACF), and the sample partial autocorrelation function (SPACF). The ACF and its behavior were already described for ARIMA models in previous sections. For an AR system, the ACF tails off exponentially rapidly, and for an MA(q) process, the ACF will cut off after lag q. As it turns out, the *order* of an AR component can be inferred from the SPACF, which we now describe in detail.

Let $\hat{\phi}_{kk}$ be the estimated coefficient of Y_{t-k} in an AR(k) model fitted to the data. The SPACF is described by a plot of $\hat{\phi}_{kk}$ versus k, the lag. Thus the lag 1 PAC is given by the regression coefficient ϕ_{11} in the AR(1) process

$$Y_t = \underline{\phi_{11}}Y_{t-1} + \varepsilon_t$$

The lag 2 PAC is given by ϕ_{22} in the AR(2) process

$$Y_t = \phi_{21}Y_{t-1} + \underline{\phi_{22}}Y_{t-2} + \varepsilon_t$$

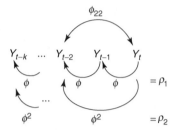

Figure 3.16 Autocorrelation and partial autocorrelation in an AR(1) process.

Continuing in this way, the lag k PAC is given by the ϕ_{kk} coefficient in the AR(k) process

$$Y_t = \phi_{k1}Y_{t-1} + \phi_{k2}Y_{t-2} + \cdots + \underline{\phi_{kk}Y_{t-k}} + \varepsilon_t$$

The ϕ_{11}, ϕ_{22}, and ϕ_{kk} coefficients in these equations form the PACF. Thus one easy way to get the SPACF is to fit AR models of increasing order to the data and keep the last estimated coefficient $\hat{\phi}_{kk}$ associated with the kth model. The interpretation of ϕ_{kk} is that it measures the correlation between Y_t and Y_{t-k} after discounting the effect of the intermediate variables Y_{t-1}, $Y_{t-2}, \ldots, Y_{t-k+1}$. For example, in an AR(1) process, Y_t and Y_{t-1} have a correlation ϕ due to the difference equation that defines the process, $Y_t = \phi Y_{t-1} + \varepsilon_t$, and there is no intermediate variable to discount. The correlation between Y_t and Y_{t-2}, however, is due to the fact that Y_{t-1} and Y_{t-2} are in turn correlated with correlation equal to ϕ. Therefore, the correlation between Y_t and Y_{t-2} (or between Y_t and Y_{t+2}) is ϕ^2. However, if we eliminate the correlation effect of the *intermediate* variable Y_{t-1}, the remaining correlation (the partial correlation) between Y_t and Y_{t-2} in an AR(1) process is zero. That is, the lag 1 correlation explains completely all the correlation at higher lags in an AR(1) process. Figure 3.16 explains this idea.

Now consider an invertible MA process. This process can be inverted into an infinite-order AR process with decreasing weights given to older Y's; thus its PACF will tail-off exponentially (geometrically). This can be seen because the SPACF can be estimated by fitting AR models of increasing order, as explained above. If the process can be represented by an AR(∞) process, the ϕ_{kk} will continue to be nonzero (although will tend to zero) as k increases. Since the MA process is invertible, the sum of coefficients will converge, which implies that at some point the coefficients will be close to zero.

For AR processes, fitting increasingly higher order AR models will yield coefficients ϕ_{kk} that will be estimating zero after lag p; thus the PACF cuts off after lag p for AR(p) processes. This will allow us to infer the order of an AR process.

To discriminate between zero values and values that are significantly different from zero in the SACF and the SPACF, we make use of their

Table 3.3   Useful Results for AR and MA Identification

	MA(q)	AR(p)	White Noise
ACF (ρ_k)	$\rho_k = 0$ for $k > q$	ρ_k tails off	$\rho_k = 0$ for all k
	$\hat{\sigma}(r_k) = \sqrt{\dfrac{1}{N}\left(1 + 2\Sigma_{v=1}^{q} r_v^2\right)}$ for $k > q$		$\hat{\sigma}(r_k) = \dfrac{1}{N}$ for all k
PACF ($\hat{\phi}_{kk}$)	ϕ_{kk} tails off	$\phi_{kk} = 0$ for $k > p$	$\phi_{kk} = 0$ for all k
		$\hat{\sigma}(\hat{\phi}_{kk}) \approx \dfrac{1}{\sqrt{N}}$ for $k > p$	$\hat{\sigma}(\hat{\phi}_{kk}) = \dfrac{1}{\sqrt{N}}$ for all k

standard errors (found by Bartlett and Quenouille). Thus we have Table 3.3. It is common practice to compare the estimated ACF and PACF against the standard errors of the white noise case. A rule of thumb is that if any r_k or $\hat{\phi}_{kk}$ is within the $\pm 2/\sqrt{N}$ band, one should conclude that the corresponding r_k or ϕ_{kk} value is zero; the difference observed in this case can be explained by noise.

3.6.1   Identification of ARMA Processes

As mentioned earlier, correct identification when both p and q are greater than 1 is difficult to achieve based only on the sample ACF and sample PACF. Two special techniques used to identify ARMA processes are the extended sample autocorrelation function (ESACF) and the smallest canonical correlation (SCAN) method, both proposed by Tsay and Tiao (1984, 1985). Some software packages [e.g., SAS (PROC ARIMA)] make it possible to use these techniques, which are outside the scope of this book.

Some Instances of ACF and PACF Functions
Figures 3.17 to 3.19 give all possible behaviors for AR(1), MA(1), AR(2), and MA(2) processes that are stationary and invertible.[9] It should be pointed out from the figures that for AR(2) processes, the ACF can oscillate according to a damped sinusoidal function if the roots of the $A(\mathcal{B})$ polynomial are complex conjugates; sinusoidal oscillations cannot occur in AR(1) processes. Similarly, the PACF of MA(2) processes can oscillate in certain cases, but this is not possible for MA(1) processes. These facts are sometimes useful in model identification.

---

[9]The power spectrum shown in the figures depicts how the frequencies that make up the process vary. See Appendix 3C for more details.

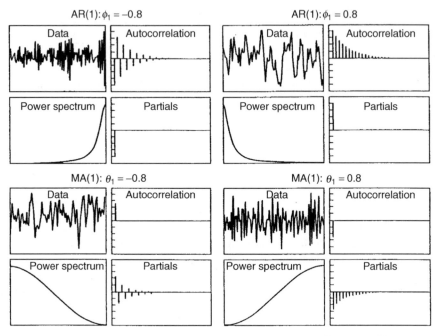

Figure 3.17 Summary of data series and theoretical properties of AR(1) and MA(1) processes. (From Makridakis et al., 1983.)

3.6.2 Identification of Nonstationary Processes

If a process is nonstationary, its SACF will fail to decay exponentially fast. If the decay is linear, not necessarily starting at 1.0, this usually indicates that the process is nonstationary. A simple plot of the time-series data can reveal whether or not the underlying process is stationary. If the process is found to be nonstationary, we proceed to difference the series by computing $\nabla y_t = y_t - y_{t-1}$. If the resulting series still appears nonstationary, we difference again, and so on. Quite often, in industrial quality control data we have $d \le 2$. Once we have obtained a stationary process by successive differencing, we proceed to identify p and q and estimate the parameters of the model.

Overdifferencing

There is always the danger of overdifferencing the series. The problem is that we will be introducing a noninvertible MA term into the model. In addition, the variance of the overdifferenced model will be inflated artificially, and this may obscure the actual structure of the data. To see how a noninvertible MA process can be introduced by overdifferencing, consider a random walk model,

$$Y_t = Y_{t-1} + \varepsilon_t$$

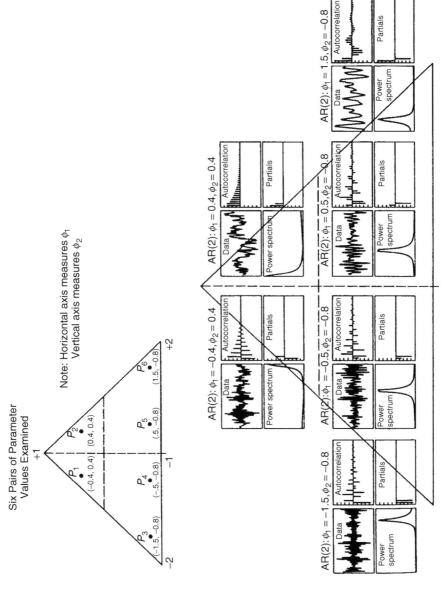

Figure 3.18 Summary of data series and theoretical properties of AR(2) processes. (From Makridakis et al., 1983.)

99

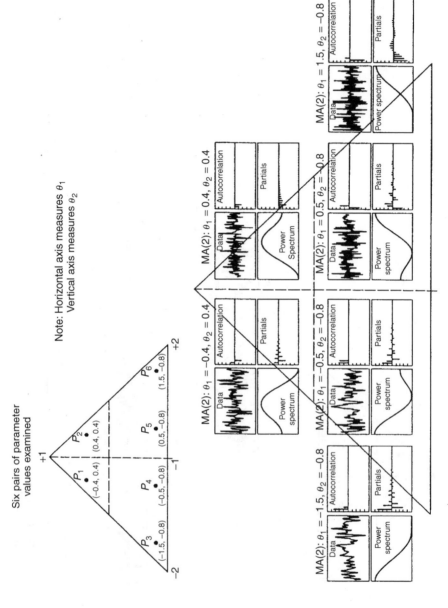

Figure 3.19 Summary of data series and theoretical properties of MA(2) processes. (From Makridakis et al., 1983.)

If we difference this process once we get a white noise process (see Section 3.4) but if for some reason we decide to difference again, we will obtain

$$\nabla^2 Y_t = \nabla \varepsilon_t = (1 - \mathcal{B}) \varepsilon_t.$$

Thus the problem of overdifferencing is that we introduce an noninvertible term, with the illogical consequence that more weight is given to older than to more recent observations. Furthermore, the variance of the overdifferenced series will be inflated. For example, in the random walk example, the variance of $\nabla^2 Y_t$ is twice that of ∇Y_t.

Several alternatives have been proposed to detect overdifferencing and thus finding the right value of d. A simple approach is to compute $\mathrm{Var}(y_t)$, $\mathrm{Var}(\nabla y_t), \ldots, \mathrm{Var}(\nabla^n y_t)$ and choose d such that this variance is a minimum, provided that the SACF of each differenced series looks stationary (exponential decay). An alternative approach is to use the *inverse autocorrelation function* (IACF). If the original process is ARMA(p, q);

$$A(\mathcal{B})\tilde{Y}_t = C(\mathcal{B}) \varepsilon_t$$

the *dual* model is defined as the ARMA(q, p) model:

$$C(\mathcal{B})\tilde{Y}_t = A(\mathcal{B}) \varepsilon_t$$

The autocorrelation function of the dual model is called the *inverse autocorrelation function* of the original series. This can be estimated using spectral analysis techniques (Cleveland, 1972; Kendall and Ord, 1990, p. 98).

If the original process is noninvertible or contains a noninvertible root due to overdifferencing, this will cause the IACF to look as the ACF of a nonstationary process (slow, linear decay). Thus we check the IACF for overdifferencing, and if we see a pattern that resembles nonstationary behavior of the dual model, we backtrack and use a degree of differencing one unit lower.

3.6.3 Box–Jenkins Approach to ARIMA Model Building

Model building is an iterative process. Figure 3.20 shows the general approach suggested by Box and Jenkins (1976). If a hypothesized ARIMA model does not provide good fit or several alternatives seems to exist, these should be tried as well. This is indicated by the dashed line on the diagram. Alternatively, if remaining autocorrelation is found in the residuals, it can be incorporated into the previously specified model. We discuss this after an example in which all diagnostic checks look correct.

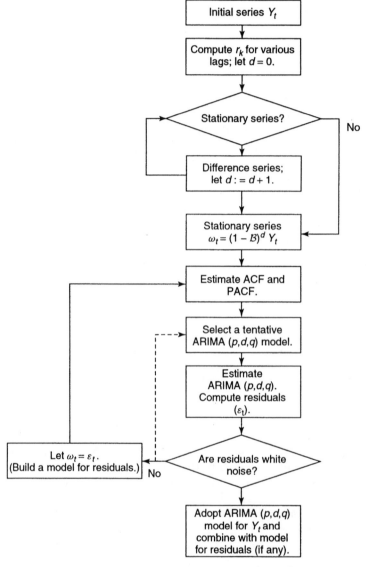

Figure 3.20   Box–Jenkins methodology for ARIMA(p, d, q) modeling.

Example 3.13: Complete ARIMA Model-Building Example. Consider the series given in Montgomery et al. (1990, Table 10.2). (The data can be found in the file *Viscosity.txt*.) The data, plotted in Figure 3.21, represent sequential readings of the viscosity of a chemical process, an important quality characteristic that is of interest to model. We conduct the analysis using the Minitab statistical software. First, we look at a plot of the data ($\{y_t\}$) and

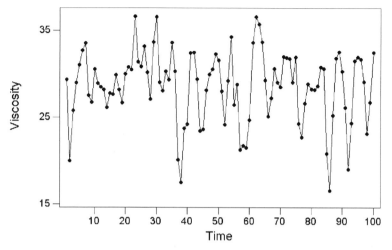

Figure 3.21 Time-series plot of the viscosity data.

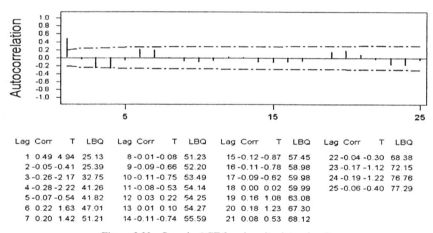

Lag	Corr	T	LBQ	Lag	Corr	T	LBQ	Lag	Corr	T	LBQ	Lag	Corr	T	LBQ
1	0.49	4.94	25.13	8	-0.01	-0.08	51.23	15	-0.12	-0.87	57.45	22	-0.04	-0.30	68.38
2	-0.05	-0.41	25.39	9	-0.09	-0.66	52.20	16	-0.11	-0.78	58.98	23	-0.17	-1.12	72.15
3	-0.26	-2.17	32.75	10	-0.11	-0.75	53.49	17	-0.09	-0.62	59.98	24	-0.19	-1.22	76.76
4	-0.28	-2.22	41.26	11	-0.08	-0.53	54.14	18	0.00	0.02	59.99	25	-0.06	-0.40	77.29
5	-0.07	-0.54	41.82	12	0.03	0.22	54.25	19	0.16	1.08	63.08				
6	0.22	1.63	47.01	13	0.01	0.10	54.27	20	0.18	1.23	67.30				
7	0.20	1.42	51.21	14	-0.11	-0.74	55.59	21	0.08	0.53	68.12				

Figure 3.22 Sample ACF for viscosity data, $d = 0$.

compute its sample autocorrelation function, which is shown in Figure 3.22. The time plot and the SACF do not show evidence of nonstationarity. The SACF decays rapidly, following damped sinusoidal-like oscillations. The sample partial autocorrelation function (SPACF) cuts off at lag 2 (see Figure 3.23). A SPACF cutting off after lag 2 and a damped sinusoidal decay in the SACF are signs of an AR(2) process, so we should proceed to fit this model.

□

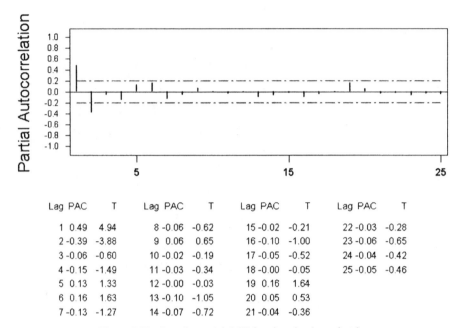

Lag PAC T | Lag PAC T | Lag PAC T | Lag PAC T

Lag	PAC	T	Lag	PAC	T	Lag	PAC	T	Lag	PAC	T
1	0.49	4.94	8	-0.06	-0.62	15	-0.02	-0.21	22	-0.03	-0.28
2	-0.39	-3.88	9	0.06	0.65	16	-0.10	-1.00	23	-0.06	-0.65
3	-0.06	-0.60	10	-0.02	-0.19	17	-0.05	-0.52	24	-0.04	-0.42
4	-0.15	-1.49	11	-0.03	-0.34	18	-0.00	-0.05	25	-0.05	-0.46
5	0.13	1.33	12	-0.00	-0.03	19	0.16	1.64			
6	0.16	1.63	13	-0.10	-1.05	20	0.05	0.53			
7	-0.13	-1.27	14	-0.07	-0.72	21	-0.04	-0.36			

Figure 3.23   Sample partial ACF for viscosity data, $d = 0$.

3.7   OFFLINE ESTIMATION

Once a tentative ARIMA(p, d, q) model has been identified, the next step is to estimate the parameters of the model. In this section we discuss *offline* methods for estimation. This means that parameter estimates are obtained once after a large set of N observations has been collected. This is in contrast with online estimation methods, discussed in Chapter 8, where the parameter estimates change as each new observation is obtained.

Maximum Likelihood and Least Squares Estimation

Suppose that we have a time series $\{y_t\}$ consisting of N observations that were generated as a realization of some ARIMA process. After differencing $d \, (\geq 0)$ times, we obtain $n = N - d$ observations of the stationary series $\{w_t\}$, where $w_t = \nabla^d y_t$. We want to estimate the parameters of the ARMA(p, q) model:

$$\tilde{w}_t - \phi\,\tilde{w}_{t-1} - \cdots - \phi_p\tilde{w}_{t-p} = -\theta_1\varepsilon_{t-1} - \theta_2\varepsilon_{t-2} - \cdots - \theta_q\varepsilon_{t-q} + \varepsilon_t$$

where $\tilde{w}_t = w_t - \mu$. Assuming that the ε's are normally distributed, their joint density function forms the likelihood function and is given by

$$p(\varepsilon_1, \varepsilon_2, \ldots, \varepsilon_n) \propto \frac{1}{\sigma_\varepsilon^n} e^{-\sum_{t=1}^n (\varepsilon_t^2 / 2\sigma_\varepsilon^2)} = \frac{1}{\sigma_\varepsilon^n} e^{-SS(\phi, \theta)/2\sigma_\varepsilon^2}$$

where $\phi' = (\phi_1, \phi_2, \ldots, \phi_p)$, $\theta' = (\theta_1, \theta_2, \ldots, \theta_q)$, and $SS(\phi, \theta)$ denotes the sum of squared errors as a function of the parameters. Maximizing the likelihood function with respect to the parameters will make the parameters as compatible as possible with the evidence in the sample. The symbol $\propto$ means is "proportional to," where the proportionality constant is not shown. Taking the logarithm, we obtain the *log-likelihood function*, which can be shown to be of the form

$$L(\phi, \theta, \sigma_\varepsilon) = f(\phi, \theta) - n \ln(\sigma_\varepsilon) - \frac{SS(\phi, \theta)}{2\sigma_\varepsilon^2}$$

where $f(\phi, \theta)$ is a function of ϕ and θ with which we do not need to be concerned for the moment. The parameter values maximizing the log-likelihodd function are the same as those maximizing the likelihood, and since the former is easier to optimize than the latter, we continue our discussion based on the log-likelihood.

The errors are estimated from the residuals obtained from given choices of ϕ and θ. The sum of squared errors (residuals) is

$$SS(\phi, \theta) = \sum_{t=1}^n \left[w_t - \hat{w}_{t|t-1}(\phi, \theta) \right]^2 = \sum_{t=1}^n \varepsilon_t^2$$

This is called the *conditional* sum of squares because its value depends on the p initial values [10] of w_t and the q initial values of ε_t. To obtain maximum likelihood estimates (MLEs) of ϕ and θ, we maximize $L(\phi, \theta, \sigma_\varepsilon)$. It turns out that for large n, the effect of the term $f(\phi, \theta)$ is almost negligible; thus, maximizing the log-likelihood function can be achieved by minimizing the sum of squares $SS(\phi, \theta)$. Thus for large n we have that

$$\left(\hat{\phi}, \hat{\theta}\right)_{MLE} \approx \left(\hat{\phi}, \hat{\theta}\right)_{OLS}$$

where OLS stands for ordinary least squares estimates.

OLS estimation can be performed in closed form if the model is a *linear statistical model*. ARIMA models are stochastic, linear difference equation models but are not always what in statistics is called a model linear in the

---

[10] Recall that ARIMA forecasts are computed recursively and need initial conditions at some point in time.

parameters, or a "linear model" for short. In the case of ARIMA models, a model is linear in the parameters if

$$\frac{\partial \varepsilon_t}{\partial \phi_i} \quad i = 1, \ldots, p \quad \text{and} \quad \frac{\partial \varepsilon_t}{\partial \theta_j} \quad j = 1, \ldots, q$$

are *not* functions of any unknown parameter. It is easy to see that all AR(p) models are linear. ARIMA(p, d, q) models where $q > 0$ are nonlinear in the parameters. Thus MA models cannot be estimated by the usual least squares closed-form estimation formula. A couple of examples will clarify these ideas.

Example 3.14: Estimation of an AR(2) Process. The model is

$$\tilde{Y}_t = \phi_1 \tilde{Y}_{t-1} + \phi_2 \tilde{Y}_{t-2} + \varepsilon_t$$

Thus

$$\varepsilon_t = \tilde{Y}_t - \phi_1 \tilde{Y}_{t-1} - \phi_2 \tilde{Y}_{t-2}$$

The derivatives with respect to the parameters are

$$\frac{\partial \varepsilon_t}{\partial \phi_i} = \tilde{Y}_{t-i} \quad i = 1, 2$$

and therefore this is a linear statistical model.

Let us see how an AR(2) model can be fitted using standard regression techniques. Avoiding initial conditions of the difference equation (something that will have little effect if the sample size n is large), a sequence of random variables defined by an AR(2) process can be written

$$\begin{pmatrix} \tilde{Y}_3 \\ \tilde{Y}_4 \\ \vdots \\ \tilde{Y}_{n-1} \\ \tilde{Y}_n \end{pmatrix} = \begin{pmatrix} \tilde{Y}_2 & \tilde{Y}_1 \\ \tilde{Y}_3 & \tilde{Y}_2 \\ \vdots & \vdots \\ \tilde{Y}_{n-2} & \tilde{Y}_{n-3} \\ \tilde{Y}_{n-1} & \tilde{Y}_n \end{pmatrix} \begin{pmatrix} \phi_1 \\ \phi_2 \end{pmatrix} + \begin{pmatrix} \varepsilon_3 \\ \varepsilon_4 \\ \vdots \\ \varepsilon_{n-1} \\ \varepsilon_n \end{pmatrix}$$

In vector notation this can be written as

$$\tilde{\mathbf{Y}} = \mathbf{Z}' \boldsymbol{\phi} + \boldsymbol{\varepsilon}$$

Therefore, the least squares estimator is given by

$$\hat{\boldsymbol{\phi}} = (\mathbf{Z}'\mathbf{Z})^{-1} \mathbf{Z}' \tilde{\mathbf{Y}} \qquad \square$$

Example 3.15: Continuation of Viscosity Data Analysis. When fitting an AR(2) model to the viscosity data of Example 3.13, we get the following output from Minitab:

ARIMA model for Viscosity

```
Final Estimates of Parameters
Type              Coef        StDev          T            P
AR   1          0.7186      0.0923        7.78       0.000
AR   2         -0.4343      0.0922       -4.71       0.000
Constant       20.5028      0.3279       62.52       0.000
Mean           28.6506      0.4583
```

```
Number of observations:   100
Residuals:       SS =  1042.82   (back forecasts excluded)
                 MS =    10.75   DF = 97
```

```
Modified Box-Pierce (Ljung-Box) Chi-Square statistic
Lag                  12           24          36           48
Chi-Square         12.2         21.1        28.3         38.0
DF                    9           21          33           45
P-Value           0.204        0.451       0.700        0.760
```

```
Correlation matrix of the estimated parameters
            1            2
  2   -0.506
  3    0.017       -0.002
```

Thus the model estimated by Mintab is

$$Y_t = 20.5028 + 0.7186Y_{t-1} - 0.4343Y_{t-2}$$

where both estimated coefficients are highly significant and are practically uncorrelated from the correlation matrix. Furthermore, the estimate of σ_ε is $\sqrt{10.75} = 3.27$. Other statistics given by Minitab are explained later.

□

The following example shows that for MA models ordinary least squares estimation cannot be used for parameter estimation.

Example 3.16: Estimation of an MA(1) Process. Consider an MA(1) process:

$$\tilde{Y}_t = -\theta_1\varepsilon_{t-1} + \varepsilon_t = (1 - \theta\mathcal{B})\varepsilon_t$$

Solving for ε_t and assuming invertibility ($|\theta| < 1$), we get

$$\varepsilon_t = \frac{\tilde{Y}_t}{1 - \theta\mathcal{B}} \tag{3.14}$$

Therefore, the derivative with respect to the parameter is

$$\frac{\partial\varepsilon_t}{\partial\theta} = \frac{\mathcal{B}\tilde{Y}_t}{(1 - \theta\mathcal{B})^2} = \frac{\mathcal{B}\varepsilon_t}{1 - \theta\mathcal{B}}$$

which is a function of the unknown parameter θ. Thus this is not a linear statistical model, and no closed formula exists for the least squares estimator. This has to be obtained numerically from minimization of the sum-of-squares function. A popular nonlinear optimization algorithm that accomplishes this is Marquardt's method.   □

Estimation of μ and σ_ε
A simple point estimate of the mean is given by $\hat{\mu} = \overline{w} = (1/n)\sum_{t=1}^{m}w_i$. In many cases, setting $\mu = 0$ ($\xi = 0$) in an ARIMA(p, d, q) process works well (Box et al., 1994, p. 226).

3.8   DIAGNOSTIC CHECKING IN TIME-SERIES MODELS

The residuals $Y_t - \hat{Y}_t = e_t$ should be white noise if the model fit is adequate. The idea is that if the model $\hat{Y}$ has captured all the autocorrelation structure in the time series, no autocorrelation should remain in the residuals. Therefore, it is recommended to check the ACF of the residuals, which should look like a white noise sequence if there are no model inadequacies. A useful statistic computed by many statistical packages (e.g., Minitab, SAS) is the Ljung–Box–Pierce statistic, defined by

$$Q = n(n + 2) \sum_{k=1}^{K} \frac{r_e(k)^2}{n - k}$$

This statistic is used in the *Portmanteau test*, which has null hypothesis H_0: $\rho_i = 0$, $i = 1, \ldots, K$. Evidently, larger values of Q are due to larger autocorrelations. The rejection rule is to reject H_0 if $Q > \chi^2_{\alpha, K-p-q}$.

Example 3.17: Continuation of Viscosity Data Analysis.   After fitting an AR(2) model to the viscosity data in Example 3.15, we proceed to analyze the model and perform diagnostic checks. A plot of the SACF of the residuals (Figure 3.24) does not reveal any remaining autocorrelation that needs to be accounted for. Also, a plot of residual versus fitted values (Figure 3.25) does not show any inadequacy. There seems to be a constant error variance and zero-centered errors. The formal test for lack of autocorrelation in the residuals is also performed by Minitab:

```
Modified Box-Pierce (Ljung-Box) Chi-Square statistic
Lag                   12        24        36        48
Chi-Square           2.2      21.1      28.3      38.0
DF                     9        21        33        45
P-Value            0.204     0.451     0.700     0.760
```

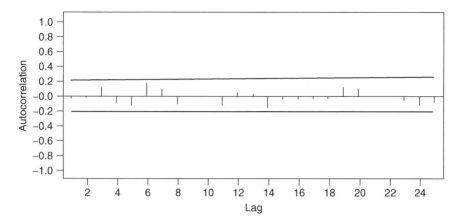

Figure 3.24 SACF of residuals of fitted AR(2) process (with 95% confidence limits for the autocorrelations).

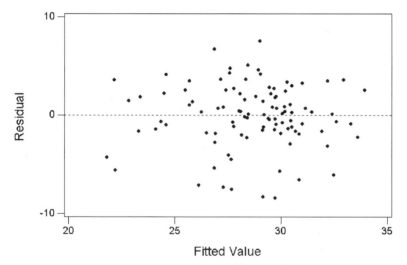

Figure 3.25 Residuals versus fitted value, AR(2) process (response is viscosity).

Minitab computes Q for lags $K = 12$ to $K = 48$. In all cases, the p-value of the test is quite large, showing that with the evidence at hand we cannot reject the hypothesis of no autocorrelations in the residuals. This and the previous graphical analysis of the sample autocorrelations of the residuals lead us to conclude that an AR(2) model is adequate.

Since the AR(2) model appears to have an adequate fit, we could use it, for example, for forecasting purposes. Figure 3.26 shows one-step-ahead MMSE forecasts for the fitted model starting at the last observation. The forecasts are computed based on the method of Section 3.5. Minitab also computes a 95% band for the forecasts.   □

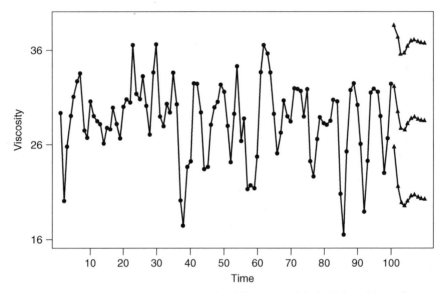

Figure 3.26   Viscosity series with one-step-ahead forecasts and their 95% confidence limits.

Use of Residuals to Modify an ARIMA(p, d, q) **Model**

If the residuals $\{\varepsilon_t\}$ show structure in the ACF other than that a white noise process would generate, the initial model has to be modified. A simple approach is to refit a different model (i.e., this usually means change p or q) and try again. Alternatively, we could fit an ARMA(p_e, q_e) model to the residuals:

$$\hat{A}_e(\mathcal{B})e_t = \hat{C}_e(\mathcal{B})\varepsilon_t \tag{3.15}$$

which, if it captures all the remaining correlation, should make the series $\{\varepsilon_t\}$ white. Then model (3.15) can be combined with the originally identified ARIMA(p, d, q) model to yield

$$\hat{A}_e(\mathcal{B})\hat{A}(\mathcal{B})\nabla^d\tilde{Y}_t = \hat{C}(\mathcal{B})\hat{C}_e(\mathcal{B})\varepsilon_t$$

which is just an ARIMA($p + p_e, d, q + q_e$) process.

3.9   USING SAS PROC ARIMA

The SAS software provides the procedure ("PROC") ARIMA, which is very powerful—not only to fit ARIMA models but also transfer function models, as will be seen in Chapter 4. A skeleton SAS program for ARIMA modeling is as follows:

```
PROC ARIMA DATA = (dataName);
IDENTIFY VAR = Y(d1,"d2,...);
ESTIMATE P = _"_"_Q_"_"_ PLOT;
END;
```

The IDENTIFY statement names the series (Y in this example) and specifies the lags at which differencing should be performed. The differences are computed by applying the operators $1 - \mathcal{B}^{d_1}, 1 - \mathcal{B}^{d_2}, \ldots$ to the original time series. Thus, if we wish to difference once (i.e., if we want to get ∇Y_t), we need to specify VAR = Y(1). If we wish to difference twice, (i.e., if we want to get $\nabla^2 Y_t$), we should specify VAR = Y(1,1). Note that specifying VAR = Y(2) will compute $(1 - \mathcal{B}^2)Y_t = Y_t - Y_{t-2}$, which clearly does not equal $\nabla^2 Y_t$. The IDENTIFY statement also prints the sample autocorrelation function (SACF), the sample partial autocorrelation function (SPACF), and the sample inverse autocorrelation function (SIACF) of the *differenced series* if any differences were requested. The ESTIMATE statement estimates the parameters of an ARMA(p, q) model for the differenced series, and the option PLOT plots the SACF, SPACF, and SIACF of the residuals.

Example 3.18. We repeat the ARIMA analysis for the viscosity data in Montgomery et al. (1990) using SAS PROC ARIMA. The input file is simply as follows (only the first four and the last observations are shown):

```
title1 'Viscosity Data';
data viscosity;
  input y;
cards;
29.33
19.98
25.76
29.00
...
32.44
;

proc arima data = viscosity;
  identify var = y nlags = 15;
  estimate p = 2 plot;
  run;
quit;
```

Figures 3.27 and 3.28 show some of the output produced by SAS in this example. The reader should compare this output with that produced by Minitab in Examples 3.13, 3.16, and 3.17.   □

Name of Variable = y

Mean of Working Series   28.5687
Standard Deviation   4.111625
Number of Observations   100

Autocorrelations

Lag	Covariance	Correlation	-1 9 8 7 6 5 4 3 2 1 0 1 2 3 4 5 6 7 8 9 1	Std Error
0	16.905457	1.00000	| |********************|	0
1	8.349453	0.49389	| . |********** |	0.100000
2	-0.837598	-.04955	| . *| .	0.121978
3	-4.471796	-.26452	| *****| .	0.122179
4	-4.785962	-.28310	| ******| .	0.127777
5	-1.217568	-.07202	| . *| .	0.133903
6	3.699606	0.21884	| . |****.	0.134290
7	3.308200	0.19569	| . |****.	0.137810
8	-0.189561	-.01124	| . | .	0.140561
9	-1.576440	-.09325	| . **| .	0.140570
10	-1.800490	-.10650	| . **| .	0.141187
11	-1.274535	-.07539	| . **| .	0.141988
12	0.528047	0.03124	| . |* .	0.142388
13	0.230977	0.01366	| . | .	0.142457
14	-1.780945	-.10535	| . **| .	0.142470
15	-2.101861	-.12433	| . **| .	0.143247

"." marks two standard errors

Inverse Autocorrelations

Lag	Correlation	-1 9 8 7 6 5 4 3 2 1 0 1 2 3 4 5 6 7 8 9 1
1	-0.55244	| ***********| . |
2	0.21916	| . |**** |
3	-0.05787	| . *| . |
4	0.05829	| . |* . |
5	0.08303	| . |** . |
6	-0.11143	| . **| . |
7	-0.01711	| . | . |
8	0.06851	| . |* . |
9	-0.01433	| . | . |
10	-0.02830	| . *| . |
11	0.07264	| . |* . |
12	-0.03958	| . *| . |
13	0.01984	| . | . |
14	0.02669	| . |* . |
15	0.01302	| . | . |

The ARIMA Procedure
Partial Autocorrelations

Lag	Correlation	-1 9 8 7 6 5 4 3 2 1 0 1 2 3 4 5 6 7 8 9 1
1	0.49389	| . |********** |
2	-0.38816	| ********| . |
3	-0.06049	| . *| . |
4	-0.14889	| .***| . |
5	0.13257	| . |***. |
6	0.16272	| . |***. |
7	-0.12657	| .***| . |
8	-0.06174	| . *| . |
9	0.06452	| . |* . |
10	-0.01858	| . | . |
11	-0.03399	| . *| . |
12	-0.00337	| . | . |
13	-0.10467	| . **| . |
14	-0.07226	| . *| . |
15	-0.02145	| . | . |

Autocorrelation Check for White Noise

To Lag	Chi-Square	DF	Pr > ChiSq	------------------------Autocorrelations------------------------
6	47.01	6	<.0001	0.494 -0.050 -0.265 -0.283 -0.072 0.219
12	54.25	12	<.0001	0.196 -0.011 -0.093 -0.107 -0.075 0.031

Conditional Least Squares Estimation

Parameter	Estimate	Standard Error	t Value	Approx Pr > |t|	Lag
MU	28.67477	0.47332	60.58	<.0001	0
AR1,1	0.70166	0.09394	7.47	<.0001	1
AR1,2	-0.40350	0.09402	-4.29	<.0001	2

Constant Estimate   20.125
Variance Estimate   11.04219
Std Error Estimate   3.322979
AIC   526.9141
SBC   534.7296
Number of Residuals   100

Figure 3.27 SAS PROC ARIMA output for the viscosity data: sample autocorrelation function and parameter estimates.

Correlations of Parameter Estimates

Parameter	MU	AR1,1	AR1,2
MU	1.000	0.024	0.014
AR1,1	0.024	1.000	−0.503
AR1,2	0.014	−0.503	1.000

Autocorrelation Check of Residuals

To Lag	Chi-Square	DF	Pr > ChiSq	-----------------------Autocorrelations-----------------------						
6	7.41	4	0.1158	−0.025	−0.027	0.077	−0.103	−0.125	0.188
12	11.86	10	0.2947	0.091	−0.115	−0.009	0.001	−0.127	0.043
18	15.60	16	0.4809	0.026	−0.150	−0.057	−0.044	−0.033	−.045
24	20.87	22	0.5285	0.111	0.080	−0.017	0.018	−0.070	−.127

Autocorrelation Plot of Residuals

Lag	Covariance	Correlation	−1 9 8 7 6 5 4 3 2 1 0 1 2 3 4 5 6 7 8 9 1	Std Error
0	11.042189	1.00000	|********************|	0
1	−0.281056	−.02545	. *| .	0.100000
2	−0.295801	−.02679	. *| .	0.100065
3	0.849051	0.07689	. |** .	0.100136
4	−1.137920	−.10305	. **| .	0.100725
5	−1.384165	−.12535	. ***| .	0.101774
6	2.074716	0.18789	. |**** .	0.103306
7	1.007433	0.09123	. |** .	0.106669
8	−1.267115	−.11475	. **| .	0.107446
9	−0.103124	−.00934	. | .	0.108665
10	0.0067976	0.00062	. | .	0.108673
11	−1.402203	−.12699	. ***| .	0.108673
12	0.478134	0.04330	. |* .	0.110147
13	0.283176	0.02564	. |* .	0.110317
14	−1.651826	−.14959	. ***| .	0.110377
15	−0.631233	−.05717	. *| .	0.112386

"." marks two standard errors

The ARIMA Procedure
Inverse Autocorrelations

Lag	Correlation	−1 9 8 7 6 5 4 3 2 1 0 1 2 3 4 5 6 7 8 9 1	
1	0.04559	. |* .	|
2	−0.09190	. **| .	|
3	−0.05503	. *| .	|
4	0.14425	. |***.	|
5	0.12944	. |***.	|
6	−0.13906	. ***| .	|
7	−0.12919	. ***| .	|
8	0.04909	. |* .	|
9	0.08485	. |** .	|
10	0.02777	. |* .	|
11	0.03612	. |* .	|
12	−0.02227	. | .	|
13	0.04355	. |* .	|
14	0.11944	. |** .	|
15	0.04845	. |* .	|

Partial Autocorrelations

Lag	Correlation	−1 9 8 7 6 5 4 3 2 1 0 1 2 3 4 5 6 7 8 9 1	
1	−0.02545	. *| .	|
2	−0.02745	. *| .	|
3	0.07560	. |** .	|
4	−0.10067	. **| .	|
5	−0.12769	. ***| .	|
6	0.17680	. |****	|
7	0.11447	. |** .	|
8	−0.10825	. **| .	|
9	−0.07064	. *| .	|
10	0.01362	. | .	|
11	−0.04700	. *| .	|
12	0.01171	. | .	|
13	−0.04581	. *| .	|
14	−0.13038	. ***| .	|
15	−0.05417	. *| .	|

Figure 3.28 SAS PROC ARIMA output for the viscosity data (*cont.*): residual diagnostics.

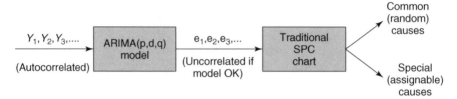

Figure 3.29   General approach for monitoring autocorrelated data using ARIMA models.

3.10   TIME-SERIES APPROACHES TO SPC OF AUTOCORRELATED DATA

A considerable number of authors (Alwan and Roberts, 1988; Montgomery and Mastrangelo 1991; Vander Weil, 1996; Lu and Reynolds, 1999) have proposed modeling the inherent autocorrelation that exists in process control data obtained in modern manufacturing. The argument is that if the autocorrelation is considered as normal for the process, once an ARIMA(p, d, q) model is fit to the data, the model becomes the in-control model. As long as new observations behave according to the model, one-step-ahead forecast errors (similar to the residuals but computed as the difference between prediction by model and actual) should form a white noise sequence. Thus the forecast errors would conform to the assumptions behind Shewhart's model and we would be able, in principle, to detect shifts and other unpredictable disturbances typical of SPC. Figure 3.29 illustrates the general approach. The main problem with this approach is that the models are just approximations, so the monitoring scheme will be sensitive to estimation errors in the parameters (Adams and Tseng, 1998; Lu and Reynolds, 1999; Jones et al., 2001).

A similar approach was proposed by Montgomery and Mastrangelo (1991), who monitor the one-step-ahead forecasts of the process on the same plot as the data, using the IMA(1, 1) as a generic model and time-varying control limits. The IMA(1, 1) model $Y_t = -\theta\varepsilon_{t-1} + \varepsilon_t$ is fit to the data by estimating the value of θ using a least squares criterion. An EWMA with parameter $\lambda = 1 - \hat{\theta}$ is then used as a generic forecasting technique for the process.

Although this approach appears to suffer from the same problems as those of the approach shown in Figure 3.29, a number of additional monitoring devices based on tracking signals have been tested extensively and seem to improve the performance of the method (Mastrangelo and Montgomery, 1995). Obviously, the method works better the closer the process is to an IMA(1, 1) process [e.g., when we have an ARMA(1, 1) process with a value of ϕ close to 1]. More recent approaches for performing SPC with autocorrelated data can be found in the "Bibliography and Comments" section.

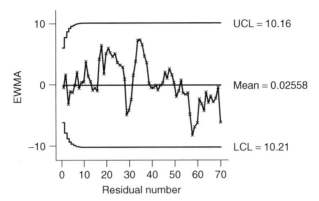

Figure 3.30 EWMA control chart applied to the residuals of the model fitted to the chemical batch process data in Example 1.7.

Example 3.19: Monitoring the Residuals of an ARMA Model. Consider the batch chemical process from Box and Jenkins (1976, series J), also described in Example 1.7. From the autocorrelation and partial autocorrelation functions, and after looking at the residuals, an ARMA(1, 1) model seems to represent the data well. The fitted model is $Y_t = 86.91 - 0.6958 Y_{t-1} + 0.03277 \varepsilon_{t-1} + \varepsilon_t$ (the $\hat{\theta}$ estimate is not very significant, but the residual sum of squares is reduced a little by inclusion of the MA term). Figure 3.30 shows an EWMA control chart applied to the residuals of the fitted model. The EWMA chart uses a value of $\lambda = 0.2$. From the chart the process seems to be in control with respect to the hypothesized ARMA(1, 1) model while the observations were collected. Online monitoring can be implemented by continuous use of the control chart to monitor future forecast errors.

□

An alternative to the previous approaches is to monitor the original autocorrelated data. Yaschin (1993) mentions that unless ρ_1 is large, shift detection in the original series is better accomplished with a CUSUM chart applied to the series itself rather than to the forecast errors of some model. Similar rationale has led some other authors to recommend use of a Shewhart $\bar{Y}$ chart in which the limits are set based on an adjusted estimator of the process variance that considers the systematic between-sample variability induced by the autocorrelation of the series. Thus, for an AR(1) process, the modified limits are set at $\hat{\mu} \pm 3\hat{\sigma}/\sqrt{1 - \hat{\phi}^2}$, where ϕ is the lag 1 autocorrelation coefficient and σ is estimated from the standard deviation of the residuals obtained after fitting an AR(1) process to the data.

A similar recommendation is given by Lu and Reynolds (1999), who applied Shewhart and EWMA charts to the residuals of a time-series model. They concluded that if the autocorrelation is low to moderate, there is no significant difference between monitoring the residuals and monitoring the observations directly using a standard SPC chart with modified limits. In fact, for large shifts, they reported that a Shewhart chart of the observations is better than a Shewhart chart on the residuals in terms of ARLs. For small shifts, an EWMA of the observations slightly outperforms an EWMA of the residuals. Lu and Reynolds (2001) consider CUSUM charts for autocorrelated data and conclude that for small levels of autocorrelation, monitoring the individual observations using a CUSUM chart performs as well as monitoring the residuals of a model using a CUSUM chart. They also indicated that EWMA charts and CUSUM charts perform similarly in the presence of autocorrelation.

Regardless of the remedial actions taken to monitor an autocorrelated process, we have to be careful in reading performance statistics related to charts for this type of process. The effect of autocorrelation on a control chart scheme depends crucially on (1) the actual model that the process really follows, (2) the model that we consider to describe in-control operation, (3) the way we estimate the in-control model, and (4) the type of disturbance that the chart is supposed to detect. Considerations 1 to 3 are particularly important because the effect of the uncertainties in the parameter estimates and in the form of the estimated model will affect the performance of the chart procedure. Lu and Reynolds (1999) report that a very large data set is necessary before the performance of an EWMA chart with estimated process parameters approaches that of a chart based on known parameters. This applies to the case when the residuals are monitored, and perhaps more important, it also applies to the case when the limits are adjusted using the parameter estimates. A related problem with autocorrelated data, sometimes ignored in practice, is that a process that has step shifts in the mean but is stable otherwise would appear as highly autocorrelated. [This has been illustrated by Wiklund (1994) with examples from the literature.]

A minor problem in the literature of SPC methods for autocorrelated data seems to be that there is no common description of what a process shift is for such processes. For example, very different ARL performance will occur if for a stationary process, assignable causes change the mean of the process $\{Y_t\}$ or if they change the mean of the error terms $\{\varepsilon_t\}$. Shift magnitudes measured as multiples of the standard deviation of the process rather than as multiples of the standard deviations of the error add to this confusion.

Finally, we point out that a completely different strategy to deal with autocorrelation is to sample less often, in this way reducing the positive autocorrelation of the data. The problem with this approach is that it may not be cost-effective to implement, since the amount of product manufactured between samples may be too large and one has to take into account that such products will not be controlled. It is also important to notice that a nonstationary process (e.g., a process with a unit root in the AR polynomial) will result in another stationary process (one with different parameters) if sampled less often (MacGregor, 1976), so this approach applies to stationary processes only. The approach utilized in the remainder of this book is to assume that there is some *controllable factor* that we can manipulate to compensate for the autocorrelation (or dynamics) of the process. For this we need first to look at how to identify and fit transfer function models.

PROBLEMS

3.1. Consider the yield data in Example 1.7. Identify and fit an ARIMA model.

3.2. Consider the machining process data in Example 1.8. Identify and fit an ARIMA model.

3.3. Boyles (2000) gives a time series for the fill weights observed for a powdered food product when a feedback controller was turned off. The data can be found in the file *BoylesData.txt*. Fit an ARIMA model to the data and do the usual model-fitting diagnostics.

3.4. The *ARMA.xls* spreadsheet simulates AR(1), AR(2), MA(1), and MA(2) processes. Every time the spreadsheet is recalculated, a new realization of the process results. Using the *ARMA.xls* spreadsheet, check the following:

(a) The effect of $\xi = 0$ in an AR(1) process

(b) The effect of $Y_0 = 0$ and $Y_0 = \mu$ on an AR(1) process

(c) The effect of $Y_0 = 0$ and $Y_0 = \mu$ on an MA(1) process

(d) The effect of $\xi = 0$ and $\xi \neq 0$ on an AR(1) process with $\phi = 1$

(e) The behavior of an AR(1) process as $\phi \to \pm 1$

(f) The behavior of an MA(1) process as $\theta \to \pm \infty$

(g) Compare the values of μ, σ, ρ_1, and ρ_2 versus their point estimates over different realizations of AR(1) and MA(1) processes.

3.5. Modify the *ARMA.xls* spreadsheet to simulate an ARMA(1, 1) process and an ARMA(2, 2) process.

3.6. Modify the *ARMA.xls* spreadsheet to compute ρ_k for lags k up to 20.

3.7. Modify the *ARMA.xls* spreadsheet to estimate $r_k = \hat{\rho}_k$ for lags k up to 20.

3.8. Show that the autocorrelations $\{\rho_k\}$ for a random walk with drift process, $Y_t = \xi + Y_{t-1} + \varepsilon_t$, $(\xi \neq 0)$, are all equal to 1.0 for all lags k.

3.9. What are the stationarity conditions of an AR(2) process in terms of ρ_1 and ρ_2? (*Hint:* Use the Yule–Walker equations of Section 3.1.)

3.10. Derive the expression for the autocorrelation function of a general MA(q) process.

3.11. Show analytically that the maximum value of $|\rho_1|$ that one can attain in an MA(1) process is 0.5.

3.12. Find the maximum values that ρ_1 and ρ_2 can attain for an invertible MA(2) process.

3.13. Derive equation (3.13). [*Hint:* Add and subtract $E_t[Y_{t+l}]$ inside the binomial term in equation (3.12).]

3.14. Show that the l-step-ahead (≥ 1) MMSE forecast for an IMA(1, 1) process is given by an EWMA of the data.

3.15. *Effect of overdifferencing on the ACF.* Suppose that you have a random walk process: $Y_t = Y_{t-1} + \varepsilon_t$, where $\{\varepsilon_t\}$ is a white noise sequence. Show that if you difference this process twice, the lag 1 autocorrelation of the resulting series is exactly -1.0.

3.16. Set up an EWMA control chart for the residuals of the model fit in Problem 3.2.

3.17. Set up a CUSUM control chart for the residuals of the model fit in Problem 3.3.

3.18. Suppose that the following additional observations were obtained in the machining process of Problem 3.2: 0.9715, 0.9750, 0.9765, 0.9801. Apply the EWMA chart you obtained in Problem 3.16 to the one-step-ahead forecast errors of these observations. What conclusion can you reach?

3.19. Show that the expression for the sample spectrum [equation (3.21)] follows from expressions (3.19) and (3.20).

BIBLIOGRAPHY AND COMMENTS

The classic book by Box and Jenkins (1970) put together a variety of results in the time series literature and proposed the ARIMA classification scheme. The original version of the book was based on a series of papers that the authors wrote in the 1960s. Other books that present very readable discussions of ARIMA models are Abraham and Ledolter (1983), Chatfield (1989), and Montgomery et al. (1990).

An interesting recent approach to monitoring an autocorrelated process is the use of ARMA charts (Jiang et al., 2001). Rather than filtering the autocorrelation using an ARMA model, a procedure that will tend to mask the changes that we wish to detect (see Appendix 3C for a frequency-domain explanation of this phenomenon), an ARMA chart seeks to amplify the signal that we want to detect by constructing an ARMA filter. Application of signal processing techniques to autocorrelated data monitoring is indeed a very fruitful area of research.

Readers interested in the frequency analysis of time series should consult the classic book by Jenkins and Watts (1968), which is still highly regarded today among electrical engineers. Chatfield (1989) and Kendall and Ord (1990) also provide excellent presentations of frequency-domain topics with emphasis on statistical time-series analysis.

APPENDIX 3A: UNIT ROOTS AND SUMMATION OPERATOR

Strictly speaking, the summation operator does not converge:

$$\nabla^{-1} = \mathcal{S} = \frac{1}{1 - \mathcal{B}} = \underbrace{1 + \mathcal{B} + \mathcal{B}^2 + \cdots}_{\text{not convergent since } |\mathcal{B}| = 1}$$

Thus we need to have an infinitely long sequence in order to apply this operator, and there is no need to define initial conditions, as the sequence has been going on "forever" since the infinite past. To avoid this conceptual difficulty, *finite summation* can be used instead. This operator is defined as

$$\mathcal{S}_m \equiv \frac{1 - \mathcal{B}^m}{1 - \mathcal{B}} = \sum_{i=0}^{m-1} \mathcal{B}^i$$

which is a finite geometric series. When applied to a sequence, it returns the sum of the last m members of the sequence. Using finite, as opposed to infinite, summation, we get the following ARIMA(p, d, q) representation:

$$A'_p(\mathcal{B}) \frac{1 - \mathcal{B}}{1 - \mathcal{B}^t} \tilde{Y}_t = C_q(\mathcal{B}) \varepsilon_t$$

where it is necessary to define $\tilde{Y}_0$ for a complete description of the process. In this book, unless otherwise specified, we follow the convention of using infinite summation and assuming infinite processes; thus there is no need to define initial conditions. The effect of initial conditions is usually minor and can be neglected for the most part. There may be situations in process control, however, where the effect of initial conditions can be crucial, a point we mention briefly in Chapter 7.

APPENDIX 3B: NONSTATIONARY MODELS THAT DRIFT

ARIMA(p, d, q) models that have $d > 0$ and $\xi \neq 0$ drift with an average slope of ξ, a term known as the *drift parameter*. These are models nonstationary in the mean, A typical example of such process is an ARIMA(0, 1, 0) process with drift, the random walk with drift (RWD) process discussed previously:

$$Y_t = Y_t + \xi + \varepsilon_t$$

Another popular drift model, equivalent to a noninvertible IMA(1, 1) process with drift, is a deterministic trend process:

$$Y_t = \xi t + \varepsilon_t$$

The main difference between the two models above is that in the random walk with drift, Var(Y_t) increases linearly in t, whereas the variance of the deterministic drift process is constant. In both cases, however, the mean is a linear function of ξ. Trying to difference the latter process will achieve stationarity in the mean, but a noninvertible unit root is introduced in the model. The correct approach to detrend such a process is to estimate ξ (perhaps using linear regression) and subtract ξt at every t. When applied to a RWD process, this detrend procedure stabilizes the mean but not the variance.

APPENDIX 3C: BRIEF OVERVIEW OF FREQUENCY ANALYSIS CONCEPTS

In this book we concentrate on analyzing, and eventually adjusting, discrete-time processes based on techniques that look at the processes as functions of time. This is called *time-domain analysis*. An alternative approach for analysis is to study the time-series processes by looking at the various frequencies with which the data oscillate. This is the *frequency-domain* or *spectral analysis* approach. Most of this book is based on time-domain techniques, but there are some places where we refer to frequency-domain concepts, primarily the spectrum of a time-series process. This appendix provides a very brief introduction to such concepts, for the sake of completeness. Interested readers seeking a deeper presentation should consult a standard reference on spectral analysis of time series, such as the Jenkins and Watts (1968) book.

Our presentation starts by defining the autocovariance and autocorrelation generating functions of a process. We then discuss the goals and main concepts of frequency analysis. Definitions of the periodogram and the sample spectrum are then given, after which the spectrum and the spectral density function are described. It is shown how the spectrum is the Fourier transform of the autocovariance function.

Autocovariance Generating Function

The autocovariance generating function of a process is defined as the z-transform[11] of the autocovariance function of the process:

$$G(z) = \mathcal{Z}(\gamma_k) = \sum_{k=-\infty}^{\infty} \gamma_k z^{-k} \qquad (3.16)$$

where z is a dummy complex variable with a range of values such that $G(z)$ is finite (i.e., a range of values such that the sum converges). The coefficients of each term in the sum give us the autocovariances, with $\sigma^2 = \gamma_0$ being the coefficient when $k = 0$, γ_1 being the coefficient when $k = \pm 1$, and so on. We first show that the autocovariance function of any process written in the *infinite AR* or linear filter form,

$$Y_t = \Psi(\mathcal{B})\varepsilon_t = \varepsilon_t + \Psi_1 \varepsilon_{t-1} + \Psi_2 \varepsilon_{t-2} + \cdots = \sum_{j=0}^{\infty} \Psi_j \varepsilon_{t-j} \qquad (3.17)$$

can be written as $G(z) = \sigma_\varepsilon^2 \Psi(z^{-1})\Psi(z)$, where we assume that $\Psi_0 = 1$. Using the autocovariance generation function it is easy to get the theoretical (population) autocovariance and autocorrelation functions of any linear process, as the example below illustrates.

To show this useful result, it turns out that for a process $\{Y_t\}$ obeying (3.17), we have that

$$\gamma_k = E[Y_t, Y_{t+k}] = E\left(\sum_{j=0}^{\infty} \Psi_j \varepsilon_{t-j} \sum_{h=0}^{\infty} \Psi_h \varepsilon_{t+k-h} \right)$$

$$= E\left(\sum_{j=0}^{\infty} \sum_{h=0}^{\infty} \Psi_j \Psi_h \varepsilon_{t-j} \varepsilon_{t+k-h} \right)$$

Since the ε's are white noise, the double sum is nonzero only when $-j = k - h$ (i.e., when $h = j + k$), so we have

$$\gamma_k = \sigma_\varepsilon^2 \sum_{j=0}^{\infty} \Psi_j \Psi_{j+k}.$$

Substituting this result into (3.16), we get

$$G(z) = \sigma_\varepsilon^2 \sum_{k=-\infty}^{\infty} \sum_{j=0}^{\infty} \Psi_j \Psi_{j+k} z^{-k}$$

---

[11] z-Transforms are described in detail in Appendix 2C.

Writing $s = j + k$ gives us

$$G(z) = \sigma_\varepsilon^2 \sum_{s=-\infty}^{\infty} \sum_{j=0}^{\infty} \Psi_j \Psi_s z^{-(s-j)} = \sigma_\varepsilon^2 \sum_{s=-\infty}^{\infty} \Psi_s z^{-s} \sum_{j=0}^{\infty} \Psi_j z^j$$

Since $\Psi_s = 0$ for $s < 0$, we finally get

$$G(z) = \sigma_\varepsilon^2 \sum_{s=0}^{\infty} \Psi_s z^{-s} \sum_{j=0}^{\infty} \Psi_j z^j = \sigma_\varepsilon^2 \Psi(z^{-1}) \Psi(z).$$

A closely related function is the autocorrelation generating function, defined as

$$G_p(z) = \frac{G(z)}{\gamma_0}.$$

Example 3.20. Consider the MA(1) process

$$Y_t = (1 - \theta \mathcal{B}) \varepsilon_t$$

The autocovariance generating function is

$$G(z) = \sigma_\varepsilon^2 C(z) C(z^{-1}) = \sigma_\varepsilon^2 (1 - \theta z^{-1})(1 - \theta z)$$
$$= \sigma_\varepsilon^2 (1 + \theta^2 - \theta z^{-1} - \theta z)$$

so we can generate

$$\gamma_0 = \text{coefficient of } z^0 = \sigma_\varepsilon^2 (1 + \theta^2)$$

$$\gamma_1 = \text{coefficient of } z^{\pm 1} = -\sigma_\varepsilon^2 \theta$$

and $\gamma_k = 0$ for $k > 1$.   □

Frequency Analysis

The goal of frequency analysis is to decompose a time series as a weighted sum of sine and cosine functions in what constitutes, more technically speaking, the Fourier transform of the series $\{Y_t\}$. Initial interest in this type of transformations applied to time series was mainly in economics and physics.[12] Fourier analysis was developed as an attempt to approximate

---

[12]A discrete periodic function Y_t is such that $Y_t = Y_{t+T}$, where T is called the *period*. A typical example of a periodic function is

$$Y_t = C \cos(\alpha t - \phi) = C \cos(2\pi f t - \phi)$$

where C gives the amplitude of the oscillations, α is the angular frequency since the sine function completes $f = \alpha/2\pi$ cycles in a time unit (the angle $\alpha t - \phi$ is measured in radians), and ϕ, called the *phase*, which determines the starting point of the function. The quantity $T = 2\pi/\alpha = 1/f$ is in time units per cycle. Graphically, T is the distance (time) between peak and peak (or between trough and trough) and the phase is the distance (time) from zero to the first peak.

deterministic continuous functions over wide intervals, as opposed to other series approximations (e.g., Taylor series) which are valid only in small regions. In our times-series context, our interest is in approximating *discrete stochastic* functions (of time).

Suppose that we have N observations of a time-series process. This means that we could fit a model with N coefficients (unknowns) such as in a Fourier series:

$$Y_t = A_0 + \sum_{j=1}^{m} \left[A_j \cos(\alpha_j t) + B_j \sin(\alpha_j t) \right]$$

where $m = (N - 1)/2$ (assuming N odd). We thus have N parameters: m A_j's, m B_j's, and A_0. We can write the Fourier series as

$$Y_t = A_0 + \sum_{j=1}^{m} \left[A_j \cos(2\pi j f_1 t) + B_j \sin(2\pi j f_1 t) \right] \qquad (3.18)$$

Here we made use of the quantities

$$\alpha_j = \frac{2\pi j}{N} = 2\pi j f_1$$

so that

$$T_j = \frac{2\pi}{\alpha_j} = \frac{N}{j}.$$

The α_j's are called the *harmonic* frequencies and are multiples of the *fundamental frequency* $f_1 = 1/N$.

Equation (3.18) shows that a discrete function of time Y_t can be expressed as a sum of sines and cosine functions whose frequencies are multiples (harmonics) of the fundamental frequency $f_1 = 1/N$ cycles per time unit (e.g., if $\Delta t = 1$ second, the units of f_1 are cycles per second) or cycles per observation, in case the time unit is defined to be equal to the time between observations. A *cycle* is defined as one complete period of a sine or cosine function.

Before proceeding to find the values of the N parameters, it is important at this point to introduce the concept of *Nyquist frequency*. Suppose that we have observations taken every hour at N discrete points in time on a process. We may ask the following question: At this sampling rate, what is the fastest oscillation, or highest frequency, that we can detect in the data? Evidently, if there are frequencies faster than 1 hour, we cannot detect them, because to do that we would need at least two observations per hour. Therefore, to get meaningful information on that process within an hour, we must take at least two observations per hour (see Figure 3.31). Thus in the ratio 1 cycle/(number of observations) we cannot lower the number of observations below 2 if

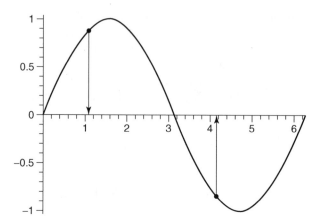

Figure 3.31 Two measurements per cycle is the minimum number of observations that will give meaningful information on the cycle.

we want to detect oscillations within 1 hour. This means that $f_m = \frac{1}{2}$ cycle per hour (or 1 cycle every 2 hours) is the largest frequency we can detect. Thus we can detect frequencies $f_1 = 1/N$, $f_2 = 2/N, \ldots f_{m-1} = (m-1)/N$, $f_m = m/N = 1/2$, but not frequencies f_j for $j > m$. In general, the Nyquist frequency equals $f_m = 1/(2\Delta t)$ cycles per time unit. Because of the Nyquist frequency bound, frequency analyses are constrained to look at frequencies in the interval $(0, \frac{1}{2})$.

If in equation (3.18) the quantities A_j and B_j are constants, we will find ourselves dealing with a nonstationary process since the mean of Y_t will vary with time. To get around this difficulty, we can think of a *random effects model* in which A_j and B_j are seen as realizations of random variables $A_{R,j} \sim N(0, \sigma_j^2)$, and $B_{R,j} \sim N(0, \sigma_j^2)$, where $A_{R,j}$ and $B_{R,j}$ are two white noise series. With this in mind, we can obtain the following least squares estimates[13] of the N parameters in (3.18):

$$A_j = \frac{2}{N} \sum_{k=1}^{N} Y_k \cos \frac{2\pi jk}{N} \qquad j = 1, 2, \ldots, m \qquad (3.19)$$

$$B_j = \frac{2}{N} \sum_{k=1}^{N} Y_k \sin \frac{2\pi jk}{N} \qquad j = 1, 2, \ldots, m \qquad (3.20)$$

and

$$A_0 = \frac{1}{N} \sum_{k=1}^{N} Y_t = \bar{Y}$$

which is simply the sample mean of Y_t. We also have that $B_0 = 0$.

---

[13] The least squares fit turns out to be perfect, so there is no need for an error term in (3.18). See Hamilton (1994) and Wilson (2001) for a derivation of this result.

Any time series may be decomposed in a set of cycles following equation (3.18). This does not imply that the series Y_t is cyclical.

Periodogram and Sample Spectrum

A measure of how strong oscillations with frequency f_i occur in the data is given by a *periodogram*, defined by

$$I(f_i) = \frac{N}{2}(A_i^2 + B_i^2) \qquad i = 1, 2, \ldots, m$$

where f_i is a harmonic and the quantity $C_i^2 = A_i^2 + B_i^2$ is an estimate of the amplitude at frequency $f_i = i/N$. A periodogram can be extended to account for amplitudes at any frequency in the continuous range $(0, 0.5)$, not only at the harmonics. From the last expression it can be shown that

$$\frac{1}{N}\sum_{k=1}^{N} Y_k^2 = \overline{Y}^2 + \frac{1}{N}\sum_{i=1}^{m} I(f_i)$$

or equivalently, that

$$\sigma^2 = \frac{1}{N}\sum_{k=1}^{n}\left(Y_k - \overline{Y}\right)^2 = \frac{1}{N}\sum_{i=1}^{m} I(f_i).$$

Therefore, the periodogram shows how the total variance of a time series is distributed between the various harmonic frequencies. This last result is known as *Parseval's theorem*.

The *sample spectrum* is usually defined as a modification of the periodogram that accounts for continuous frequencies, not only the harmonics, and that turns out to be related to the sample autocovariance function $\hat{\gamma}_k = c_k$. The sample spectrum is

$$I(f) = \frac{N}{2}\left(A_f^2 + B_f^2\right) \qquad 0 \le f \le \tfrac{1}{2}$$

Direct substitution of expressions (3.19) and (3.20) in the sample spectrum definition and use of mathematical induction can be made to show that the sample spectrum reduces to

$$I(f) = 2\left[c_0 + 2\sum_{k=1}^{N-1} c_k \cos\left(2\pi f k\right)\right]. \qquad (3.21)$$

This indicates that the sample spectrum is the Fourier cosine transform of the sample autocovariance function.

Power Spectrum

The *power spectrum*, or simply the *spectrum* of a stochastic process, is defined as the limit

$$p(f) = \lim_{N \to \infty} E[I(f)] = 2\left[\gamma_0 + 2\sum_{k=1}^{\infty} \cos(2\pi fk)\right] \qquad 0 \le f \le \tfrac{1}{2}$$

Note that the sum is infinite, since this is a population characteristic, not a function based on a sample of N observations like the sample spectrum. In particular, Parseval's theorem shows that

$$\int_0^{1/2} p(f)\, df = \gamma_0 = \sigma_Y^2$$

Thus the spectrum shows how the total variance of a stochastic process is distributed across all possible frequencies, in analogy with the periodogram. An equivalent function given in terms of the autocorrelations is the *spectral density function*, defined as

$$g(f) = \frac{p(f)}{\gamma_0} = 2\left[1 + 2\sum_{k=1}^{\infty} \rho_k \cos(2\pi fk)\right] \qquad 0 \le f \le \tfrac{1}{2}$$

which is actually a probability density function since $g(f) \ge 0$ and $\int_0^{1/2} g(f)\, df = 1$.

Notice that

$$\frac{g(f)}{2} = \sum_{k=-\infty}^{\infty} \rho_k e^{-i\alpha k} = \sum_{k=-\infty}^{\infty} \rho_k z^{-k} = G_p(z)$$

Therefore, the spectral density function is proportional to the autocorrelation generating function of a process [i.e., $G_p(z) = g(f)/2$]. This implies that for stationary ARMA processes $\phi(\mathcal{B})Y_t = \theta(\mathcal{B})\varepsilon_t$, we have that

$$\frac{g(f)}{2} = \frac{\sigma^2 \theta(z)\theta(z^{-1})}{\gamma_0 \phi(z)\phi(z^{-1})}$$

Example 3.21. Consider a white noise process $Y_t = \varepsilon_t$. We then have that $\rho_0 = 1$, $\rho_k = 0$ for all integers $k \ne 0$, $\theta(z) = 1$, and $\phi(z) = 1$. The spectral density function is

$$\frac{g(f)}{2} = \frac{\sigma^2}{\sigma^2} = 1 \qquad \text{for } 0 \le f \le \tfrac{1}{2}$$

This shows that the total variability of a white noise process is distributed equally among all possible frequencies. $\qquad \square$

One might be tempted to estimate the spectrum with the sample spectrum, since from the definition of spectrum, the sample spectrum $I(f)$ is an asymptotically unbiased estimator of the spectrum $p(f)$, that is,

$$\lim_{N \to \infty} E[I(f)] = p(f)$$

However, $I(f)$ is not a consistent estimator of $p(f)$, that is, $\text{Var}[I(f)]$ does not go to zero as N increases. The effect of this is that values of $I(f)$ obtained from a single realization of $\{Y_t\}$ will vary widely and will be difficult to interpret. In practice, some form of *smoothing* takes place and this results in consistent estimates. A smooth estimate of the spectrum is

$$\widehat{p(f)} = 2\left[\lambda_0 \hat{\gamma}_0 + 2 \sum_{k=1}^{M} \lambda_k \hat{\gamma}_k \cos(2\pi fk) \right]$$

If $\lambda_k = 1$ and $M = N - 1$, we get $I(f)$, which is not a consistent estimator. Since the estimates $\hat{\gamma}_k$ have larger variances as k increases, we can use weights λ_k that decrease with increasing k. For similar reasons, it makes sense to truncate the sum at some point $M < N - 1$. For example, a set of weights commonly used in practice was proposed by Tukey:

$$\lambda_k = \tfrac{1}{2}\left(1 + \cos\frac{\pi k}{M} \right) \qquad k = 0, 1, \ldots, M$$

Typical values for M are around 20. The smaller M is, the smaller the variance of $\widehat{p(f)}$, but the larger its bias. A typical guideline is $M \approx 2\sqrt{N}$. Chatfield (1989) gives a very readable description of this and other spectrum smoothing methods.

Linear Filters
The discussion about SPC for autocorrelated data is related to the idea of the spectrum of a linear filter. Consider a time series Y_t created as a weighted average of another series X_t as follows:

$$Y_t = \sum_{j=0}^{\infty} w_j X_{t-j} = W(\mathcal{B}) X_t \tag{3.22}$$

where $W(\mathcal{B}) = 1 + w_1 \mathcal{B} + w_2 \mathcal{B}^2 + \cdots$. As mentioned before, such a weighted average is called a *linear filter*. The function $W(e^{-i\alpha}) = \sum_{j=0}^{\infty} w_j e^{-i\alpha j}$ is called the *frequency transfer function* of the filter. Suppose that X_t follows in turn a stationary process such as

$$X_t = \sum_{j=0}^{\infty} \phi_j \varepsilon_{t-j} = \Psi(\mathcal{B}) \varepsilon_t$$

where $\Psi(\mathcal{B}) = 1 + \phi\mathcal{B} + \phi_2\mathcal{B}^2 + \cdots$ and ε_t is a white noise sequence. If $W(\mathcal{B})$ has weights such that $\sum_{j=0}^{\infty} |w_j| < \infty$ and $\Psi(\mathcal{B})$ has weights such that $\sum_{j=0}^{\infty} |\phi_j| < \infty$, then Y_t is a stationary stochastic process. Then the spectral density function of Y_t is given by

$$g_y(f) = 2G_{p,y}(z) = (\text{constant})G_{p,x}(z)G_{p,\varepsilon}(z)$$

$$= (\text{constant})W(z)W(z^{-1})\Psi(z)\Psi(z^{-1})$$

Therefore, by choosing the constants w_j in (3.22), we can manipulate which frequencies of X_t are observed in Y_t. This is a powerful idea with application in SPC for autocorrelated data.

Example 3.22 (Kendall and Ord, 1990). Suppose that we have the process

$$X_t = \phi X_{t-1} + \varepsilon_t \quad \text{or} \quad (1 - \phi\mathcal{B}) X_t = \varepsilon_t$$

which is filtered using

$$Y_t = X_t - a X_{t-1} \quad \text{or} \quad Y_t = (1 - a\mathcal{B})X_t.$$

Therefore, we have that

$$G_{p,\varepsilon}(z) = \frac{1}{1 - \phi z} \times \frac{1}{1 - \phi z^{-1}} \sigma_\varepsilon^2$$

$$G_{p,x}(z) = \sigma_x^2(1 - az)(1 - az^{-1})$$

Therefore, the filter has spectral density:

$$g_y(f) = (\text{constant})\frac{(1 - a z^{-1})(1 - a z)}{(1 - \phi z^{-1})(1 - \phi z)}$$

or

$$g_y(f) = (\text{constant})\frac{1 - a(z + z^{1-}) + a^2}{1 - \phi(z + z^{-1}) + \phi^2}$$

which, since $z = e^{i2\pi f}$, results in

$$g_y(f) = (\text{constant})\frac{1 - a(e^{i2\pi f} + e^{-i2\pi f}) + a^2}{1 - \phi(e^{i2\pi f} + e^{-i2\pi f}) + \phi^2}$$

$$= (\text{constant})\frac{1 - 2a\cos(2\pi f) + a^2}{1 - 2\phi\cos(2\pi f) + \phi^2}.$$

Table 3.4   Effect of the Values of a on the Numerator of $g_y(f)$ in the Linear Filter of Example 3.22

a	$f \to 0$	$f \to \frac{1}{2}$
$a \to 1$	$\to 0$	$\to 4$
$a \to -1$	$\to 4$	$\to 0$

This means that as $a \to \phi$, the spectral density gets flatter. With $a = \phi$ we get that Y_t is white noise, and its spectral density reflects that. By manipulating the value of a in the invertibility range ($|a| < 1$), we can analyze the effect on the observed series Y_t. From Table 3.4 it can be seen that if $a \to 1$ only the high frequencies are observed in Y_t, thus the linear filter is a *high-pass filter*. Similarly, if $a \to -1$, we end up with a *low-pass filter* since only the lower frequencies are observed in Y_t.   □

The idea of SPC for autocorrelated data is to use a high-pass filter that will filter out positive autocorrelation (corresponding to low frequencies) while letting us "see" the shifts. The problem is that sudden, infrequent shifts will also be filtered out, so detecting them in the residuals of a fitted linear filter is very difficult unless the shift magnitude is very large.

In Chapter 4 we also make use of these results when we *prewhiten* the input to make its spectrum flatter, since the autocorrelation in X_t (the controllable factor) may mask the cross-correlation between X and Y.

CHAPTER 4

Transfer Function Modeling

Transfer functions describe the dynamical relation, if any, between a controllable factor and a quality characteristic. Some notions of transfer function models were discussed in Chapter 2 in relation to difference equation models and the stability of a dynamic process. In this chapter we describe transfer functions in further detail. Identification of transfer function models in the presence of noise is an important practical problem, which, as mentioned before, is the basis for finding good adjustment policies. The types of dynamic processes studied in this chapter have their main area of application in continuous (chemical) processes, where "inertial elements" exist in the process. Discrete-part manufacturing processes can usually be modeled by the ARIMA models described in Chapter 3 with the addition of a trivial transfer function which simply indicates that the full effect of changing the controllable factor is observed immediately in the quality characteristic.

In this chapter we discuss mainly what in the control engineering literature has come to be known as the Box–Jenkins transfer function model. However, we also discuss other forms of a transfer function model, some of them useful for deriving optimal controllers, such as the ARMAX formulation. In Appendix 4A we show additional material with reference to the relations between discrete- and continuous-time dynamic models and the state-space approach to modeling dynamic processes. State-space analysis offers a different modeling view that complements the polynomial transfer function models described in the main part of this chapter. State-space models of ARIMA and transfer function models are valuable for estimation purposes through the use of Kalman filters, a topic discussed in Chapter 8.

4.1 DESCRIPTION OF TRANSFER FUNCTION SYSTEMS

In this section we look at a dynamic system as a black box (Figure 4.1) by modeling the relationships between its input and its output without recourse to the inner workings of the system or process [multiple-input, multiple-output (MIMO) systems are discussed in Chapter 9]. Models that describe a

Figure 4.1 Dynamic process.

process internally, the so-called *state-space models*, are presented in Appendix 4A. Transfer functions described in the main part of this chapter are based on input–output data only and will make recourse to difference equations that lead to equations with polynomials in the backshift operator. For this reason, in the control engineering literature, transfer function models are sometimes called *polynomial models* to distinguish them from state-space models.

Suppose that observations $(\{X_t\}, \{Y_t\})$ are available at equidistant points in time $t = 0, 1, \ldots$. Our goal is to find the dynamic relationship between $\{X_t\}$ and $\{Y_t\}$. Later, these two sequences are considered stochastic processes, but for the moment we treat them as deterministic sequences of variables. In later chapters it will be assumed that X_t is under our control, so it will be called the *controllable factor* and Y_t will be the quality characteristic we wish to control. In this chapter they are simply the input and the output (or response), respectively. Let us assume that the process under study is stable, as discussed in Section 2.4. If X_t is fixed at a constant level for all time periods t, say at a numerical value of 1.0, then after a transient period Y_t will eventually reach a steady-state level $Y_\infty = g$. Since the process is assumed linear, for other constant values X at which we can fix X_t we will obtain a steady-state response:

$$Y_\infty = gX$$

where g is called the *steady-state gain*. The gain models the change in the steady-state output that will eventually be obtained by a unit change in the input. Evidently, this is a static model, of the type obtained after conducting designed experiments by means of utilizing regression techniques. For example, when running a DOE in a chemical process, it is usually necessary to wait until the process "settles" in order to measure the response. This means that the experimenter is interested in the long-run, or steady-state, response of the system, disregarding the transient phase.

In process control applications, interest is usually in bringing the process back to target rapidly, and this implies that the dynamic response, as evidenced in the transient phase, should be modeled. In such a case we will be interested in knowing what the process response is when the input is *changed* from time to time. Since the system is assumed linear, it will have a response equal to some linear combination of *all* the present and past inputs. That is, we have the polynomial model

$$Y_t = v_0 X_t + v_{t-1} X_{t-1} + \cdots = H(\mathcal{B}) X_t \tag{4.1}$$

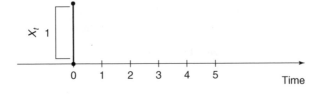

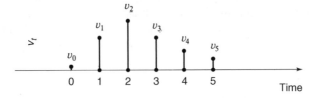

Figure 4.2   Unit impulse response function.

where the $H(\mathcal{B})$ polynomial has an infinite order and is defined as the transfer function of the system.

A simple instance of a changing input, useful for defining a transfer function, is the unit impulse or pulse input:

$$X_t = \begin{cases} 1 & \text{if } t = 0 \\ 0 & \text{if } t \neq 0 \end{cases}$$

The weights $v_0, v_1, \ldots$ are called the *impulse response function*, a connotation that can be understood by considering Figure 4.2. From the figure we see that

$$Y_0 = v_0 X_0 = v_0(1) = v_0$$

$$Y_1 = v_0 X_1 + v_1 X_0 = v_1$$

$$Y_2 = v_0 X_2 + v_1 X_1 + v_2 X_0 = v_2$$

$$Y_3 = v_3$$

$$\vdots$$

Thus we see that the response at each point in time equals the weights $\{v_j\}$, and because of this, a transfer function is defined as the response of the system to a unit impulse.

Example 4.1: First-Order Dynamical Process. Suppose that in a process the deviations from target obey the transfer function

$$Y_t = \frac{0.5}{1 - 0.2\mathcal{B}} X_{t-2}$$

where observations are taken every 20 minutes. This can be written in difference equation form as $Y_t = 0.2Y_{t-1} + 0.5X_{t-2}$. Suppose that we have $Y_0 = 0$ and that at time zero we introduce a unit pulse (i.e., we make $X_0 = 1$, $X_t = 0$, $t \neq 0$). Then we have that

$$Y_0 = 0 = v_0$$
$$Y_1 = 0.2Y_0 + 0.5X_{-1} = 0 = v_1$$
$$Y_2 = 0.2Y_1 + 0.5X_0 = 0.5(1) = 0.5 = v_2$$
$$Y_3 = 0.2Y_2 + 0.5X_1 = 0.2(0.5) = 0.1 = v_3$$
$$Y_4 = 0.2Y_3 + 0.5X_2 = 0.2(0.1) = 0.02 = v_4$$
$$Y_5 = 0.2Y_4 + 0.5X_3 = 0.2(0.02) = 0.004 = v_5$$
$$\vdots$$

Thus the deviations from target Y, and the weights $\{v_j\}$ converge to zero as $t \to \infty$. This is an instance of a first-order dynamical process in which there are two time periods of input–output *delay*. That is, the effect of a change in X is first observed after $2(20) = 40$ minutes, thus we have $Y_0 = Y_1 = 0$ but $Y_2 \neq 0$. If we sample more often, the delay time may change, of course. □

Stability
A transfer function process as described by equation (4.1) is stable if and only if

$$\sum_{j=0}^{\infty} |v_j| < \infty$$

This means that eventually, the weights v_t will decrease to zero or will equal zero for large t, implying that the effect of past inputs on the future response is eventually zero or negligible after some time. The steady-state gain g is easy to obtain by fixing $X_t = 1$ for all discrete t, so

$$Y_t = Y_\infty = \sum_{j=0}^{\infty} v_j = g$$

In general for any constant input X, $Y_\infty = gX$.

Examples of Transfer Functions for Discrete-Part and Continuous Manufacturing Processes

The modeling methodology shown in the next few sections is very general and powerful, allowing us to model input–output relations of a wide variety of manufacturing processes, but it is complicated. Fortunately, for the majority of manufacturing processes, simple instances of transfer functions will provide an adequate representation of the dynamic behavior of the process.

Discrete-Part Manufacturing Processes

Consider first the case of a discrete-part manufacturing process, for example, machining of metal parts. The controllable factor will be typically the machine set point (i.e., the dimension we aim the machine at to perform a cut on the part). Suppose that we change the set point from part to part. Whatever setting X_{t-1} we have before producing part number t, this setting will affect only part number t, so we have

$$Y_t = \beta X_{t-1}$$

where Y_t is the observed deviation from target (actual dimension minus nominal) and β is the process gain which for a well-calibrated machine will be very close to 1.0. The process is subject to disturbances, due primarily to wearing off in the cutting tool but also due to variations in raw materials, thermal effects on the machine tool, and other factors. Thus we actually observe

$$Y_t = \beta X_{t-1} + N_t$$

where the noise term N_t will typically follow a low-order ARIMA process with drift like those described in Appendix 3B. This transfer function model is known as a *pure unit delay* or *responsive model* in which the dynamic behavior of the $\{Y_t\}$ series is due only to *disturbance dynamics* (noise), not to *process dynamics* (i.e., there is no dynamical relation between X and Y).

Continuous Manufacturing Processes

Consider, instead, a continuous manufacturing process such as a chemical process in which product flows through a series of pipes and tanks. Typical controllable factors are temperatures, pressures, flow rates, or amounts of certain ingredients that are mixed in order to achieve desired quality characteristics. The relation between the quality characteristic Y and the controllable factors X will usually be governed by a differential equation that models the delays and inertial elements (pipes, tanks, etc.) of such a manufacturing process. If Y and X are measured at discrete points in time, the effect of a single change in X at time t will be spread over several subsequent values of Y, eventually making Y reach a steady-state value. As it turns out in practice, many continuous manufacturing processes will be well approximated by either a first- or a second-order transfer function. The first-order transfer function is

$$Y_t = \frac{b_0}{1 - a_1 \mathcal{B}} X_{t-k}$$

where b_0 and a_1 are parameters and k is the input–output delay. In some other processes, a second-order transfer function

$$Y_t = \frac{b_0}{1 - a_1 \mathcal{B} - a_2 \mathcal{B}^2} X_{t-k}$$

will be necessary. In either case, Y_t will be observed in the presence of disturbances due to raw material changes, fluctuating environmental conditions, and similar factors, so an ARIMA disturbance N_t can be added to either of these two models. Thus, whenever process dynamics are suspected, these two models should be entertained as candidates due to their simplicity. Evidently, there will be more complicated processes that require more involved transfer function models, such as models with *numerator dynamics*.[1] Identifying and fitting these relatively more complicated linear transfer functions is explained in the remaining sections of this chapter. Two excellent references on modeling linear dynamic chemical processes with a view on process control are Seborg et al. (1989) and Ogunnaike and Ray (1994). The statistical modeling methodology presented in this chapter follows Box et al. (1994).

Transfer Functions as Ratios of Polynomials

Model (4.1) contains an infinite number of parameters. A more "parsimonious" representation, useful for estimation purposes when the responses are measured in the presence of error, is the rational form, introduced in Section 2.4. This is based on the linear, time-invariant, discrete difference equation

$$Y_t = a_1 Y_{t-1} + a_2 Y_{t-2} + \cdots + a_r Y_{t-r} + b_0 X_{t-k} - b_1 X_{t-k-1} - \cdots - b_s X_{t-k-s}$$

or

$$A_r(\mathcal{B})Y_t = B_s(\mathcal{B})X_{t-k} = B(\mathcal{B})\mathcal{B}^k X_t$$

[1] The first- and second-order model names refer to denominator dynamics, that is, a transfer function where it is the denominator that has a nonzero order polynomial in the backshift operator. We point out that nonlinear and even chaotic behavior is possible in complicated processes, and such behavior will not be captured by even the more complicated linear transfer function models explained in this chapter. Readers interested in nonlinear, discrete-time dynamic models should consult the book by Pearson (1999).

where $A_r(\mathcal{B})$ is a polynomial in $\mathcal{B}$ of order r with first element[2] equal to 1 and $B_s(\mathcal{B})$ is a polynomial of order s with first element equal to b_0. The time k is called the *input–output delay.*

If the two representations of a transfer function, the impulse response form and the rational form, are indeed equivalent, we must have

$$H(\mathcal{B}) = \frac{B_s(\mathcal{B})\,\mathcal{B}^k}{A_r(\mathcal{B})}$$

which is called an (r, s, k) *transfer function* by Box and Jenkins. From the rational representation on the right-hand side, we know (see Section 2.4) that for stability, $A_r(\mathcal{B}) = 0$ must have all roots outside the unit circle. For a stable system, the steady-state gain can be obtained from the rational form of a transfer function by setting $X_t = 1$ for all discrete t. We get

$$g(1 - a_1 - a_2 - \cdots - a_r) = (b_0 - b_1 - \cdots - b_s)$$

so we have

$$\lim_{t \to \infty} Y_t = Y_\infty = \frac{b_0 - b_1 - \cdots - b_s}{1 - a_1 - a_2 - \cdots - a_r} = g$$

a result that is related to the *final value theorem* of z-transforms (see Appendix 2C). In general, $Y_\infty = gX$ for any fixed input X.

Example 4.2. For the transfer function in Example 4.1,

$$Y_t = \frac{0.5}{1 - 0.2\mathcal{B}} X_{t-2}$$

we have that if we set $X_t = 1$ for all periods t, then $Y_\infty = 0.5/(1 - 0.2) = 0.625 = g$, which is the long-run (or asymptotic) gain of this process. Note how $\sum_{j=0}^{\infty} v_j$ converges to $g = 0.625$. This implies that the transfer function is stable.   □

4.2   IDENTIFICATION OF TRANSFER FUNCTION PROCESSES

Our first objective is to find the values of the orders of the denominator and numerator polynomials, r and s, and the input–output delay k that best describes a process. For doing this, the impulse response form (4.1) is very useful, as advocated by Box and Jenkins. As mentioned earlier, if the

---

[2]A common terminology, used in control theory, is to say that $A_r(\mathcal{B})$ is a *monic polynomial* and that $B_s(\mathcal{B})$ is a *nonmonic polynomial.*

two representations of $H(\mathcal{B})$ are equivalent, we must have the polynomial equality

$$v_0 + v_1\mathcal{B} + v_2\mathcal{B}^2 + \cdots = \frac{\left(b_0 - b_1\mathcal{B} - \cdots - b_s\mathcal{B}^s\right)\mathcal{B}^k}{1 - a_1\mathcal{B} - \cdots - a_r\mathcal{B}^r}$$

Analyzing this relation provides useful guidelines for the identification of r, s, and k. Note that the following equation results[3]:

$$\left(1 - a_1\mathcal{B} - \cdots - a_r\mathcal{B}^r\right)\left(v_0 + v_1\mathcal{B} + \cdots\right) = \left(b_0 - b_1\mathcal{B} - \cdots - b_s\mathcal{B}^s\right)\mathcal{B}^k \quad (4.2)$$

For this equation to hold, we must have equal terms on both sides of the equality for all powers j of $\mathcal{B}$. Thus, equating coefficients that multiply like powers of $\mathcal{B}$, we can find the relation between the impulse response weights, v_j, and the coefficients in the $A(\mathcal{B})$ and $B(\mathcal{B})$ polynomials. For example, consider the kth power of $\mathcal{B}$. Equating terms that contain $\mathcal{B}^k$ on each side, we get

$$\left(v_k - a_1 v_{k-1} - a_2 v_{k-2} - \cdots - a_r v_{k-r}\right)\mathcal{B}^k = b_0\mathcal{B}^k$$

from which we get

$$v_k = a_1 v_{k-1} + a_2 v_{k-2} + \cdots + a_r v_{k-r} + b_0.$$

Proceeding this way, we get the following:

$$v_j = \begin{cases} 0 & \text{for } j < k \\ a_1 v_{j-1} + a_2 v_{j-2} + \cdots + a_r v_{j-r} + b_{j-k} & j = k, k+1, \ldots, k+s \\ a_1 v_{j-1} + a_2 v_{j-2} + \cdots + a_r v_{j-r} & j > k+s \end{cases} \quad (4.3)$$

The weights are grouped according to the three different equations they obey. Note that in the first group there are k weights, in the second group there are $s + 1$ weights, and in the third group there is an infinite number of weights.

The first equation in (4.3) follows because there are no powers lower than k on the right hand side of (4.2). Thus, starting from $j = 0$, it is easy to see that $v_0 = 0$, then $v_1 = 0$, and so on, until weight $v_{k-1} = 0$. Similarly, the last equation in (4.3) follows because there is no right-hand-side power in (4.2) that is greater than $k + s$. However, contrary to the case when $j < k$, the starting point when powers are zero on the right-hand side is preceded by a nonzero power at time $j = k + s$. Thus, for $j > k + s$, the weights v_j obey

---

[3]This equation is related to the Diophantine identity used to divide polynomials, discussed in Chapter 5.

the rth-order difference equation

$$A_r(\mathcal{B})v_j = 0 \tag{4.4}$$

which requires r initial values $v_{k+s-(r-1)}, v_{k+s-r}, \ldots, v_{k+s-1}, v_{k+s}$. Thus r out of the $s + 1$ weights of the second group in (4.3) will be weights that initialize this difference equation. The remaining $s + 1 - r$ weights will not follow any fixed pattern since they depend only on the actual values of the a_i and b_i coefficients. If the system is stable, the difference equation (4.4) for $j > k + s$ will always show decay, either exponential [in case $A(\mathcal{B})$ has all roots real] or as a damped sinusoidal function in case $A(\mathcal{B})$ has complex roots. Note that this is identical to the behavior of the autocorrelation function ρ_k of AR(p) models since for AR(p) models the autocorrelation obeys the same difference equation as the process. Mathematically speaking, ARIMA models can be thought of as transfer functions where the input is a white noise process.

In summary, the weights v_j of a transfer function model of order (r, s, k) can be classified as follows:

1. k weights $v_0, v_1, \ldots, v_{k-1}$ equal to zero.
2. $s + 1 - r$ weights $v_k, v_{k+1}, \ldots, v_{k+s-r}$ following no fixed pattern [no such weights exist if $s < r$, in which case the first $r - s$ initial values of (4.4) will be zero].
3. r weights $v_{k+s-r+1}, \ldots, v_{k+s-1}, v_{k+s}$ the starting values for (4.4), which do not follow any special pattern.
4. Weights $v_j, j > k + s$, following equation (4.4). These will show exponential or sinusoidal decay.

Based on these observations, while identifying transfer function models it is very useful to keep in mind the following general facts:

- If there is decay in the weights ($r > 0$), it starts from the weight v_{k+s+1} (i.e., $|v_{k+s+1}|$ is the largest value in a decaying sequence of remaining weights). The decay can be exponential or sinusoidal, depending on the roots of the characteristic equation.
- If we restrict ourselves to cases where $r \leq 2$, sinusoidal decay is possible only for a second-order dynamical system ($r = 2$).
- Weights v_0 through v_{k-1} are always zero. The first nonzero weight is v_k.

Knowing the typical behavior of the impulse response $\{v_j\}$, we can identify r, s, and k from observing the impulse response of a system. In the type of quality control applications we are concerned with, the impulse response will always be observed with noise, and identification in such conditions is feasible only if the noise level is moderate compared to the signal we wish to identify. Table 4.1 illustrates typical behaviors of the impulse response function $\{v_j\}$ for different values of r and s when the delay is 3 time units ($k = 3$). In practical applications, r and s are usually less than or equal to 2. The graphs in the table also show the typical response to a unit step function.

Models with $r = 0$
If $s = 0$, the output is a single lagged value $v_k = b_0$ (Table 4.1, case 1). This case is referred to in the control literature as a *pure delay system*. If in addition, $k = 0$, we have a static, not a dynamic, system.

   If $s > 0$ and $r = 0$, there will be $s + 1$ nonzero weights $v_k = b_0$, $v_{k+1} = -b_1, \ldots, v_{k+s} = -b_s$ following no fixed pattern (Table 4.1, cases 2 and 3). This case is referred to in the control literature as the *input* or *denominator dynamics* case.

Models with $r = 1$
If $s = 0$, the weights exhibit exponential decay starting at $v_k = b_0$, which is the starting value for the difference equation $v_j = a_1 v_{j-1}$ (Table 4.1 case 4). Since $s < r$, there are no weights without a fixed pattern. This is a first-order dynamical system, and many processes can be modeled with it. If $a_1 < 0$, this system has an impulse response that oscillates, but not in a sinusoidal manner. Sinusoidal oscillations are possible only in second- and higher-order systems.

   If $s > 0$, there will be s ($= s + 1 - r$) weights v_j starting at $j = k$ that do not follow any pattern, followed by exponential decay from weight v_{k+s}. In this case, the weight v_{k+s} provides the single initial value for the difference equation (4.4) (Table 4.1, cases 5 and 6).

Models with $r = 2$
If $s = 0$, the weights v_j ($j \geq k$) behave according to the solution to

$$v_j = a_1 v_{j-1} + a_2 v_{j-2} \tag{4.5}$$

with the initial values v_k and $v_{k-1} = 0$ (Table 4.1, case 7). Since $s < r$, there are no weights that have no fixed pattern. Note that the decay starts from weight v_k.

   If $s = 1$, the weights obey (4.5) with initial values v_k and v_{k+1}. Since $s < r$, there are no weights that have no fixed pattern. Decay starts from weight v_{k+1} (Table 4.1, case 8).

   Finally, if $s = 2$, the weights v_{k+1} and v_{k+2} provide initial values for (4.5). Also, in this case ($r = s$) there is a single weight (since $s + 1 - r = 1$) that does not follow any pattern. The decay starts from weight v_{k+2} and can be exponential or sinusoidal, according to the roots of (4.5) (Table 4.1, case 9).

   In practice, process engineering insight must be used to help determine the right values of r, s, and k. This, together with an iterative identification–estimation–diagnostic checking procedure will typically result in useful transfer function models.

Table 4.1 Typical Impulse and Step Response for Some Common Transfer Function Models

(r, s, k)	Model	Typical Impulse and Step Response[a]	Values Used in Graph
1. $(0, 0, 3)$	$Y_t = b_0 X_{t-3}$		$Y_t = X_{t-3}$
2. $(0, 1, 3)$	$Y_t = (b_0 + b_1 \mathcal{B}) X_{t-3}$		$Y_t = (0.5 + 0.5\mathcal{B}) X_{t-3}$
3. $(0, 2, 3)$	$Y_t = (b_0 + b_1 \mathcal{B} + b_2 \mathcal{B}^2) X_{t-3}$		$Y_t = (0.25 + 0.5\mathcal{B} + 0.25\mathcal{B}^2) X_{t-3}$
4. $(1, 0, 3)$	$Y_t = \dfrac{b_0}{1 - a_1 \mathcal{B}} X_{t-3}$		$Y_t = \dfrac{0.5}{1 - 0.5\mathcal{B}} X_{t-3}$
5. $(1, 1, 3)$	$Y_t = \dfrac{b_0 + b_1 \mathcal{B}}{1 - a_1 \mathcal{B}} X_{t-3}$		$Y_t = \dfrac{0.25 + 0.25\mathcal{B}}{1 - 0.5\mathcal{B}} X_{t-3}$

6. (1, 2, 3) $\quad Y_t = \dfrac{b_0 + b_1 B + b_2 B^2}{1 - a_1 B} X_{t-3}$

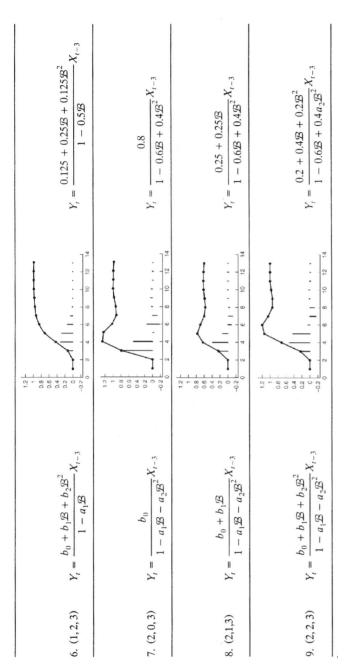

$$Y_t = \dfrac{0.125 + 0.25B + 0.125B^2}{1 - 0.5B} X_{t-3}$$

7. (2, 0, 3) $\quad Y_t = \dfrac{b_0}{1 - a_1 B - a_2 B^2} X_{t-3}$

$$Y_t = \dfrac{0.8}{1 - 0.6B + 0.4B^2} X_{t-3}$$

8. (2,1,3) $\quad Y_t = \dfrac{b_0 + b_1 B}{1 - a_1 B - a_2 B^2} X_{t-3}$

$$Y_t = \dfrac{0.25 + 0.25B}{1 - 0.6B + 0.4B^2} X_{t-3}$$

9. (2, 2, 3) $\quad Y_t = \dfrac{b_0 + b_1 B + b_2 B^2}{1 - a_1 B - a_2 B^2} X_{t-3}$

$$Y_t = \dfrac{0.2 + 0.4B + 0.2B^2}{1 - 0.6B + 0.4a_2 B^2} X_{t-3}$$

^{a}Bars, impulse; lines, step.

4.2.1 Transfer Function Models with Added Noise

Disturbances normally enter into the process and usually are not controllable, in contrast with X_t, which usually is controllable. To model the disturbances, Box and Jenkins (1976) proposed to add a noise term N_t that follows an ARIMA(p, d, q) model of the transfer function model:

$$Y_t = \frac{B(\mathcal{B})}{A(\mathcal{B})} X_{t-k} + N_t$$

where

$$N_t = \frac{C(\mathcal{B})}{D(\mathcal{B})} \varepsilon_t$$

and $D(\mathcal{B})$ may have one or more roots on the unit circle. In this way we obtain the model known in the control engineering literature as the *Box–Jenkins* (BJ) *model*:

$$Y_t = \frac{B(\mathcal{B})}{A(\mathcal{B})} X_{t-k} + \frac{C(\mathcal{B})}{D(\mathcal{B})} \varepsilon_t \tag{4.6}$$

which can be represented as in Figure 4.3.

4.2.2 Identification of Transfer Function Models When Noise Is Present

Identification by means of directly observing the impulse response function is feasible only if the amount of noise relative to the transfer function dynamics is low, as assumed in control engineering. Unfortunately, this does not occur in quality control applications in which the measurements of the product quality characteristics are related to process controllable factors. If the noise is large relative to the contribution of X_t, the controllable factor is probably not important in describing the process. If noise is present, statistical analysis based on the *cross-correlation function* is useful for identification purposes. The identification techniques described in this section, due to Box and Jenkins (1974), assume that the input X_t is not being manipulated based on

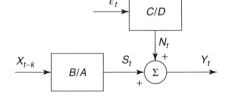

Figure 4.3 Box–Jenkins transfer function plus noise model.

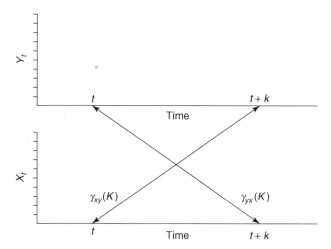

Figure 4.4 Cross-covariance functions.

values of the output Y_t. In other words, it is assumed there is no feedback control mechanism in operation.[4]

Let $X_t = \nabla^d X_t'$ and $Y_t = \nabla^d Y_t'$ be two stationary stochastic processes, where Y_t' and X_t' are the original, possibly nonstationary input and output processes. Their *cross-covariance* at lag j is defined as

$$\gamma_{xy}(j) = E\big[(X_t - \mu_x)(Y_{t+j} - \mu_y)\big] \qquad j = 0, 1, 2, \ldots$$

The cross covariance between Y and X is defined as

$$\gamma_{yx}(j) = E\big[(Y_t - \mu_y)(X_{t+j} - \mu_x)\big] \qquad j = 0, 1, 2, \ldots$$

Figure 4.4 illustrates the meaning of $\gamma_{xy}(k)$ and $\gamma_{yx}(k)$. Note that $\gamma_{xy}(j) \neq \gamma_{yx}(j)$. However, since

$$\gamma_{xy}(j) = \gamma_{yx}(-j) = E\big[(Y_t - \mu_y)(X_{t-j} - \mu_x)\big] \qquad (4.7)$$

we need only consider one of these functions (usually, γ_{xy} is analyzed). We used property (4.7) when we analyzed the autocovariance function of ARMA processes in Section 3.3.

The *cross correlation* between X and Y at lag j is defined as

$$\rho_{xy}(j) = \frac{\gamma_{xy}(j)}{\sigma_x \sigma_y} \qquad j = 0, \pm 1, \pm 2, \ldots$$

---

[4]An alternative and probably simpler approach to identifying and fitting single-input, single-output transfer functions is based on multivariate time-series techniques, and is described in Example 9.2. The multivariate approach provides a simple way to detect if the data were collected under open-loop (no feedback control) conditions.

Since $\rho_{xy}(j) \neq \rho_{yx}(j)$, the cross correlation is not symmetric around zero (it is an odd function).

Estimation of $\gamma_{xy}(j)$ and $\rho_{xy(j)}$
If n pairs of observations (X_t, Y_t) are available, an estimate of $\gamma_{xy}(j)$ is given by

$$c_{xy}(j) = \hat{\gamma}_{xy}(j) = \begin{cases} \dfrac{1}{n}\Sigma_{t=1}^{n-j}(X_t - \bar{X})(Y_{t+j} - \bar{Y}) & j = 0, 1, 2, \ldots \\ \dfrac{1}{n}\Sigma_{t=1}^{n+j}(Y_t - \bar{Y})(X_{t-j} - \bar{X}) & j = 0, -1, -2, \ldots \end{cases}$$

$$(4.8)$$

The formula for the estimates for negative lags comes directly from property (4.7).

The sample cross-correlation function is given by

$$r_{xy}(j) = \hat{\rho}_{xy}(j) = \frac{c_{xy}(j)}{S_x S_y} \qquad j = 0, \pm 1, \pm 2, \ldots$$

If X_t and Y_t are not correlated, it was shown by Bartlett (1955) that

$$\mathrm{Var}\big[r_{xy}(j)\big] \approx \frac{1}{n} \sum_{i=-\infty}^{i=\infty} \rho_x(i)\rho_y(i) \qquad \text{for } j = 0, 1, 2, \ldots$$

From this it follows that if two series are not cross-correlated and one is white noise (perhaps after prewhitening, as described below), then

$$\mathrm{Var}\big[r_{xy}(j)\big] = \mathrm{Var}\big[r_{xy}(-j)\big] \approx \frac{1}{n-j}.$$

Thus the cross-correlations will vary around zero with standard deviation approximately equal to $(n-j)^{-1/2}$ if X_t and Y_t are not cross-correlated and one is white.

Prewhitening for Identification of Transfer Functions
From the above it is clear that the correlation within each series will mask the correlation between the two series, which is of interest for identifying transfer function models. Thus it will make sense to try to make one of the two series white. Such a procedure is called *prewhitening*.

Since X_t is stationary, we can build an ARMA model for it of the form

$$A_\alpha(\mathcal{B})X_t = C_\alpha(\mathcal{B})\alpha_t$$

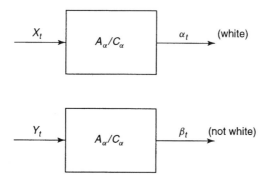

Figure 4.5 Prewhitening the input and applying the prewhitening filter to both input and output.

where α_t is white noise. Thus, applying the same filter to the output,

$$\frac{A_\alpha(\mathcal{B})}{C_\alpha(\mathcal{B})}Y_t = \beta_t$$

where β_t is not necessarily white noise. Figure 4.5 illustrates the two filtering operations we have conducted so far. The BJ transfer function can then be written as

$$\frac{A_\alpha(\mathcal{B})}{C_\alpha(\mathcal{B})}Y_t = \frac{B(\mathcal{B})}{A(\mathcal{B})}\mathcal{B}^k\frac{A_\alpha(\mathcal{B})}{C_\alpha(\mathcal{B})}X_t + \frac{A_\alpha(\mathcal{B})}{C_\alpha(\mathcal{B})}N_t$$

or

$$\beta_t = H(\mathcal{B})\alpha_t + \varepsilon_t^*$$

where $\varepsilon_t^* = [A_\alpha(\mathcal{B})/C_\alpha(\mathcal{B})]N_t$ is *colored noise*. Multiplying both sides by α_{t-j} and taking expectations yields,

$$E[\alpha_{t-j}\beta_t] = E[\alpha_{t-j}H(\mathcal{B})\alpha_t] + E[\alpha_{t-j}\varepsilon_t^*]$$

or

$$\gamma_{\alpha\beta}(j) = E[\alpha_{t-j}v_j\mathcal{B}^j\alpha_t] + E[\alpha_{t-j}\varepsilon_t^*]$$

This last step follows since α_t is white noise as long as autocorrelation goes, but it is clearly related to β_t, so the expected value on the left is not zero. The first term on the right follows since $E[\alpha_{t-j}\alpha_{t-l}] = 0$ for $j \neq l$. The last term is zero because ε_t^* is not a function of α_t, and ε_t^* and α_t are uncorrelated.[5] Thus

$$v_j = \frac{\gamma_{\alpha\beta}(j)}{\sigma_\alpha^2} = \frac{\rho_{\alpha\beta}(j)\sigma_\beta}{\sigma_\alpha} \qquad j = 0, 1, 2, \ldots$$

$$\hat{v}_j = \frac{r_{\alpha\beta}(j)s_\beta}{s_\alpha} \qquad j = 0, 1, 2, \ldots$$

(4.9)

This is the estimated impulse response function used to identify (r, s, k). Notice that since $\rho_{\alpha\beta}(j) = v_j\sigma_\alpha/\sigma_\beta$, the estimates $\hat{\rho}_{\alpha\beta}(j) = r_{\alpha\beta}(j)$ will be more reliable as σ_α increases. Thus the input should vary enough to let us "see" the input–output effects v_j.

Identification of the Noise Model
An estimate of the noise disturbance of the series can be obtained from

$$\hat{N}_t = Y_t - \hat{H}(\mathcal{B})X_t$$

where $\hat{H}(\mathcal{B})$ contains the weights $\hat{v}_j$. If a rational transfer function is available, the noise can be estimated from

$$\hat{N}_t = Y_t - \frac{\hat{B}(\mathcal{B})}{\hat{A}(\mathcal{B})}X_{t-k}$$

The SACF and SPACF of $\hat{N}_t$ are used to identify an ARIMA(p, d, q) noise model.

Offline Estimation of Transfer Function Models
Once a set of pairs of observations (X_t, Y_t) is obtained (size set of at least 100 pairs is usually recommended), and an (r, s, k) and ARIMA(p, d, q) identified, the next step is to estimate the parameters of these models. Denote the model parameters by $\mathbf{B}' = (b_0, b_1, \ldots, b_s)$, $\mathbf{A}' = (a_1, a_2, \ldots, a_r)$, $\mathbf{C}' = (c_1, c_2, \ldots, c_p)$ and $\mathbf{D}' = (d_1, d_2, \ldots, d_q)$. These are estimated by minimizing the conditional sum of squares:

$$SS(\mathbf{A}, \mathbf{B}, \mathbf{C}, \mathbf{D}) = \sum_{t=1}^{n} \varepsilon_t^2(\mathbf{A}, \mathbf{B}, \mathbf{C}, \mathbf{D}|k, \mathbf{X}_0, \mathbf{Y}_0, \boldsymbol{\varepsilon}_0)$$

---

[5] If the data were collected during the closed-loop operation of a linear feedback controller, the term $E[\alpha_{t-j}\varepsilon_t^*]$ will not be zero, and the identification method shown here breaks down. We discuss this problem in more detail in Section 8.5. See also Box and MacGregor (1974, 1976) and del Castillo (2001).

where $\mathbf{X}_0$, $\mathbf{Y}_0$, and $\boldsymbol{\varepsilon}_0$ denote the initial values of the difference equation. The solution to this minimization is a good approximation of the MLE values if the errors are normally distributed. The initial values can be "back-forecasted" (see Box et al., 1994) or simply neglected if n is large (and get the unconditional sum of squares). The sum of squares function will typically be nonlinear in the parameters. Marquardt's algorithm is frequently used for the minimization.

Example 4.3: Semiconductor Manufacturing Process. Chemical mechanical polishing (CMP) is an important processing step in the manufacture of semiconductors. The goal of this process is essentially to make the silicon wafers "flat" by polishing them. In a CMP machine, wafers are pressed down onto a rotating plate that has a polishing pad. A slurry is added to the pad to increase abrasion. *Planarizing* the wafers is essential to be able to add more circuits to the wafers. The two goals of this process are the nonuniformity of the wafer (a measure of the variability of the thickness of the wafer) and the removal rate of silicon oxide. Some of the controllable factors are the down-force pressure, rotation speed of the plate, amount of slurry, and the polishing time. In this example, let us consider removal rate (in Amstrongs per minute) to be the quality characteristic of interest and the *platen speed* to be the controllable factor (in rpm). Removal rate is a response that manufacturers wish to maximize, since evidently "time is money" and this is a measure of how fast the job is done. To identify the input–output transfer function, all other controllable factors were fixed at nominal values. The platen speed was varied randomly every 10 wafers within the interval (22, 28). The input–output data of this example can be found in the file *CMP.txt*, and the time-series data are plotted in Figure 4.6. The computations were performed both in SAS and with Matlab's Systems ID toolbox. (We discuss how to use these software packages in Sections 4.4 and 4.5.)

The first step is to prewhiten the input series. Figure 4.7 shows the autocorrelation and partial autocorrelation function of the controllable factor. Since the platen speed was varied only a few times over the 100 observations (wafers), its mean wanders with time, so it is nonstationary, as can be seen from the plots. This nonstationarity behavior would make estimation of the cross-correlation difficult. Thus, in this case, a simple first-order difference is enough as a prewhitening filter. The removal rate series was also differenced. Figure 4.8 shows the impulse response function $\{\hat{v}_j\}$ estimated from the prewhitened input and output. The graph indicates that the CMP process is simply a *pure unit delay process*, that is, there are no process dynamics and the delay is one time unit (here the time unit is simply the run or batch number). This was expected, as a change in the controllable factors affects the output at the next run of wafers. The dynamic behavior of this process comes from the wearing off that occurs in the polishing pads, which is modeled by a nonstationary disturbance. Fitting the model

$$\nabla Y_t = b_0 \nabla X_{t-1} + N_t$$

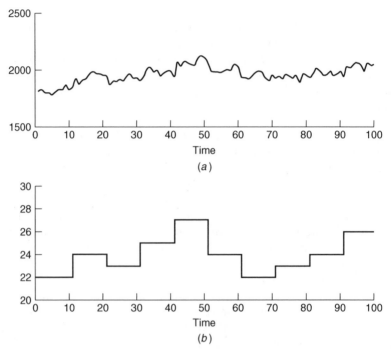

Figure 4.6   (*a*) Removal rate and (*b*) platen speed time series for 100 runs, Example 4.3.

to the differenced input and output gives an estimate $\hat{b}_0 = 32.15$. The sample autocorrelation and partial autocorrelation functions of the residuals obtained from $Y_t - \hat{Y}_t$, where $\hat{Y}_t = Y_{t-1} + 32.15\nabla X_{t-1}$, are shown in Figure 4.9. The sample autocorrelation function clearly cuts off after lag 1, while the first two lags of the sample partial autocorrelation function are significant. This indicates as a tentative disturbance model in MA(1). Fitting the model

$$\nabla Y_t = b_0 \nabla X_{t-1} + (1 - \theta \mathcal{B}) \, \varepsilon_t$$

to the data, we obtain $\hat{B}_0 = 33.86$ and $\hat{\theta} = 0.7061$. Both estimates are highly significant. At this point, diagnostic checks on the model should be performed, as explained later. The fitted model so far is

$$Y_t = 33.86 X_{t-1} + \frac{1 - 0.7061\mathcal{B}}{1 - \mathcal{B}} \varepsilon_t, \qquad \varepsilon_t \sim (0, 30.19^2)$$

which corresponds to a pure unit delay or responsive process with an IMA(1, 1) disturbance.   □

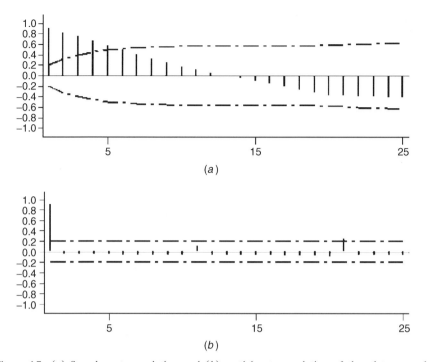

Figure 4.7 (*a*) Sample autocorrelation and (*b*) partial autocorrelation of the platen speed, Example 4.3.

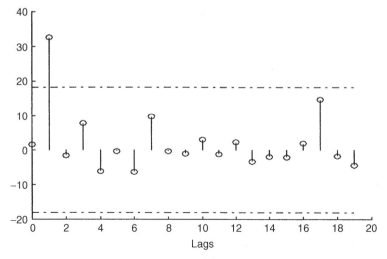

Figure 4.8 Estimated impulse response function $\{\hat{v}_j\}$ between the differenced platen speed and the differenced removal rate, Example 4.3.

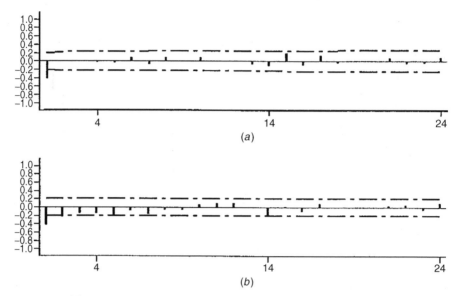

Figure 4.9 (*a*) Autocorrelation and (*b*) partial autocorrelation of the residuals of the prelimi-nary model $\nabla Y_t = 32.15 \nabla X_{t-1} + N_t$, Example 4.3.

4.2.3 Other Input–Output Dynamic Models

The Box–Jenkins transfer function (4.6) model has the advantage that it has a nice signal plus noise interpretation and avoids common terms that might give additional complexity. This is not the only form for an input–output dynamical model. In particular, a popular model in the econometrics litera-ture that will be referred to in later chapters is the *ARMAX* (ARMA exogenous variable) *model*:

$$A(\mathcal{B})Y_t = B(\mathcal{B})X_t + C(\mathcal{B})\varepsilon_t$$

Here the exogenous variable is X_t and the endogenous variable is Y_t. This form is simpler to manipulate for controller design purposes. Going from a BJ model to the ARMAX format is a simple algebraic step, thus models will be fitted in BJ form but will be manipulated[6] starting from an ARMAX form.

4.3 DIAGNOSTIC CHECKING IN TRANSFER FUNCTION MODELS

Denote the residuals of the fitted model by $e_t = Y_t - \hat{Y}_t$. Suppose that the true process is

$$Y_t = H(\mathcal{B})X_t + \Psi(\mathcal{B})\varepsilon_t$$

[6] Primarily to derive optimal controllers in Chapter 5.

where $\{\varepsilon_t\}$ is a white noise sequence. However, suppose that we fit the model

$$Y_t = H_0(\mathcal{B})X_t + \Psi_0(\mathcal{B})\varepsilon_{0,t}.$$

Then the errors of the model are

$$\varepsilon_{0,t} = \frac{H(\mathcal{B}) - H_0(\mathcal{B})}{\Psi_0(\mathcal{B})}X_t + \frac{\Psi(\mathcal{B})}{\Psi_0(\mathcal{B})}\varepsilon_t$$

so we can make the following remarks:

- If there is no system−model mismatch, $\varepsilon_{0,t} = \varepsilon_t$ and the $\varepsilon_{0,t}$ will not be autocorrelated or cross-correlated with the X's (or with the α's).
- If the transfer function is *correct* $[H(\mathcal{B}) = H_0(\mathcal{B})]$ but the noise model is incorrect $[\Psi_o(\mathcal{B}) \neq \Psi(\mathcal{B})]$,

$$\varepsilon_{0,t} = \frac{\Psi(\mathcal{B})}{\Psi_0(\mathcal{B})}\varepsilon_t$$

and the $\varepsilon_{0,t}$'s will be autocorrelated but not cross-correlated with the X's. Thus if r_{ee} shows structure but r_{ae} does not, the noise model is probably wrong.

- If the transfer function is *incorrect* (even if the noise is OK),

$$\varepsilon_{0,t} = \frac{H(\mathcal{B}) - H_0(\mathcal{B})}{\Psi_0(\mathcal{B})}X_t + \varepsilon_t$$

Thus the $\varepsilon_{0,t}$ values will be autocorrelated and cross-correlated with the X's. To check this condition, if both r_{ee} and r_{ae} show structure, the transfer function is incorrect.

An overall check for lack of structure in the first K autocorrelations of the residuals is given by the chi-square statistic:

$$Q_1 = m(m+2)\sum_{j=1}^{K}\frac{r_{ee}^2}{m-j}$$

where $m = n - \max(r, s+k) - p$. This statistic, under H_0: $\rho_{ee}(j) = 0$ for $j = 1, \ldots, K$, follows a χ_{K-p-q}^2 distribution. Evidently, the rejection region is the right tail of the distribution, since large values will be due to large autocorrelations. Similarly, an overall check for H_0: $\rho_{ae}(j) = 0$ for $j =$

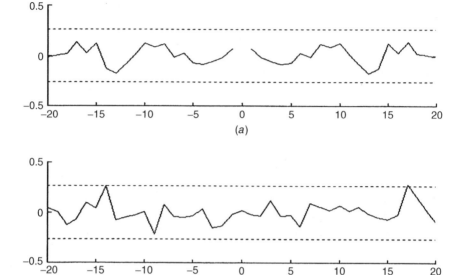

Figure 4.10 (*a*) Sample autocorrelation of the residuals, r_{ee}; (*b*) sample cross-correlation between prewhitened input and residuals, $r_{\alpha e}$, Example 4.4.

$1, \ldots, K$ is to compute

$$Q_2 = m(m + 2) \sum_{j=1}^{K} \frac{r_{\alpha e}^2}{m - j}$$

which under H_o follows a $\chi^2_{K+1-(r+s+1)}$ distribution.

Example 4.4: Diagnostic Checking for Semiconductor Manufacturing Process.
In the CMP process of Example 4.3, the tentative model fitted was $\nabla Y_t = 33.86 \nabla X_{t-1} + (1 - 0.7061\, \mathcal{B})\varepsilon_t$. Figure 4.10 shows plots of the sample autocorrelation function of the residuals [i.e., a plot of $r_{ee}(k)$ versus. k], and a plot of the estimated cross-correlations between the prewhitened input and the residuals, [i.e., a plot of $r_{\alpha e}(k)$ versus. k]. In both cases, no significant correlations can be seen indicating that both the transfer function and the disturbance have been identified adequately. Overall checks for lack of correlation can be performed from the Q_1 and Q_2 statistics referred to earlier. Using SAS, we found that $Q_1 = 19.67$ up to lag $K = 24$, for a p-value equal to 0.6618. Thus there is no evidence to reject the hypothesis which says that residuals are uncorrelated. Also using SAS, we find that $Q_2 = 18.84$ up to lag $K = 23$, with a p-value of 0.7105. This indicates that there is no

evidence to conclude there is a significant cross-correlation left between the input and the residuals. Based on these statistics and the graphs, we can conclude that the removal rate model is adequate.   □

4.4 OFFLINE IDENTIFICATION AND ESTIMATION USING SAS PROC ARIMA

SAS PROC ARIMA provides a very powerful tool for fitting not only ARIMA models but also Box–Jenkins transfer function models. The procedure follows quite closely Box and Jenkins's methodology. Assuming that we call the input x and the output y, a typical session for transfer function modeling would look as follows:

```
PROC ARIMA data = seriesName;
/* Identify and fit a prewhitening filter for the input */
identify var = x;
estimate p = -- q = --;
/* Identify and fit TF and noise */
identify var = y(d1,d2,...) crosscorr = (x(d1,d2,...));
estimate p = -- q = -- input = (k$(L1,L2,...)/(L1,L2,...) X);
run;
quit;
```

As explained in Chapter 3, the terms `d1,d2` and so on, are used to difference the input and output series (i.e, the operators $1 - \mathcal{B}^{d1}$, $1 - \mathcal{B}^{d2}$, etc., are computed). For example, to specify a second-order difference (∇^2) applied to a variable called Y, we would enter `var = Y(1,1)`. The delay is entered in the input statement (`k`). The terms `(L1,L2,...)/(L1,L2,...)` are used to enter the orders of the $B(\mathcal{B})$ and $A(\mathcal{B})$ polynomials in the Box–Jenkins model (4.6). The cross-correlation function will be that of the prewhitened input and the prewhitened output. SAS automatically differences both input and output with the filter that was fitted to the input. The prewhitened series are only used for computation of the cross-correlation. The original series are used in subsequent identify/estimate statements. SAS computes the Q_2 diagnostic statistic described before only if a prewhitening filter is constructed for the input series using a pair of identify/estimate statements, as shown above. The Q_1 diagnostic statistic is always computed when fitting ARIMA and transfer function models.

Fitting models with nonstationary noise is quite common but a little tricky in SAS. The key is to difference both input and output in such a way that the differences cancel except for the noise denominator, as shown in the following example.

Example 4.5: Fitting a Transfer Function with Additive IMA(1, 1) Noise Using SAS. Suppose that we wish to fit a $(r, s, k) = (1, 0, 2)$ transfer function with IMA(1, 1) noise. Then we enter the following statements, inserted after prewhitening the input:

```
identify var = y(1)  crosscorr = (x(1));
estimate p = 0  q = 1  input = (2$ / (1)  x);
run;
```

This fits the model

$$\nabla Y_t = \frac{b_0 \nabla}{1 - a_1 \mathcal{B}} X_{t-2} + (1 - c\mathcal{B})\varepsilon_t$$

which reduces to the desired form,

$$Y_t = \frac{b_0}{1 - a_1 \mathcal{B}} X_{t-2} + \frac{1 - c\mathcal{B}}{\nabla}\varepsilon_t. \qquad \square$$

Example 4.6: Transfer Function Analysis of a Polymerization Process Using SAS. Consider a continuous polymerization process in which it is of interest to model the polymer viscosity as a function of changes in the reactor temperature.[7] Let the output variable Y_t be a melt index, a measure of polymer viscosity, and let X_t be the average temperature of the reactor. Samples are taken every 2 hours, so X denotes the average temperature over a 2-hour period. The file *Polymer.txt* contains the input–output time-series data, which were plotted in Figure 4.11 using Matlab.

The first step is to build a model for the input. From the sample autocorrelation function of the input, it is evident that the input needs to be differenced. In SAS we do this with the statement

```
proc arima data = Capilla;
identify var = X(1)  nlags = 10;
run;
```

A first-order difference transforms the input into a white noise sequence, as seen in the autocorrelation plots of the differenced input. For SAS to treat a simple difference as a prewhitening filter, we estimate the model with the statement

```
estimate noconstant;
run;
```

[7]This example is based on a paper by Capilla et al. (1999).

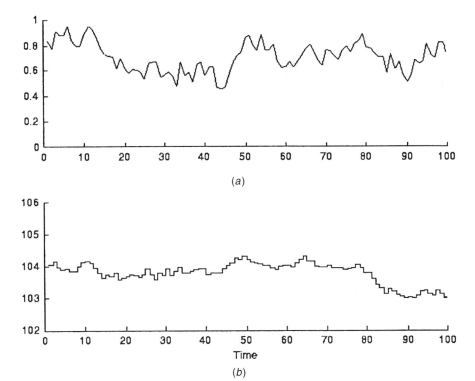

Figure 4.11 Input and output series, Example 4.6. (*a*) Melt index (output); (*b*) temperature (input).

Next, we want to estimate the cross-correlation between the prewhitened input and output, and we wish that any further modeling be made on ∇Y_t and ∇X_t. Thus we specify

```
identify var = Y(1) crosscorr = X(1) nlags = 10;
estimate input = (1$(1) X) plot noconstant;
run;
```

The cross-correlation function is shown in Figure 4.12. It is clear this is a numerator dynamics process with two significant lags and delay equal to one period; thus $k = 1$ and $s = 1$. The input statement specified above generates preliminary parameter estimates for such a process.

To build a disturbance model, the sample autocorrelation and sample partial autocorrelation functions of the residuals of the preliminary model are needed, and these are shown in Figure 4.13. As can be seen from the estimated autocorrelation and partial autocorrelation plots of the residuals, a good tentative disturbance model is an MA(1), since the SACF cuts off after lag 1 and the SPACF tails off.

Crosscorrelations

Lag	Covariance	Correlation	−1 9 8 7 6 5 4 3 2 1 0 1 2 3 4 5 6 7 8 9 1
−1	−0.0011354	−.14140	| ***| . |
0	−0.0007442	−.09268	| **| . |
1	0.0025944	0.32309	| . |****** |
2	0.0025131	0.31297	| . |****** |
3	−0.0000623	−.00775	| . | . |
4	6.93666E-6	0.00086	| . | . |
5	−0.0001903	−.02370	| . | . |
6	0.0012451	0.15506	| . |*** . |
7	−0.0011894	−.14812	| ***| . |
8	−0.0004194	−.05223	| *| . |
9	0.00073335	0.09133	| . |** . |
10	−0.0006342	−.07898	| **| . |

"." marks two standard errors

Crosscorrelation Check Between Series

To Lag	Chi-Square	DF	Pr > ChiSq	− − − − − − − − − − − − Crosscorrelations − − − − − − − − − − − −
5	20.94	6	0.0019	−0.093 0.323 0.313 −0.008 0.001 −0.024

Conditional Least Squares Estimation

Parameter	Estimate	Standard Error	t Value	Approx Pr > |t|	Lag	Variable	Shift
NUM1	0.22592	0.06349	3.56	0.0006	0	X	1
NUM1,1	−0.23179	0.06345	−3.65	0.0004	1	X	1

Variance Estimate	0.004502
Std Error Estimate	0.067096
AIC	−246.864
SBC	−241.715
Number of Residuals	97

* AIC and SBC do not include log determinant.

Figure 4.12 Sample cross-correlation and preliminary parameter estimates obtained using SAS, Example 4.6.

The complete transfer model with MA(1) noise is fitted with the statement

```
estimate q = 1 input = (1(1) X) plot noconstant;
run;
```

Figure 4.14 shows the parameter estimates of the complete transfer function and disturbance model fitted thus far, together with some diagnostic information. The fitted model is

$$\nabla Y_t = 0.1866 \nabla X_{t-1} + 0.23016 \nabla X_{t-2} - 0.46438 \varepsilon_{t-1} + \varepsilon_t$$

or

$$Y_t = (0.1886 + 0.23016 \mathcal{B}) X_{t-1} + \frac{1 - 0.46438 \mathcal{B}}{1 - \mathcal{B}} \varepsilon_t$$

Autocorrelation Check of Residuals

To Lag	Chi–Square	DF	Pr > ChiSq	— — — — — — — — — — Autocorrelations — — — — — — — — — — — —					
6	17.22	6	0.0085	−0.378	−0.035	0.030	0.120	−0.106	−0.018
12	19.33	12	0.0809	0.067	−0.040	0.085	−0.077	0.010	−0.001
18	24.84	18	0.1295	0.092	−0.017	0.012	0.111	−0.149	0.056
24	31.51	24	0.1395	−0.064	0.077	−0.094	0.102	0.083	−0.126

Autocorrelation Plot of Residuals

Lag	Covariance	Correlation	−1 9 8 7 6 5 4 3 2 1 0 1 2 3 4 5 6 7 8 9 1	Std Error
0	0.0045018	1.00000	| |********************|	0
1	−0.0017018	−.37802	| ********| . |	0.101535
2	−0.0001582	−.03513	| . *| . |	0.115133
3	0.00013702	0.03044	| . |* . |	0.115244
4	0.00054026	0.12001	| . |** . |	0.115326
5	−0.0004786	−.10631	| . **| . |	0.116607
6	−0.0000823	−.01827	| . | . |	0.117602
7	0.00030330	0.06737	| . |* . |	0.117631
8	−0.0001806	−.04011	| . *| . |	0.118028
9	0.00038388	0.08527	| . |** . |	0.118169
10	−0.0003458	−.07681	| . **| . |	0.118801

"." marks two standard errors

Partial Autocorrelations

Lag	Correlation	−1 9 8 7 6 5 4 3 2 1 0 1 2 3 4 5 6 7 8 9 1
1	−0.37802	| ********| . |
2	−0.20771	| ****| . |
3	−0.07819	| . **| . |
4	0.11813	| . |** . |
5	−0.00034	| . | . |
6	−0.04614	| . *| . |
7	0.02242	| . | . |
8	−0.02596	| . *| . |
9	0.10756	| . |** . |
10	−0.00080	| . | . |

Crosscorrelation Check of Residuals with Input X

To Lag	Chi–Square	DF	Pr > ChiSq	— — — — — — — — — — — Crosscorrelations — — — — — — — — — — —					
5	3.47	4	0.4819	−0.003	0.018	0.007	−0.035	−0.002	0.186
11	8.04	10	0.6249	−0.122	−0.001	0.066	−0.141	0.036	−0.085
17	11.28	16	0.7921	0.001	−0.061	−0.094	−0.041	0.139	0.019
23	15.72	22	0.8296	0.030	−0.078	0.043	0.107	−0.159	0.024

Figure 4.13 Autocorrelation and partial autocorrelations of residuals estimated using SAS, Example 4.6.

Conditional Least Squares Estimation

Parameter	Estimate	Standard Error	t Value	Approx Pr > |t|	Lag	Variable	Shift
MA1,1	0.46438	0.09413	4.93	<.0001	1	Y	0
NUM1	0.18666	0.05851	3.19	0.0019	0	X	1
NUM1,1	−0.23016	0.05769	−3.99	0.0001	1	X	1

Variance Estimate	0.00372

Correlations of Parameter Estimates

Variable	Parameter	Y MA1,1	X NUM1	X NUM1,1
Y	MA1,1	1.000	−0.217	0.010
X	NUM1	−0.217	1.000	0.426
X	NUM1,1	0.010	0.426	1.000

Autocorrelation Plot of Residuals

Lag	Covariance	Correlation	−1 9 8 7 6 5 4 3 2 1 0 1 2 3 4 5 6 7 8 9 1	Std Error
0	0.0037198	1.00000	| |********************|	0
1	−0.0000566	−.01523	| . | . |	0.101535
2	−0.0000843	−.02266	| . | . |	0.101558
3	0.00029305	0.07878	| . |** . |	0.101610
4	0.00043114	0.11590	| . |** . |	0.102238
5	−0.0002682	−.07209	| . *| . |	0.103584
6	−0.0000794	−.02134	| . | . |	0.104100
7	0.00022621	0.06081	| . |* . |	0.104145
8	0.00001609	0.00432	| . | . |	0.104510
9	0.00025627	0.06889	| . |* . |	0.104512
10	−0.0001446	−.03887	| . *| . |	0.104979

Partial Autocorrelations

Lag	Correlation	−1 9 8 7 6 5 4 3 2 1 0 1 2 3 4 5 6 7 8 9 1
1	−0.01523	| . | . |
2	−0.02289	| . | . |
3	−0.07814	| . |** . |
4	0.11850	| . |** . |
5	−0.06566	| . *| . |
6	−0.02602	| . *| . |
7	0.04029	| . |* . |
8	−0.00273	| . | . |
9	0.09233	| . |** . |
10	−0.04501	| . *| . |

Crosscorrelation Check of Residuals with Input X

To Lag	Chi-Square	DF	Pr > ChiSq	Crosscorrelations					
5	3.76	4	0.4395	−0.004	0.015	0.014	−0.030	−0.017	0.194
11	7.24	10	0.7028	−0.049	−0.024	0.076	−0.123	−0.027	−0.108
17	12.94	16	0.6771	−0.056	−0.088	−0.145	−0.099	0.105	0.082
23	16.54	22	0.7883	0.047	−0.069	0.006	0.118	−0.125	−0.029

Figure 4.14 Final parameter estimates and diagnostics (including Q_2) obtained using SAS, Example 4.6.

with $\hat{\sigma}_{\varepsilon}^2 = 0.0037$. Note that the noise term corresponds to an IMA(1, 1) model.

The SACF and PACF of the residuals do not indicate any remaining autocorrelation, and the Q_2 statistic equals 16.24 up to lag $K = 23$. The corresponding p-value for the overall null hypothesis of no autocorrelations for the first 23 lags is 0.7883. Thus there is no evidence to reject, and we conclude that the model is appropriate.   □

4.5 OFFLINE IDENTIFICATION AND ESTIMATION USING MATLAB'S SYSTEMS IDENTIFICATION TOOLBOX

Matlab's Systems ID Toolbox also provides nice capabilities to fit Box–Jenkins transfer function models.[8] Assuming the input and output time series are stored in an $n \times 2$ matrix S, with the first column being the output and the second column being the input,[9] the function

```
cra(S)
```

automatically prewhitens the input and output [fitting an AR(10) model to the input] and plots the impulse response function (the $\hat{v}_j$'s) with 99% significance limits. The same AR(10) prewhitening filter is applied to the output series. If a lower AR model is desired to prewhiten the input, you should enter

```
cra(S,M,na)
```

where na is the degree of the AR model desired and M is the number of lags for which we want the impulse response to be estimated. Using

```
cra(S,M,na,2)
```

will plot the estimated cross-correlation and the estimated impulse response function, which differ by a constant, as shown by equation (4.9). If one wishes to difference a series, the command

```
diff(X)
```

---

[8] A full account of this toolbox is described by its author, L. Ljung, in *System Identification Toolbox User's Guide*, The Math Works Inc., Natick, Massachusetts, 1995. Matlab version 11 and the Systems ID Toolbox release 4 were used in this book. A new Systems ID toolbox release 5 is now available (Ljung, 2001).
[9] Systems ID release 5 requires the input–output data to be stored instead in an iddata object; see Ljung (2001) for more information.

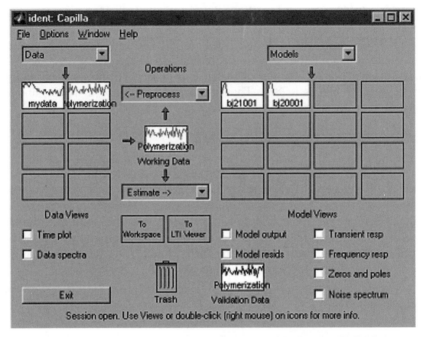

Figure 4.15 The ident graphical user interface, Matlab's Systems ID Toolbox.

will compute the first-order difference of the time series stored in vector X. This is useful before analyzing and fitting transfer function models, perhaps as part of prewhitening.

Entering the name of the function ident starts Matlab's identification GUI, which contains capabilities for time-series plotting, model fitting (including Box–Jenkins models and many others), and autocorrelation and cross-correlation plots of residuals.[10] Figure 4.15 shows the GUI.

To use the ident GUI, a user must first specify the vectors that contain the input and output data. This is done by selecting "Data → Import." The user then enters the names of the vectors containing input and output data. An icon associated with the input just imported into the GUI will appear on the left-hand side.

From the cross-correlation (impulse response) function, the user should select the orders of the models to fit. Matlab Systems ID Toolbox uses the notation

$$Y_t = \frac{B}{F} X_{t-k} + \frac{C}{D} \varepsilon_t$$

[10]All the capabilities of the ident GUI can also be performed by specific commands entered from the main Matlab window. See Ljung (1995, 2001) for details.

to denote the four polynomials in a Box–Jenkins transfer function with noise model ("BJ" according to Matlab's Systems ID Toolbox). To fit this model, select "Estimate → Parametric model → BJ." Specify the orders of the models by entering the orders of the four polynomials in the following order, followed at the end by the delay (k):

$$n_B \quad n_C \quad n_D \quad n_F \quad k$$

The program will name the model according to the orders of the polynomials and the delay. For example, in Matlab a *bj11101 model* is a Box–Jenkins model of the form

$$Y_t = b_0 X_{t-1} + \frac{1 + c_1 \mathcal{B}}{1 + d_1 \mathcal{B}} \varepsilon_t$$

Notice that for Matlab, a B polynomial of order 1 is what we have called before (more properly) a polynomial of order zero. Also notice the signs of the parameters in the polynomials: They are all positive, contrary to the Box and Jenkins convention that we have followed in this book.

Once a model is fitted, an icon will appear on the right-hand side associated with the fitted model (e.g., a "bj11101" icon will appear). Then one can plot the autocorrelation of the residuals and the cross-correlations between the residuals and the input. To look at residuals, be sure to use the same working file as validation file, as indicated by the corresponding GUI icons [ident allows us to validate (i.e., look at residuals against other series, not necessarily the series used to fit the model)]. To get the parameter estimates, double-click on the icon of a fitted model and select present. The parameter estimates will appear on Matlab's main window environment.

Example 4.7: Transfer Function Modeling Using Matlab's Systems ID Toolbox. Consider the polymerization process analyzed in Example 4.6. The input and output appear nonstationary (see Figure 4.11), so we difference both input and output using the diff command. Then, using the cra command, we get the estimated impulse response function $\{\hat{v}_j\}$ of Figure 4.16. It is evident that we are dealing with a $(r, s, k) = (0, 1, 1)$ transfer function between ∇Y_t and ∇X_t:

$$\nabla Y_t = \left(b_0 + b_1 \mathcal{B} \right) \nabla X_{t-1} + N_t$$

We enter the ident GUI and import the differenced series. We initially estimate the model bj20001. The sample autocorrelations of the residuals of this model (Figure 4.17) indicate that the noise is probably a MA(1).[11] Thus,

---

[11] Note that the Matlab Systems ID Toolbox does not estimate the partial autocorrelation function.

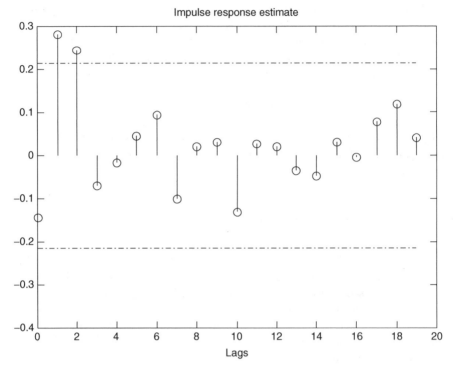

Figure 4.16 Impulse response function estimated with Matlab's Systems ID Toolbox, Example 4.7.

we estimate the bj21001 model, which according to the residuals autocorrelation and the cross-correlation between residuals and input (Figure 4.18) shows no remaining structure, so we conclude that this model represents the data adequately.

Finally, we present the information related to the fitted bj21001 model, and we obtain the listing below. The numbers on the second row of each matrix correspond to the standard errors of the parameter estimates:

```
The polynomial coefficients and their standard deviations are

B =
                0           0.1794          0.25087
                0           0.058219        0.056942

C  =
                1           -0.46477
                0           0.092943
```

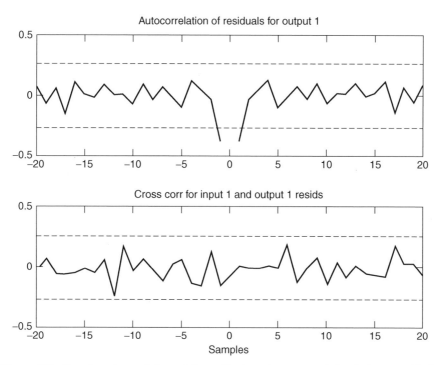

Figure 4.17 Sample autocorrelation of the residuals and cross-correlation between residuals and input for the bj20001 model, estimated with Matlab's Systems ID Toolbox, Example 4.7.

The number of trailing zeros in the B matrix indicates the input–output delay. The final model is

$$\nabla Y_t = (0.1794 + 0.25087\mathcal{B})\, \nabla X_{t-1} + (1 - 0.4647\mathcal{B})\, \varepsilon_t$$

which is very similar to what we obtained using SAS. The numerical differences are due to minor differences in the estimation routines used by these packages. □

PROBLEMS

4.1. Build a transfer function for the semiconductor manufacturing data in Example 4.3 using (**a**) using SAS PROC ARIMA; (**b**) Matlab's Systems ID Toolbox.

4.2. Build a transfer function for the polymerization process data of Example 4.6 using Matlab's Systems ID Toolbox.

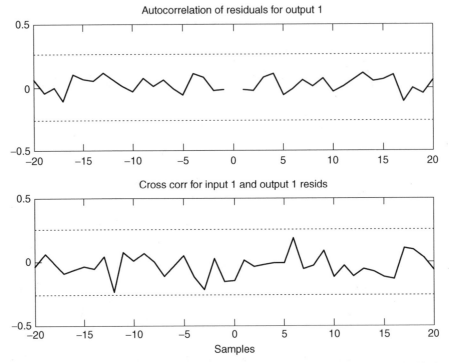

Figure 4.18 Sample autocorrelation of the residuals and cross-correlation between residuals and input for the bj21001 model, estimated with Matlab's Systems ID Toolbox, Example 4.7.

4.3. Consider the gas furnace data (series J) in Box and Jenkins, which can be found in the file *BJ-J.txt*. Build a transfer function for CO_2 in the outlet as a function of the input gas rate. Use **(a)** SAS PROC ARIMA, **(b)** Matlab's System's ID Toolbox.

4.4. Write the SAS statements needed to identify and fit a $(2, 1, 2)$ transfer function with IMA$(1, 1)$ noise.

4.5. Write the SAS statements needed to identify and fit a $(0, 0, 1)$ transfer function with ARIMA$(1, 1, 1)$ noise.

4.6. A set of 100 input–output observations gives the sample cross-correlations shown below:

Lag, k	0	1	2	3	4	5	6	7	8	9	10
$r_{\alpha\beta}(k)$	0.098	0.012	0.15	0.34	0.22	0.096	0.023	-0.04	0.05	-0.024	0.01

What tentative transfer model (r, s, k) would you suggest?

4.7. A set of 100 input–output observations gives the sample cross-correlations shown below.

Lag, k	0	1	2	3	4	5	6	7	8	9	10
$r_{g\alpha\beta}(k)$	-0.088	-0.068	0.297	-0.590	0.318	0.135	0.001	-0.097	0.183	-0.127	0.03

Give a tentative (r, s, k) transfer function model.

4.8. Consider the estimated transfer function

$$\nabla Y_t = \frac{0.5 - 0.1\mathcal{B}}{1 - 0.2\mathcal{B} - 0.3\mathcal{B}} \nabla X_{t-2} + (1 - 0.6\mathcal{B})\varepsilon_t$$

Find the estimated steady-state process gain.

4.9. *Simulation of step response.* Consider the transfer function fitted in Example 4.6 for the polymerization data. Suppose that the controllable factor is the step function $X_t = 1$, for $t \geq 1$, $X_t = 0$ for $t < 0$. Simulate and plot the response for 30 realizations of the process under such an input sequence. (*Suggestion*: Use a spreadsheet program.)

4.10. Repeat Problem 4.9 for the transfer function fitted in Example 4.3.

4.11. Repeat Problem 4.9 for the transfer function you fitted to the Box–Jenkins data in Problem 4.3.

4.12. *Simulation of impulse response.* Consider the transfer function fitted in Example 4.6 for the polymerization data. Suppose that the controllable factor is the *impulse* function $X_0 = 1$ and $X_t = 0$ for $t \neq 0$. Simulate and plot the response for 30 realizations of the process under such an input sequence. (*Suggestion*: Use a spreadsheet program.)

4.13. Repeat Problem 4.12 for the transfer function fitted in Example 4.3.

4.14. Repeat Problem 4.12 for the transfer function you fitted to the Box–Jenkins data in Problem 4.3.

4.15. *Known steady-state gains.* In some chemical processes, the steady-state gain g of a transfer function may be known from theoretical or experimental knowledge. How can one use this knowledge in the estimation of the transfer function parameters?

4.16. *Steady model.* Consider the steady model described in Example 4.14, Appendix 4A. Show that if the parameter θ of an IMA(1, 1) process satisfies equation (4.10), the first difference $(\nabla(Y_t))$ of an IMA(1,1) and the steady model have the same autocorrelation function.

4.17. Write the model

$$Y_t = a \, Y_{t-1} + bX_{t-2} - c\varepsilon_{t-1} + \varepsilon_t$$

in state-space form. Verify that your answer has the same form as the model above.

BIBLIOGRAPHY AND COMMENTS

The main part of this chapter is based heavily on Box et al. (1994). The two types of models, polynomial models and state-space models, are described and contrasted in most stochastic control books (e.g., Åström, 1970; Lewis, 1996; Åström and Wittenmark, 1997). The book by Ljung (1999) contains an extensive discussion on many issues related to identifying and estimating stochastic systems and should be consulted, in particular by anyone interested in using Matlab's Systems ID Toolbox (Ljung, 1995, 2001).

R. Kalman developed the state-space approach, proposed the celebrated Kalman filter and solved many of the main problems in estimation, filtering, and control. The Kalman filter was first presented in Kalman (1960). Kalman was a student of Ragazzini at Columbia in the mid-1950s. Ragazzini is probably one of the most successful engineering Ph.D. advisors ever, as many other celebrities of control theory studied for their Ph.D. degree with him as well (among others, E. I. Jury, developer with Tsypkin of z-transforms, and L. Zadeh, who years later developed fuzzy sets). Åström and Wittenmark (1997) provide a nice historical development of the main contributors in the field of stochastic control.

APPENDIX 4A: BASICS OF STATE-SPACE MODELING

All the dynamic models discussed thus far are external to the process in the sense that this is seen as a black box. The model characterizes the relationships between input and output. An alternative description of a dynamic process is based on *internal* models, which, on the contrary, describe all internal couplings among the input, output, and a set of internal variables, usually not directly measurable, called the *state variables*. In these models, knowledge about the first principles of the process (physics, chemistry) is usually described as a set of differential equations in the state variables that lead to the *state-space formulation*.

The *state* of a process is a set of numbers, the state variables, such that knowledge of these numbers, the input, and the equations describing the process dynamics provide the future state and output of the process. *State variables* are internal variables that determine the future behavior of the process given the input and the current state of the process. More formally, a set of state variables $\mathbf{Z}' = (z_1, z_2, \ldots, z_n)$ is a set such that knowledge of the initial state (Z_0) and of the input X_t $(t > 0)$ is sufficient to determine future values of the outputs and state variables. The state-space is simply the n-dimensional space of state variables. We hasten to point out that state variables do not need to have a physical meaning. In aerospace and mechanical engineering systems, the state usually has a physical meaning (e.g., the position of some object), but in empirical modeling of industrial processes this does not need to be so.

The state-space formulation consists of two equations. The first, the state or system equation, describes the dynamics of the state variables using differential or difference equations. It is usually assumed that the inputs to the process, X_t, affect the state variables through this equation. The state variables are usually assumed not to be observable or measurable directly. Instead, the outputs of the process, Y_t, are assumed directly observable. The observation, or measurement, equation links the process output vector with the state vector. In what follows, we drop the boldface convention to denote vectors.

In this appendix we first show how continuous-time state-space models can be transformed by sampling into discrete-time difference equation models. We then review some fundamental concepts of state-space models. Finally, we show how to transform a discrete-time state-space model into a transfer function model of the type described in this chapter and how to get a state-space formulation from a transfer function model.

A state-space formulation is most useful for the recursive estimation of the parameters of a model using Kalman filters, a topic discussed in Chapter 8. Estimation of multivariate models is also better done in state space, as discussed in Chapter 9.

Discrete-Time Modeling of Continuous Processes
A deterministic continuous-time process can be modeled in state-space form by a set of differential equations describing the state dynamics and a vector observation equation[12]:

$$\frac{dZ}{dt} = \Phi Z(t) + \Gamma X(t)$$

$$Y(t) = CZ(t)$$

where Z is a $(n \times 1)$ vector, where n is the number of state variables; X is a $(r \times 1)$ vector of inputs; and Y is a $(p \times 1)$ vector of outputs or process responses. If the process is sampled periodically at times $t_k = k\Delta t$, the state of the process at the discrete points t_k is given by the solution to the state

---

[12] Unless stated otherwise, in this appendix we consider the general case in which there are several responses of interest in the process.

equation, obtained from applying Laplace transforms:

$$Z_{t_{k+1}} = e^{\Phi(t_{k+1} - t_k)} Z_{t_k} + \int_{t_k}^{t_{k+1}} e^{\Phi(t_{k+1} - s)} \, ds \, \Gamma X_{t_k}$$

which can be written as

$$Z_{t_{k+1}} = AZ_{t_k} + BX_{t_k}.$$
$$Y_{t_k} = CZ_{t_k}$$

If we assume that $t_{k+1} - t_k = \Delta t$ is a constant (i.e., if sampling is at equidistant points in time), the last equations can be written as

$$Z_{t+1} = AZ_t + BX_t$$
$$Y_t = CZ_t$$

where $A = e^{\Phi \Delta t}$ and $B = \int_0^{\Delta t} e^{\Phi s} \, ds \, \Gamma$. This is *zero-order hold* (ZOH) *sampling*. It assumes that X_t does not change between sampling instants; that is, the input is held at the value set at the beginning[13] of period t.

Example 4.8: First-Order Process. Consider the state equation corresponding to a first-order dynamical process:

$$\frac{dZ}{dt} = \alpha Z + \beta X$$

Applying ZOH sampling, we obtain $A = e^{\alpha \Delta t}$ and $B = \int_0^{\Delta t} e^{\alpha \Delta t} \, ds \, \beta = (\beta / \alpha)(e^{\alpha \delta t} - 1)$. Therefore, the discrete-time representation is:

$$Z_{t+1} = e^{\alpha \Delta t} Z_t + \frac{\beta}{\alpha} (e^{\alpha \delta t} - 1) X_t$$
$$= e^{\alpha \Delta t} Z_t + (1 - e^{\alpha \Delta t}) X_t$$
$$= (1 - \lambda) Z_t + \lambda X_t$$

which has the EWMA form with weight $\lambda = 1 - e^{\alpha \Delta t}$ and the observation equation is simply $Y_t = CZ_t$.   □

Solution to the State Equation

Discrete-time state equations are usually given by first-order linear difference equations that can be solved easily. Suppose that the state-space description

---

[13] Note how no second-order information (i.e., derivatives with respect to time) is used to derive the discrete representation of the process under ZOH. For other approaches to sampling, see Franklin et al. (1998).

is

$$Z_{t+1} = AZ_t + BX_t + w_t$$
$$Y_t = CZ_t + R_t$$

where w_t and R_t are two given vectors that usually model disturbances. To solve the state equation, we have that, by repeated substitution,

$$Z_{t+2} = AZ_{t+1} + BX_{t+1} + w_{t+1} = A^2Z_t + ABX_t + A\,w_t + BX_{t+1} + w_{t+1}$$
$$Z_{t+3} = A^3Z_t + A^2BX_t + A^2w_t + ABX_{t+1} + Aw_{t+1} + BX_{t+2} + w_{t+2}$$
$$\vdots$$

Let $t = 0$ be the initial time and consider the state after some arbitrary number of steps k after. We then have

$$Z_k = A^kZ_0 + \sum_{j=0}^{k-1} A^{k-j-1}\left(BX_j + w_j\right)$$

The first term on the right represents the transient response due to the initial condition. Notice that an AR(1) process has the same solution since it is a first-order (stochastic) difference equation model. Thus, if Y is a scalar, a value A less than 1 in absolute value guarantees the stability of the process. If Y is a vector (i.e., if $n > 1$), the absolute value of the eigenvalues of the A matrix (called the *transition matrix*) need to be smaller than 1 in absolute value for stability.

Controllability, Observability, and Minimality
Some notions from state-space theory that are used widely are controllability and observability, first introduced by Kalman. We define these notions for a discrete-time process. We then introduce the concept of minimality, useful to define the set of state variables. The questions answered by Kalman were: (1) What conditions are needed to be able to steer a process from a given initial state to a given final state? and (2) How can we determine the state of the process from input–output data? These questions are answered by the notions of controllability and observability, respectively.

Controllability
A system is *controllable* if it is possible to find a sequence of inputs such that any arbitrary state F can be reached from any initial state I in finite time. That is, a process is controllable if $0 < j < \infty$ such that

$$Z_{t+1} = AZ_t + BX_t \qquad t = 0, 1, \ldots, j - 1$$

with $Z_0 = I$ and $Z_j = F$. This is called *reachability* by some authors.

To obtain the conditions for a process to be controllable, consider the state at sample n, where n is the dimension of Z. This is obtained by repeated substitution as follows:

$$
\begin{aligned}
Z_n &= A Z_{n-1} + X_{n-1} = A(AZ_{n-2} + BX_{n-2}) + BX_{n-1} \\
&= A^2 Z_{n-2} + ABX_{n-2} + BX_{n-1} \\
&\;\;\vdots \\
&= A^n Z_0 + A^{n-1}BX_0 + A^{n-2}BX_1 + \cdots + BX_{n-1} \\
&= A^n Z_0 + W_c \mathcal{X}
\end{aligned}
$$

where W_c is the $n \times nr$ controllability matrix

$$
W_c = [B \quad AB \;\cdots\; A^{n-1}B]
$$

(this is a square matrix if $r = 1$) and

$$
\mathcal{X} = [X_{n-1} X_{n-2} \;\cdots\; X_0]
$$

is a $nr \times 1$ vector. Using the initial and final states, we find that a process is controllable if we can find a vector χ such that

$$
F - A^n I = W_c \mathcal{X}
$$

which implies that a process is controllable if and only if rank$(W_c) = n$, a fact proved by Kalman. Note that if $r = 1$ (single-input case), this says that W_c must be invertible.

Example 4.9. Consider the scalar first-order process given in Example 4.8: $Z_{t+1} = AZ_t + BX_t$ with $A = e^{\alpha\Delta t}$ and $B = (\beta/\alpha)(e^{\alpha\delta t} - 1)$. We have $n = 1$, $r = 1$, and therefore

$$
W_c = B = \frac{\beta}{\alpha}(e^{\alpha\Delta t} - 1)
$$

and $\chi = X_0$. Therefore,

$$
X_0 = \frac{F - e^{\alpha\Delta t}S}{(\beta/\alpha)(e^{\alpha\Delta t} - 1)}
$$

and the process is controllable if $\beta \neq 0$.   □

Example 4.10. Consider the state equation

$$
Z_{t+1} = \begin{bmatrix} 2 & 0 \\ 0 & 2 \end{bmatrix} Z_t + \begin{bmatrix} 2 \\ 2 \end{bmatrix} X_t
$$

so we have that $n = 2$ and $r = 1$ (two state variables, one input) and

$$W_c = [B \quad AB] = \begin{bmatrix} 2 & 4 \\ 2 & 4 \end{bmatrix}$$

and since $\text{rank}(W_c) = 1 < n = 2$, the process is not controllable. □

It is perhaps worth mentioning that the Matlab Control Systems Toolbox function `ctrb` computes the controllability matrix.

Observability
The notion of observability relates to being able to determine exactly (or to observe) the initial state vector Z_0 from values of inputs and outputs that occur afterward. By iterative substitution in the observation equation we have that

$$Y_0 = CZ_0$$
$$Y_1 = CZ_1 = CAZ_0 + CBX_0$$
$$Y_2 = CZ_2 = CAZ_1 + CBX_1 = A^2CZ_0 + CABX_0 + CBX_1$$
$$Y_3 = CZ_3 = A^3CZ_0 + CA^2BZ_0 + CABX_1 + CBX_2$$
$$\vdots$$
$$Y_{n-1} = CZ_{n-1} = A^{n-1}CZ_0 + CB\sum_{t=0}^{n-2} A^{n-2-t}X_t$$

where the last term on the right is a function of the inputs only, which are given. If we define the $np \times n$ matrix

$$W_0 = \begin{bmatrix} C \\ CA \\ CA^2 \\ \vdots \\ CA^{n-1} \end{bmatrix}$$

the $np \times 1$ vector

$$\mathcal{Y} = \begin{bmatrix} Y_0 \\ Y_1 \\ \vdots \\ Y_{n-1} \end{bmatrix}$$

and the $np \times 1$ vector

$$\mathcal{X} = \begin{bmatrix} 0 \\ CBX_0 \\ CABX_0 + CBX_1 \\ \vdots \\ CB\sum_{t=0}^{n-2} A^{n-2-t} X_t \end{bmatrix}$$

we can find Z_0 if the system of equations

$$\mathcal{Y} - \mathcal{X} = W_0 Z_0$$

can be solved which implies that a process is observable if and only if rank(W_0) = n. The matrix W_0 is called the *observability matrix*.

Example 4.11. For the first-order scalar process of Example 4.9 we have that $n = 1$ and $p = 1$, and the observation equation is $Y_t = CZ_t$. Then $W_0 = C$, so we have $W_0 Z_0 = Y_0$ or $Z_0 = Y_0/C$, so the process is observable if $C \neq 0$. □

Example 4.12. Consider the process discussed in Example 4.10. Here $n = 2$, $p = 1$ (two state variables, one response) and

$$Z_{t+1} = \begin{bmatrix} 2 & 0 \\ 0 & 2 \end{bmatrix} Z_t + \begin{bmatrix} 2 \\ 2 \end{bmatrix} X_t$$

and suppose that the observation equation is $Y_t = CZ_t$ with $C = [1, 1]$. Then

$$W_0 = \begin{bmatrix} C \\ CA \end{bmatrix} = \begin{bmatrix} 1 & 1 \\ 2 & 2 \end{bmatrix}$$

and $\mathcal{Y} = Y_0$. Thus rank(W_0) = $1 < n = 2$ and the process is not observable in the sense of Kalman (we have that $W_0 Z_0 = \mathcal{Y}$ has no solution). □

The Matlab Control Systems Toolbox function obsv computes the observability matrix.

Minimality
If a process is controllable and observable, it can be proved that the state vector is of minimal dimension. The reverse implication is also true: A state-space model with minimal state vector dimension is observable and controllable. Thus, if a process is not observable or not controllable, it can be reexpressed in terms of a lower dimensional (and minimal) state vector such that the resulting formulation is controllable and observable.

Obtaining the Transfer Function from a State-Space Formulation
To obtain a transfer function [in the way described in the main body of this chapter (i.e., as a ratio of polynomials in the backshift operator)] from a state-space model, we need to eliminate the state variables. Once we have the transfer function of a process, we can make use of results discussed elsewhere in this book, such as finding control algorithms. Consider the state equation

$$Z_{t+1} = \mathcal{F}Z_t = AZ_t + BX_t$$

where $\mathcal{F}$ is the forward shift operator. Then $(\mathcal{F}I - A)Z_t = BX_t$ and $Y_t = CZ_t = C(\mathcal{F}I - A)^{-1}BX_t$. Thus the transfer function is given by either

$$H^*(\mathcal{F}) = C(\mathcal{F}I - A)^{-1}B$$

in the forward shift operator or

$$H(\mathcal{B}) = C(I - \mathcal{B}A)^{-1}\mathcal{B}B$$

in the backshift operator.

Example 4.13. Suppose that we have the state equation

$$Z_{t+1} = \begin{bmatrix} a_1 & 1 \\ a_2 & 0 \end{bmatrix} Z_t + \begin{bmatrix} b_0 \\ -b_1 \end{bmatrix} X_t$$

where X_t is simply a scalar $(r = 1)$ and the observation equation is $Y_t = [1 \quad 0]Z_t$. Therefore, we have that

$$H^*(\mathcal{F}) = [1 \quad 0] \begin{pmatrix} \mathcal{F} - a_1 & -1 \\ -a_2 & \mathcal{F} \end{pmatrix}^{-1} \begin{bmatrix} b_0 \\ -b_1 \end{bmatrix}$$

$$= \frac{b_0\mathcal{F} - b_1}{\mathcal{F}^2 - a_1\mathcal{F} - a_2} = \frac{b_0\mathcal{B} - b_1\mathcal{B}^2}{1 - a_1\mathcal{B} - a_2\mathcal{B}^2}$$

$$= \frac{b_0 - b_1\mathcal{B}}{1 - a_1\mathcal{B} - a_2\mathcal{B}^2}\mathcal{B} = H(\mathcal{B})$$

which corresponds to an $(r, s, k) = (2, 1, 1)$ transfer function using the notation of Section 4.1. Note that the delay is one period. If we had, instead,

$B' = (0, b_0)$, we would have obtained

$$H(\mathcal{B}) = \frac{b_0}{1 - a_1\mathcal{B} - a_2\mathcal{B}^2} \mathcal{B}^2$$

a $(2, 0, 2)$ transfer function. Similarly, a $(2, 0, 1)$ transfer function is obtained using $B' = (b_0, 0)$. □

State-Space Models with Noise

The discrete state-space formulation with noise is given by

$$Z_{t+1} = AZ_t + BX_t + Nw_t$$
$$Y_t = CZ_t + v_t$$

where $w_t \sim (0, R_w)$ is called the *process noise* (this is assumed white, not necessarily normal). The covariance matrix R_w is $n \times n$. The term $v_t \sim (0, R_v)$ is called the *measurement noise* (or *error*). The errors w_t and v_t may be correlated (i.e., $E[w_t v_t] = \gamma_{wv} = R_{wv}$). This model can be put into transfer function form as follows:

$$Y_t = H_1(\mathcal{B})X_t + H_2(\mathcal{B})w_t + v_t$$

where, as we saw before,

$$H_1(\mathcal{B}) = C(I - \mathcal{B}A)^{-1} \mathcal{B}B$$
$$H_2(\mathcal{B}) = C(I - \mathcal{B}A)^{-1} \mathcal{B}N$$

Example 4.14: Steady Model. A quite useful state-space model in the area of time series is the steady model, introduced by West and Harrison (1997). This model was used in Chapter 1 to discuss when SPC and EPC are needed. The model is

$$Z_t = Z_{t-1} + w_t \qquad w_t \sim (0, \sigma_w^2)$$
$$Y_t = Z_t + v_t$$

with $v_t \sim (0, \sigma_v^2)$. Here all quantities are scalars. From the previous discussion, we have that $H_1(\mathcal{B}) = 0$ since this is not a controlled process, but

$$H_2(\mathcal{B}) = (1 - \mathcal{B})^{-1} \mathcal{B}$$

since in this case $A = C = N = 1$. Thus the corresponding transfer function description is

$$Y_t = \frac{w_{t-1}}{1 - \mathcal{B}} + v_t$$

It can be shown (see the Problems) that if the signal-to-noise ratio SN $= \sigma_w^2/\sigma_v^2$ satisfies

$$\theta = \tfrac{1}{2}\left(\text{SN} - \sqrt{4\text{SN} + \text{SN}^2}\right) + 1 \qquad (4.10)$$

where θ is the MA parameter of the IMA(1, 1) model $Y_t = Y_{t-1} - \theta\varepsilon_{t-1} + \varepsilon_t$, the steady model and a differenced IMA(1, 1) have the same autocorrelation function. Note that the steady process can be understood as a random walk process that we can observe only with measurement error. Also, notice how if $\sigma_w^2 = 0$, the steady model reduces to Shewhart's SPC model. □

Transforming a Transfer Function Model into State-Space Form

It is sometimes useful to transform a transfer function model into state-space form. This is particularly true for estimating the parameters of the transfer function. One way in which the transformation can be accomplished is as follows[14]: Suppose that we write our model in ARMAX form:

$$A(\mathcal{B})Y_t = B(\mathcal{B})X_{t-k} + C(\mathcal{B})\varepsilon_t$$

where

$$A(\mathcal{B}) = 1 - a_1\mathcal{B} - a_2\mathcal{B}^2 - \cdots - a_{n_a}\mathcal{B}^{n_a}$$

$$B(\mathcal{B}) = b_0 - b_1\mathcal{B} - \cdots - b_{n_b}\mathcal{B}^{n_b}$$

$$C(\mathcal{B}) = 1 - c_1\mathcal{B} - \cdots - c_{n_c}\mathcal{B}^{n_c}$$

If the number of responses is greater than 1 ($p > 1$), this is a multivariate ARMAX process (see Chapter 9) in which we assume that all input–output couplings have the same delay of k time periods. What follows applies only to the case of one controllable factor and one response. For the multiple-input, multiple-output case, see Chapter 9.

---

[14] In this case, the transformation is not unique.

A single-input, single-output state-space formulation[15] is given by

$$
Z_{t+1} = \begin{pmatrix} a_1 & 1 & 0 & \cdots 0 \\ a_2 & 0 & 1 & \cdots & 0 \\ \vdots & \vdots & \vdots & \ddots & \vdots \\ a_{n-1} & 0 & 0 & \cdots & 1 \\ a_n & 0 & 0 & \cdots & 0 \end{pmatrix} Z_t + \begin{bmatrix} 0 \\ 0 \\ \vdots \\ 0 \\ b_0 \\ -b_1 \\ \vdots \\ -b_{n_b} \\ 0 \\ \vdots \\ 0 \end{bmatrix} X_t + \begin{bmatrix} 1 \\ -c_1 \\ \vdots \\ -c_{n_c} \\ 0 \\ \vdots \\ 0 \end{bmatrix} \varepsilon_{t+1}
$$

$$
= AZ_t + BX_t + N\varepsilon_{t+1} \tag{4.11}
$$

where $n = \max\{n_a, n_b + k, n_c + 1\}$, using zeros in case no such coefficients exist for a particular polynomial. The B and N vectors are $n \times 1$. B has $k - 1$ rows of zeros, followed by $n_b + 1$ rows of coefficients followed by $n - n_b - k$ zeros provided that $n - n_b - k > 0$. N has $n_c + 1$ rows of coefficients followed by $n - n_c - 1$ rows of zeros provided that $n - n_c > 0$. The observation equation is given simply by

$$
Y_t = (1 \quad 0 \quad 0 \quad \cdots \quad 0) Z_t = CZ_t
$$

To prove this result, we need to compute $H_1(\mathcal{B}) = \mathcal{B}C(I - \mathcal{B}A)^{-1}B$ and $H_2(\mathcal{B}) = \mathcal{B}C(I - \mathcal{B}A)^{-1}N$ to check that we indeed end up with the ARMAX model. In either case we need to compute $\mathcal{B}C(I - \mathcal{B}A)^{-1} = (\mathcal{B} \ 0 \ \cdots \ 0)(I - \mathcal{B}A)^{-1}$, so we only need to compute the first row of the inverse of $I - \mathcal{B}A$. Now we have that

$$
L = I - \mathcal{B}A = \begin{pmatrix} 1 - a_1\mathcal{B} & -\mathcal{B} & 0 & \cdots & 0 \\ -a_2\mathcal{B} & 1 & -\mathcal{B} & \cdots & 0 \\ \vdots & \vdots & \vdots & \ddots & \vdots \\ -a_{n-1}\mathcal{B} & 0 & 0 & \cdots & -\mathcal{B} \\ -a_n\mathcal{B} & 0 & 0 & \cdots & 1 \end{pmatrix}
$$

[15] This representation is not unique. It assumes that the controllable factor X_t is given and not a part of the state vector. Different formulations where values of X_t are included in the state vector are discussed in Chapter 9 with application to empirical modeling of multivariate processes.

Since the inverse of a matrix L equals $\text{adj}(L)/|L|$ and the first row in the adjoint matrix is made of the cofactors[16] $L_{11}, L_{21}, \ldots, L_{n1}$, it can be seen that the cofactors we seek are $L_{11} = 1$, $L_{21} = \mathcal{B}$, $L_{31} = \mathcal{B}^2$, $L_{41} = \mathcal{B}^3, \ldots$, and $L_{n1} = \mathcal{B}^{n-1}$. Recall that the determinant of L is expanded by cofactors as $|L| = \sum_{j=1}^{n} l_{1j} L_{ij}$ where l_{1j} is the jth entry in the first column of matrix L. Then we have that the determinant is $|L| = 1 - a_1\mathcal{B} - a_2\mathcal{B}^2 - \cdots - a_n\mathcal{B}^n$. Therefore,

$$H_1(\mathcal{B}) = \frac{\mathcal{B}(1,\ \mathcal{B},\ \mathcal{B}^2, \ldots,\ \mathcal{B}^{n-1})B}{|L|} = \frac{b_0\mathcal{B}^k - b_1\mathcal{B}^{k+1} - \cdots - b_{n_b}\mathcal{B}^{n_b+k}}{1 - a_1\mathcal{B} - a_2\mathcal{B}^2 - \cdots - a_n\mathcal{B}^n}$$

$$= \frac{b_0 - b_1\mathcal{B} - b_2\mathcal{B}^2 - \cdots - b_{n_b}\mathcal{B}^{n_b}}{1 - a_1\mathcal{B} - a_2\mathcal{B}^2 - \cdots - a_n\mathcal{B}^n}\mathcal{B}^k$$

Similarly,

$$H_2(\mathcal{B}) = \frac{1 - c_1\mathcal{B} - c_2\mathcal{B}^2 - \cdots - c_{n_c}\mathcal{B}^{n_c}}{1 - a_1\mathcal{B} - a_2\mathcal{B}^2 - \cdots - a_n\mathcal{B}^n}$$

In either of $H_1(\mathcal{B})$ or $H_2(\mathcal{B})$ we use zeros for denominator coefficients that do not exist in the ARMAX formulation. The ARMAX model is then given by:

$$Y_t = \frac{B(\mathcal{B})}{A(\mathcal{B})}X_{t-k} + \frac{C(\mathcal{B})}{A(\mathcal{B})}\varepsilon_t$$

Example 4.15. Consider the following ARMAX model:

$$\left(1 - a_1\mathcal{B} - a_2\mathcal{B}^2\right)Y_t = b_0 X_{t-2} + \left(1 - c_1\mathcal{B}\right)\varepsilon_t$$

We have that $n = \max(2, 2, 2) = 2$ and $k = 2$. To transform to state space, we follow equation (4.11) and get

$$Z_{t+1} = \begin{pmatrix} a_1 & 1 \\ a_2 & 0 \end{pmatrix} Z_t + \begin{pmatrix} 0 \\ b_0 \end{pmatrix} X_t + \begin{pmatrix} 1 \\ -c_1 \end{pmatrix}\varepsilon_{t+1}$$

and $Y_t = (1, 0)\, Z_t$. To corroborate that we have the desired ARMAX model, look at the first state variable, call it S_1:

$$S_{1,t+1} = a_1 S_{1,t} + S_{2,t} + \varepsilon_{t+1}$$

---

[16] Recall that the adjoint of a matrix is the *transpose* of the matrix of cofactors. The cofactors, in turn, are given by $L_{ij} = (-1)^{i+j}|M_{i,j}|$, where $M_{i,j}$ is the (i, j)th minor of L.

The second state variable is

$$S_{2,t+1} = a_2 S_{1,t} + b_0 X_t - c_1 \varepsilon_{t+1}$$

so we have that

$$S_{1,t+1} = a_1 S_{1,t} + a_2 S_{1,t-1} + b_0 X_{t-1} - c_1 \varepsilon_t + \varepsilon_{t+1}$$

or

$$S_{1,t} = a_1 S_{1,t-1} + a_2 S_{1,t-2} + b_0 X_{t-2} - c_1 \varepsilon_{t-1} + \varepsilon_t$$

which has exactly the same form as the desired ARMAX format. Thus we can see that the first state variable is simply Y_t, that is, the state vector is given by

$$Z_t = \begin{pmatrix} Y_t \\ S_{2,t} \end{pmatrix}$$

Then, from $Y_t = (1, 0) Z_t$, we retrieve the ARMAX model.   □

CHAPTER 5

Optimal Feedback Controllers

In this chapter we consider a practical problem central to this book: how to adjust a process with the assurance that our adjustments are, in some well-defined sense, "optimal" for the quality characteristic of the product we wish to control. In other words, we want to find a control rule, or *controller*, for a single-input, single-output process that is optimal for our purposes. We focus on optimal *feedback* controllers. The *feedback principle* can be stated as follows: *Increase* the controllable variable when the response is smaller than the set point; *decrease* the controllable variable when the response is larger than the set point. Actually, this is the *negative* feedback principle. Consider Figure 5.1 where $Y = T - Q$ denotes deviations from set point or target, where Q is the quality characteristic and T is its target. Thus we have that:

- If $Y > 0$ (i.e., if $T > Q$), *increase* X.
- If $Y < 0$ (i.e., if $T < Q$), *decrease* X.

Thus X moves in the opposite (negative) direction than Q. Sometimes the minus sign is included in the control equation:

$$X_t = -f(Y_t, Y_{t-1}, \ldots, X_{t-1}, \ldots)$$

and the deviations from target are defined as $Y_t = Q_t - T$ instead. The resulting block diagram is neater (see Figure 5.2).

In the latter case, some authors prefer to eliminate the target or set point, as in Figure 5.3. This is appropriate if the target T does not change with time, as frequently happens in quality control. All three forms are equivalent.

In this chapter we present the general derivation for how to obtain a minimum variance [or minimum mean square error (MMSE)] feedback controller. Although sometimes impractical, MMSE controllers provide considerable intuition and can easily be modified to get more robust alternatives. In this chapter we also examine what is the best policy if there are adjust-

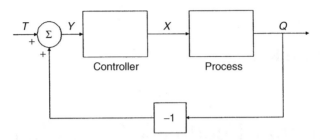

Figure 5.1   First form of a negative feedback controller.

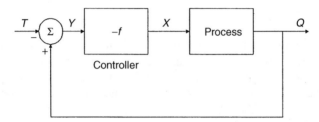

Figure 5.2   Second form of a negative feedback controller.

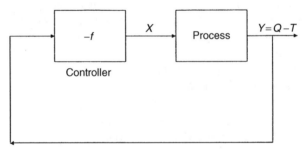

Figure 5.3   Third equivalent form of a negative feedback controller for a constant target.

ment costs. Adjustment techniques necessary due to errors during setup are useful for discrete-part manufacturing systems and are presented in Section 5.5.

5.1   MINIMUM VARIANCE (MINIMUM MSE) CONTROLLERS

By themselves, control strategies that *only* seek to minimize the variability of the quality characteristic, disregarding any cost involved in doing so, are impractical. However, as will be shown, this type of control rule is easy to modify to achieve other goals as well, such as keeping the variability of the adjustments below a certain level. Furthermore, the derivation of minimum variance control strategies provides a great deal of insight into the type of adjustment problems we are concerned with. A couple of simple examples will help illustrate the type of control laws we wish and will give insight on how to obtain them.

Example 5.1: Process with No Dynamics. Suppose that the process under study can be described by

$$Q_t = \mu + bX_{t-1} + \varepsilon_t$$

where Q_t is the quality characteristic. The value of μ is the mean quality characteristic if we do not adjust. Assume that μ is a known quantity.[1] Let us assume, as we do more formally later, that our interest is to minimize the mean square error (MSE) of the quality characteristic by manipulating the controllable factor. The minimum MSE control law is obtained by setting $E[Q_t] = T$, which achieves a MSE equal to the minimum possible variance of σ_ε^2. Thus the control law is simply

$$X_t = \frac{T - \mu}{b} \tag{5.1}$$

This implies that the *adjustments*, ∇X_t, always equal zero. If $X_0 \neq (T - \mu)/b$, only one adjustment (∇X_1) is needed. If $\mu = T$, no adjustments are needed. This is in accordance to Deming's funnel experiment assumptions (Section 1.4), where only if the funnel was originally off target is a single adjustment necessary to aim the funnel on target, after which no further adjustments are necessary.

Figure 5.4 shows that the minimum variance law (5.1) is a *feedforward*, not a feedback, controller (i.e., the values of the quality characteristic are not used in the computation of the level of the controllable factor). Instead, the

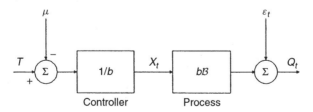

Figure 5.4   Feedforward controller, Example 5.1.

[1]A more interesting adjustment problem occurs when μ is unknown and not necessarily equal to the target T. This is addressed in Section 5.5 with relation to setting up a machine that at startup can be off target by some unknown "offset" $d = \mu - T$.

changes in set point or target (T) are used by the controller, so we "forward" this information to the controller. □

Example 5.2: Process with Simple Dynamics. As a second instance of MMSE control, suppose now that the process to be regulated is

$$Y_t + a\,Y_{t-1} = bX_{t-1} + c\varepsilon_{t-1} + \varepsilon_t$$

where it is assumed that $|c| < 1$ (invertibility) and where Y_t now denotes deviation from target or set point. At time t, we need to decide on a control action X_t such that

$$Y_{t+1} = -a\,Y_t + bX_t + c\varepsilon_t + \varepsilon_{t+1}. \tag{5.2}$$

Thus Y_{t+1}, the next deviation from target, depends on X_t. We want to choose X_t to minimize MSE(Y_{t+1}). Now, the mean square error[2] is

$$\mathrm{MSE}(Y_{t+1}) = \mathrm{Var}(Y_{t+1}) + E[Y_{t+1}]^2$$

As will be shown, to minimize MSE we need to make $E[Y_{t+1}] = 0$; thus it turns out that the minimum MSE equals the minimum possible variance. Thus, at optimality, from (5.2) we get

$$\mathrm{MSE}(Y_{t+1}) = \mathrm{Var}(Y_{t+1}) \geq E[\varepsilon_{t+1}^2] = \sigma_\varepsilon^2$$

Also from (5.2), we get the MMSE one-step-ahead forecast[3]:

$$\hat{Y}_{t+1|t} = -aY_t + bX_t + c\varepsilon_t$$

from where, solving for X_t, we obtain

$$X_t = \frac{aY_t - c\varepsilon_t}{b} \tag{5.3}$$

This is, of course, not usable immediately because we need to estimate the error. We can estimate the error ε_t by the one-step-ahead prediction error $e_t = Y_t - \hat{Y}_{t|t-1} = Y_t$ since we are setting $E_t[Y_{t+1}] = \hat{Y}_{t+1|t} = 0$ for all t (recall that Y denotes deviation from target). Then we get that

$$X_t = \frac{a - c}{b}Y_t \tag{5.4}$$

---

[2] This should more appropriately be called mean square *deviation*, since we are not computing square deviations from a "true" population value (as when computing the MSE of a parameter estimate) but deviations from a *desired* target of zero. We use the term *MSE* since it is widely used in the area of time-series control.

[3] See Section 3.5.

If we substitute (5.4) into (5.2), we get the *closed-loop output equation* of the process:

$$Y_{t+1} = -cY_t + c\varepsilon_t + \varepsilon_{t+1}$$

There are two ways of analyzing this equation: from a transient point of view and from an asymptotic point of view. If we use backshift operators, we look at the asymptotic behavior of the process:

$$(1 + c\mathcal{B})Y_{t+1} = (1 + c\mathcal{B})\varepsilon_{t+1} \Rightarrow Y_t = \varepsilon_t$$

Thus, in the long run, the minimum variance of σ_ε^2 is achieved. The transient can be considered not by using backshift operators but, instead, by solving the closed-loop difference equation assuming some initial condition:

$$Y_{t+1} - \varepsilon_{t+1} + c(Y_t - \varepsilon_t) = 0$$

which is a first-order difference equation. If the initial condition of the equation is $Y_0 - \varepsilon_0 = K$, the solution is (Goldberg, 1986, p. 86)

$$Y_t - \varepsilon_t = Kc^t$$

Since $|c| < 1$, $Y_t \to \varepsilon_t$ as $t \to \infty$. Figure 5.5 illustrates this behavior and Figure 5.6 shows the controller operating in closed loop. $\square$

Before proceeding further, let us define more precisely the goals of the controllers we are seeking and the type of processes on which they will operate.

1. *Process dynamics.* We assume that the dynamics of the process can be represented by a transfer function of the form

$$Y_t = \frac{B_1(\mathcal{B})}{A_1(\mathcal{B})}X_{t-k}$$

where k is the input–output delay. We assume that both polynomials are of the same order m, although if they are not, we can always use trailing zeros as coefficients in the smaller-order polynomial.

2. *Environment (noise dynamics).* Since we assume a linear model, by the principle of superposition we can represent all disturbances by a single disturbance ε_t acting on the output. The noise process will then be assumed to be

$$N_t = \frac{C_1(\mathcal{B})}{D(\mathcal{B})}\varepsilon_t$$

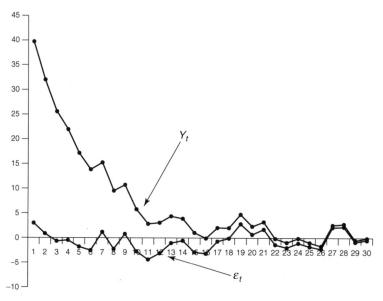

Figure 5.5 Transient response provided by MMSE controller, Example 5.2. In the graph, $K = 40$, $c = 0.85$, and $\sigma_\varepsilon = 2$ were used.

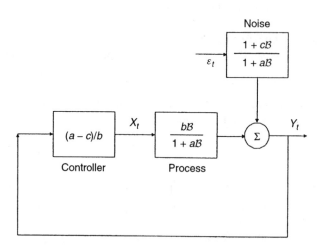

Figure 5.6 Block diagram for Example 5.1 (MMSE controller shown).

where $\{\varepsilon_t\}$ is a white noise sequence with $\varepsilon_t \sim (0, \sigma_\varepsilon^2)$. The process and environment can be written as a Box–Jenkins transfer function plus noise model:

$$Y_t = \frac{B_1(\mathcal{B})}{A_1(\mathcal{B})} X_{t-k} + \frac{C_1(\mathcal{B})}{D(\mathcal{B})} \varepsilon_t \tag{5.5}$$

For deriving optimal control laws, we will use the equivalent ARMAX form[4]:

$$A_1(\mathcal{B}) D(\mathcal{B}) Y_t = B_1(\mathcal{B}) D(\mathcal{B}) X_{t-k} + C_1(\mathcal{B}) A_1(\mathcal{B}) \varepsilon_t$$

which can be reduced to the form we will use:

$$A(\mathcal{B}) Y_t = B(\mathcal{B}) X_{t-k} + C(\mathcal{B}) \varepsilon_t$$

3. *Admissible control rules.* X_t is admissible if it is of the form

$$X_t = f(Y_t, Y_{t-1}, Y_{t-2}, \ldots; X_{t-1}, X_{t-2}, \ldots)$$

That is, a control rule is admissible if it is a function only of past values of the response and past values of the controllable factor.

4. *Problem statement.* Find an admissible control law such that the mean square error (MSE) of Y_t is a minimum. As shown in Examples 5.1 and 5.2 and as discussed more generally below, to get MMSE control it is required to set $E[Y_t] = 0$, thus the MMSE of Y_t equals the minimum variance of Y_t. This explains the differences in terminology between the time-series literature (where this problem is described as MMSE control) and the control engineering literature (where it is usually described as minimum variance control).

Ideal Controller
It is instructive to look at what would be the "ideal" or perfect controller and compare it with the MMSE controller derived later in this chapter. Suppose that we have an ARMA process description:

$$A(\mathcal{B}) Y_t = B(\mathcal{B}) X_{t-k} + C(\mathcal{B}) \varepsilon_t$$

or

$$Y_t = \frac{B(\mathcal{B})}{A(\mathcal{B})} X_{t-k} + \frac{C(\mathcal{B})}{A(\mathcal{B})} \varepsilon_t$$

---

[4]See Section 4.2.

where Y denotes deviation from target. To set Y_t to zero, it is tempting to try making

$$\frac{B(\mathcal{B})}{A(\mathcal{B})} \mathcal{B}^k X_t = -\frac{C(\mathcal{B})}{A(\mathcal{B})} \varepsilon_t$$

or

$$X_t = -\frac{C(\mathcal{B})}{B(\mathcal{B})} \mathcal{J}^k \varepsilon_t \tag{5.6}$$

where $\mathcal{J}$ is the forward shift operator. Control law (5.6) is not possible to implement because we would have to know some future disturbances ε_{t+j}, $0 < j \leq k$. If we could use (5.6), we would end up with the closed-loop equation $Y_t = \varepsilon_t$, which would always provide minimum variance. However, the controller is not practical and a minimum variance of σ_ε^2 is not always achievable in general. The ideal controller points to the necessity of predicting the disturbance, a topic we discuss next.

MMSE Forecasting

The optimal control law (5.3) was obtained by making the one-step-ahead forecast equal to zero (which is the "target" for Y_t):

$$\hat{Y}_{t+1|t} = -aY_t + bX_t + c\varepsilon_t = 0$$

so that the closed-loop equation yields

$$Y_{t+1} = \varepsilon_{t+1}$$

and we obtain a minimum variance of σ_ε^2 in this case. If the process were left uncontrolled, that is, the level of the controllable factor were left fixed at the "baseline" value of zero ($X_t = 0$), the one-step-ahead forecast would be

$$\hat{Y}_{t+1|t} = -aY_t + c\varepsilon_t \tag{5.7}$$

Comparing (5.7) with (5.3), we see that *the minimum variance controller is selected such that it cancels out with the one-step-ahead forecast.* As will be seen, to obtain minimum variance (or MMSE) controllers, we need an MMSE forecast of the process assuming that we do not adjust. In Chapter 3, a heuristic description of how to obtain MMSE forecasts of uncontrolled ARIMA processes was given; here we look at optimal forecasting in more detail and relate forecasting with adjustment.

Consider process (5.2) with $X_t = 0$ for all time periods:

$$Y_{t+1} = \frac{1 + cB}{1 + aB}\varepsilon_{t+1}$$

Adding and subtracting aB in the numerator, we get

$$Y_{t+1} = \underbrace{\varepsilon_{t+1}}_{\text{future error}} + \underbrace{\frac{c - a}{1 + aB}\varepsilon_t}_{\text{current and previous errors}} \tag{5.8}$$

From model (5.2) under the assumption that $X_t = 0$, we get that

$$\varepsilon_t = \frac{1 + aB}{1 + cB}Y_t$$

and substituting this into (5.8), we obtain

$$Y_{t+1} = \varepsilon_{t+1} + \frac{c - a}{1 + cB}Y_t.$$

Thus, at the current time t, we have expressed the next observation Y_{t+1} as a function only of current and past observations and a future error. Let $\hat{Y}$ denote a function of $Y_t, Y_{t-1}, Y_{t-2}, \ldots$ that forecasts future deviations from target. We wish to find $\hat{Y}$ such that

$$\text{MSE}(\hat{Y}_{t+1}) = E_t\left[(Y_{t+1} - \hat{Y})^2\right]$$

is a minimum. Working out the algebra yields

$$E_t\left[(Y_{t+1} - \hat{Y})^2\right] = E_t\left[\left(\varepsilon_{t+1} + \frac{c - a}{1 + cB}Y_t - \hat{Y}\right)^2\right]$$

$$= E_t\left[\varepsilon_{t+1}^2 + 2\varepsilon_{t+1}\left(\frac{c - a}{1 + cB}Y_t - \hat{Y}\right) + \left(\frac{c - a}{a + cB}Y_t - \hat{Y}\right)^2\right]$$

$$= E_t\left[\varepsilon_{t+1}^2\right] + E\left[\left(\frac{c - a}{1 + cB}Y_t - \hat{Y}\right)^2\right]$$

since the cross term

$$2E\left[\varepsilon_{t+1}\left(\frac{c - a}{1 + cB}Y_t - \hat{Y}\right)\right] = 0$$

given that ε_{t+1} is independent of $Y_t, Y_{t-1}, \ldots$. Thus we have that

$$E\left[\left(Y_{t+1} - \hat{Y}\right)^2\right] \geq E\left[\varepsilon_{t+1}^2\right] = \sigma_\varepsilon^2$$

with equality when

$$\hat{Y} = \frac{c - a}{1 + c\mathcal{B}}Y_t \equiv E_t[Y_{t+1}] = \hat{Y}_{t+1|t}$$

which is what we would obtain if the rules in Section 3.5 were used instead. The one-step-ahead forecast error or "innovation" is

$$\hat{Y}_{t+1|t} = Y_{t+1} - \hat{Y}_{t+1|t} = \varepsilon_{t+1}$$

where we use a different notation to distinguish the forecast errors from the residuals e_t. The forecast errors are called *innovations* because, if they were zero, it would mean that there is nothing new to learn from the data at time t, so any nonzero forecast error can be thought of giving us new information about our forecasting model in relation to the underlying process.

Two-Step-Ahead Forecast
Assuming again that $X_t = 0$ in process (5.2), at time $t + 2$ we have that

$$Y_{t+2} = \frac{1 + c\mathcal{B}}{1 + a\mathcal{B}}\varepsilon_{t+2} = \varepsilon_{t+2} + \frac{c - a}{1 + a\mathcal{B}}\varepsilon_{t+1}$$

in analogy with equation (5.8). Adding and subtracting $(c - a)a\,\mathcal{B}$ in the numerator on the right-hand side, we get

$$Y_{t+2} = \underbrace{\varepsilon_{t+2} + (c - a)\varepsilon_{t+1}}_{\text{future errors}} + \underbrace{\frac{a(a - c)}{1 + a\mathcal{B}}\varepsilon_t}_{\text{current and past errors}}. \qquad (5.9)$$

Since $\varepsilon_t = (1 + a\mathcal{B})/(1 + c\mathcal{B})Y_t$, we have that

$$Y_{t+2} = \varepsilon_{t+2} + (c - a)\varepsilon_{t+1} + \frac{a(a - c)}{1 + c\mathcal{B}}Y_t \qquad (5.10)$$

which shows Y_{t+2} as a function of current and past observations and of two future errors (looking at the process from time t, which is "now"). We wish to get a forecast function $\hat{Y}$ such that it minimizes

$$E_t\left[\left(Y_{t+2} - \hat{Y}\right)^2\right] = E_t\left[\varepsilon_{t+2}^2\right] + (c - a)^2 E_t\left[\varepsilon_{t+1}^2\right] + E_t\left[\left(\frac{a(a - c)}{1 + c\mathcal{B}}Y_t - \hat{Y}\right)^2\right]$$

where we proceeded similarly to before. Then we note that

$$E_t\left[\left(Y_{t+2} - \hat{Y}\right)^2\right] \geq E\left[\varepsilon_{t+2}^2\right] + (c - a)^2 E\left[\varepsilon_{t+1}^2\right] = \left[1 + (c - a)^2\right]\sigma_\varepsilon^2$$

where we have an equality when

$$\hat{Y} = \frac{a(a - c)}{1 + c\mathcal{B}}Y_t = E_t[Y_{t+2}] = \hat{Y}_{t+2|t}.$$

Again, this last expression is what we would obtain if the rules for computing forecasts described in Section 3.5 were applied. Thus the two-step-ahead MMSE forecast is given by

$$\hat{Y}_{t+2|t} = -c\hat{Y}_{t+1|t-1} + a(a - c)Y_t$$

Note that from (5.10) the two-step-ahead forecast error is

$$\tilde{Y}_{t+2|t} = Y_{t+2} - \hat{Y}_{t+2|t} = \varepsilon_{t+2} + (c - a)\varepsilon_{t+1} = \left(1 + (c - a)\mathcal{B}\right)\varepsilon_{t+2}$$

which is a moving average process of first order.

General k-Step-Ahead Forecast
Consider a general ARIMA process

$$A(\mathcal{B})Y_t = C(\mathcal{B})\varepsilon_t$$

where $A(\mathcal{B})$ and $C(\mathcal{B})$ are of order n (trailing zeros can be used in the lower-ordered polynomial) and where we allow $A(\mathcal{B})$ to have zeros on the unit circle. The MMSE k-step-ahead forecast minimizes

$$\text{MSE}(\hat{Y}_{t+1}) = E_t\left[\left(Y_{t+k} - \hat{Y}_{t+k|t}\right)^2\right].$$

At time t we want to forecast Y_{t+k} so we have

$$Y_{t+k} = \frac{C(\mathcal{B})}{A(\mathcal{B})}\varepsilon_{t+k}. \tag{5.11}$$

The variables $\varepsilon_t, \varepsilon_{t-1}, \ldots$ can be estimated from the residuals and $\varepsilon_{t+1}, \varepsilon_{t+2} \ldots$ are unknown random variables at time t. We separate between these two groups of error terms by writing (5.11) as

$$\frac{C(\mathcal{B})}{A(\mathcal{B})} \varepsilon_{t+k} = F(\mathcal{B}) \varepsilon_{t+k} + \mathcal{B}^k \frac{G(\mathcal{B})}{A(\mathcal{B})} \varepsilon_{t+k} \tag{5.12}$$

where we introduce two new polynomials:

$$F(\mathcal{B}) = 1 + f_1 \mathcal{B} + \cdots + f_{k-1} \mathcal{B}^{k-1} \qquad \text{(order is } k - 1\text{)}$$

$$G(\mathcal{B}) = g_0 + g_1 \mathcal{B} + \cdots + g_{n-1} \mathcal{B}^{n-1} \qquad \text{(order is } n - 1\text{)}$$

Equation (5.11) can then be rewritten as

$$Y_{t+k} = \underbrace{F(\mathcal{B}) \varepsilon_{t+k}}_{\text{future errors at time } t} \underbrace{\frac{G(\mathcal{B})}{A(\mathcal{B})} \varepsilon_t}_{\text{current } (t) \text{ and past errors}} \tag{5.13}$$

Before we proceed any further, we pause for an example of how this partition of errors is made.

Example 5.3: Partition of Errors Needed in Forecasting. Consider the process in Example 5.2. We have from equation (5.9) that

$$Y_{t+2} = \varepsilon_{t+2} + (c - a) \varepsilon_{t+1} + \frac{a(a - c)}{1 + a\mathcal{B}} \varepsilon_t$$

Thus in the notation just introduced, we have $k = 2$ and $n = 1$, so

$$F(\mathcal{B}) = 1 + (c - a)\mathcal{B} \quad \left[\text{i.e., } f_1 = c - a, \text{ and } F(\mathcal{B}) \text{ is of order } k - 1 = 1\right]$$

$$G(\mathcal{B}) = a(a - c) \qquad \left[\text{i.e., } g_0 = a(a - c), \text{ and } G(\mathcal{B}) \text{ is of order } n - 1 = 0\right]$$

□

Looking back at equation (5.11), we notice that the second term can be estimated through the residuals, computed from

$$\varepsilon_t = \frac{A(\mathcal{B})}{C(\mathcal{B})} Y_t.$$

Substituting this into (5.13), we get

$$Y_{t+k} = F(\mathcal{B}) \varepsilon_{t+k} + \frac{G(\mathcal{B})}{C(\mathcal{B})} Y_t$$

Let $\hat{Y}$ be any forecasting function of current and past observations Y_t. Then

$$E_t\left[\left(Y_{t+k} - \hat{Y}\right)^2\right] = E\left[\left(F(\mathcal{B})\varepsilon_{t+k} + \frac{G(\mathcal{B})}{C(\mathcal{B})}Y_t - \hat{Y}\right)^2\right]$$

$$= E\left[\left(F(\mathcal{B})\varepsilon_{t+k}\right)^2\right] + 2\,E\left[\left(F(\mathcal{B})\varepsilon_{t+k}\right)\left(\frac{G(\mathcal{B})}{C(\mathcal{B})}Y_t - \hat{Y}\right)\right]$$

$$+ E\left[\left(\frac{G(\mathcal{B})}{C(\mathcal{B})}Y_t - \hat{Y}\right)^2\right]$$

where the cross term is zero given that future errors are independent of current and past observations. Thus we must have that

$$E\left[\left(Y_{t+k} - \hat{Y}\right)^2\right] \geq E\left[\left(F(\mathcal{B})\varepsilon_{t+k}\right)^2\right] = \left(1 + f_1^2 + f_2^2 + \cdots + f_{k-1}^2\right)\sigma_\varepsilon^2$$

where we have an equality if

$$\hat{Y} = \frac{G(\mathcal{B})}{C(\mathcal{B})}Y_t = E_t[Y_{t+k}] = \hat{Y}_{t+k|t}. \tag{5.14}$$

This constitutes the MMSE forecast k periods ahead. The forecast errors or innovations are

$$\tilde{Y}_{t+k|t} = Y_{t+k} - \hat{Y}_{t+k|t} = F(\mathcal{B})\varepsilon_{t+k}$$

a moving average process of order $k - 1$. Therefore, the variance of these forecast errors is

$$\mathrm{Var}\left[\left(\tilde{Y}_{t+k|t} - \hat{Y}\right)^2\right] = \left(1 + f_1^2 + f_2^2 + \cdots + f_{k-1}^2\right)\sigma_\varepsilon^2.$$

As can be seen, the key to obtaining the MMSE forecasts is to find the G and F polynomials. These are obtained by equating coefficients of like powers of $\mathcal{B}$ in the *Diophantine identity*:

$$C(\mathcal{B}) = F(\mathcal{B})A(\mathcal{B}) + \mathcal{B}^k G(\mathcal{B}) \tag{5.15}$$

The Diophantine identity[5] represents the long division of C by A, which yields a quotient F with remainder $\mathcal{B}^k G/A$. The procedure of obtaining the

---

[5]Diophantus ($\approx$ A.D. 300) found this expression for the division of polynomials.

coefficients of G and F is analogous to the procedure used in Section 4.2 to obtain the impulse response function. The main difference is that both the C and A polynomials are of finite order and there is a remainder. In the derivations of Section 4.2, the polynomial $H(\mathcal{B})$ was assumed to be infinite order, so there is no need for an explicit remainder. This equation is central in the derivation of MMSE control schemes. For this reason we delay an illustration of its use until after MMSE controllers are explained.

5.1.1 General Solution to the Minimum Variance (MMSE) Control Problem

A general solution to the minimum MSE control problem is presented here. As shown, the resulting controllers are not very practical in that they require considerable control effort. However, these controllers are very easy to modify to obtain a more acceptable performance.

Consider a generic Box–Jenkins transfer function model with environmental noise [i.e., equation (5.5)] and suppose that we look at the decision we need to make at time t, that is, consider the situation from the point of view of selecting X_t:

$$Y_{t+k} = \frac{B(\mathcal{B})}{A(\mathcal{B})} X_t + \frac{C(\mathcal{B})}{A(\mathcal{B})} \varepsilon_{t+k}$$

where k is the system delay. The second term on the right contains unknown errors $\varepsilon_{t+1}, \ldots, \varepsilon_{t+k}$ and past errors that can be estimated from residuals $(\varepsilon_t, \varepsilon_{t-1}, \ldots)$. Thus we write this expression as

$$Y_{t+k} = \frac{B(\mathcal{B})}{A(\mathcal{B})} X_t + \underbrace{F(\mathcal{B}) \varepsilon_{t+k}}_{\text{future errors at time } t} + \underbrace{\frac{G(\mathcal{B})}{A(\mathcal{B})} \varepsilon_t}_{\text{current and past errors}} \qquad (5.16)$$

where, as before, F is of order $k - 1$ and G is of order $n - 1$. Here $n = \max\{n_A, n_B, n_C\}$, using trailing zeros where necessary. The coefficients of F and G are obtaining from equating terms of like powers of $\mathcal{B}$ in the Diophantine identity (5.15).

From the definition of the ARMAX model,

$$A(\mathcal{B}) Y_t = B(\mathcal{B}) X_{t-k} + C(\mathcal{B}) \varepsilon_t$$

we can write Y_{t+k} as a function of the X's and Y's only by using

$$\varepsilon_t = \frac{A(\mathcal{B})}{C(\mathcal{B})} Y_t - \frac{B(\mathcal{B})}{C(\mathcal{B})} \mathcal{B}^k X_t$$

Substituting this equation into (5.16), we get

$$Y_{t+k} = \frac{B}{A}X_t + F\varepsilon_{t+k} + \frac{G}{C}Y_t - \frac{BG}{AC}\mathcal{B}^k X_t$$

$$= F\varepsilon_{t+k} + \frac{G}{C}Y_t + \left(\frac{B}{A} - \frac{BG}{AC}\mathcal{B}^k\right)X_t$$

$$= F\varepsilon_{t+k} + \frac{G}{C}Y_t + \left(\frac{B(C - \mathcal{B}^k G)}{AC}\right)X_t$$

where we omit the backshift operator in the polynomials for simplicity of notation. Using the Diophantine identity in the numerator of the last term simplifies the expression and we get

$$Y_{t+k} = F\varepsilon_{t+k} + \frac{G}{C}Y_t + \frac{BF}{C}X_t. \tag{5.17}$$

Suppose that T_y denotes the target for Y_t. This will, in general, be zero, since Y_t denotes deviations from target. To solve the MMSE control problem, we want to minimize

$$\mathrm{MSE}(Y_{t+k}) = E_t\left[(Y_{t+k} - T_y)^2\right]$$

$$= \underbrace{E_t\left[(Y_{t+k} - E_t[Y_{t+k}])^2\right]}_{\mathrm{Var}(Y_{t+k})} + \underbrace{(T_y - E_t[Y_{t+k}])^2}_{\text{squared bias}}$$

where the last term is the squared bias, or the squared *offset* the quality characteristic will have. Now, from equation (5.17), we have that

$$E_t[Y_{t+k}] = \frac{G}{C}Y_t + \frac{BF}{C}X_t$$

where the first term on the right is $E_t[Y_{t+k}|X_t = 0] = \hat{Y}_{t+k|t,\,X_t=0}$, the k-step-ahead forecast *if the process were uncontrolled* [equation (5.14)]. Therefore, since $T_y = 0$,

$$\mathrm{MSE}(Y_{t+k}) = E_t[Y_{t+k}^2] = E_t\left[(Y_{t+k} - E_t[Y_{t+k}])^2\right] + E_t[Y_{t+k}]^2$$

$$= E_t\left[(F\varepsilon_{t+k})^2\right] + E_t[Y_{t+k}]^2 \geq \left(1 + f_1^2 + \cdots + f_{k-1}^2\right)\sigma_\varepsilon^2$$

where we have an equality if we choose X_t such that it makes $E_t[Y_{t+k}] = 0$, that is, we choose X_t such that

$$-\frac{G}{C}Y_t = \frac{BF}{C}X_t$$

from where we obtain the *minimum variance or MMSE control law*:

$$X_t = -\frac{G(\mathcal{B})}{B(\mathcal{B})F(\mathcal{B})}Y_t \tag{5.18}$$

This can be referred to either as a minimum *variance* controller (Åström, 1970) or as a minimum MSE controller (Box et al., 1994) since

$$\min\{\text{MSE}(Y_{t+k})\} = \min\{\text{Var}(Y_{t+k})\} = \text{Var}(F\varepsilon_{t+k}).$$

An important feature of the MMSE control law is that it is based on the *separation principle*, which states that the control law can be thought of as two separate tasks:

1. Forecast where the process will be k periods ahead if it is left uncontrolled; that is, find $E_t[Y_{t+k}|X_t = 0] = \hat{Y}_{t+k|t,\,X_t=0} = (G/C)\,Y_t$.
2. Find a control action that cancels with the prediction in task 1 above; that is, find X_t such that (omitting backshift notation):

$$-\frac{G}{C}Y_t = \frac{BF}{C}X_t$$

This step makes the k-step-ahead prediction of the process if we *do control the process*, $\hat{Y}_{t+k|t} = E_t[Y_{t+k}]$, equal to zero.

Figure 5.7 illustrates the closed-loop operation of the process under a MMSE controller.

By substituting equation (5.18) into the process description, we obtain the closed-loop output equation:

$$Y_{t+k} = F\varepsilon_{t+k} + \frac{G}{C}Y_t + \frac{BF}{C}\left(-\frac{G}{BF}\right)Y_t = F\varepsilon_{t+k}$$

Thus

$$Y_t = F(\mathcal{B})\varepsilon_t$$

showing that *the closed-loop output of an MMSE controlled process follows an MA(k − 1) process*, where k is the number of periods of input−output delay.

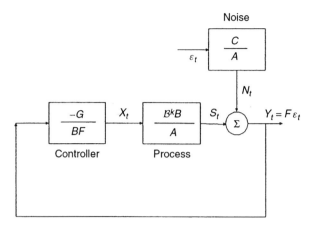

Figure 5.7   Block diagram of an MMSE controller applied to a generic BJ transfer function.

This implies that the output will be stationary[6], but it will *not* be uncorrelated in general. Furthermore, we have that in closed-loop operation,

$$\text{MSE}(Y_{t+k}) = \text{Var}(Y_t) = \sigma_\varepsilon^2 \sum_{i=0}^{k-1} f_i^2$$

where $f_0 = 1$. Under MMSE control, the output will equal the forecast error [i.e., $Y_t = \tilde{Y}_t = F(\mathcal{B})\varepsilon_t$]. Examples that illustrate how to obtain MMSE controllers for given transfer function models are given in Section 5.2. First we discuss the MMSE control rule in more detail.

Comparison with the Ideal Controller
The ideal controller that would attempt to cancel the disturbances perfectly was introduced at the beginning of this section. It is quite illustrative to compare the MMSE control law (5.18) with the ideal controller. Consider the Box–Jenkins transfer function process model:

$$Y_{t+k} = \frac{B}{A}X_t + \frac{C}{A}\varepsilon_{t+k}$$

Instead of making

$$\frac{B}{A}X_t = -\frac{C}{A}\varepsilon_{t+k} \qquad \text{(ideal controller)}$$

which is impossible since we need to know future disturbances, the MMSE

---

[6]Although not a very robust stationarity, as shown later.

controller partitions the last term:

$$\frac{B}{A}X_t + \frac{C}{A}\varepsilon_{t+k} = \frac{B}{A}X_t + F\varepsilon_{t+k} + \frac{G}{A}\varepsilon_t$$

where the last term contains errors that are estimable from the residuals. The MMSE controller then makes

$$\frac{B}{A}X_t = -\frac{G}{A}\varepsilon_t$$

That is, *the MMSE controller only cancels out with that part of the disturbance that has occurred by time t and can be estimated from residuals.* By doing this, the closed-loop equation is

$$Y_{t+k} = F\varepsilon_{t+k}$$

From this it is evident that only when $k = 1$ (unit delay) can the MSE equal the lower bound of σ_ε^2. The MMSE control law is obtained by noting that since

$$\varepsilon_{t+k} = \frac{Y_{t+k}}{F} \Rightarrow \varepsilon_t = \frac{Y_t}{F}$$

then

$$\frac{B}{A}X_t = -\frac{G}{AF}Y_t \Rightarrow X_t = -\frac{G}{BF}Y_t$$

which is the same as equation (5.18).

Testing for Minimum Variance

If a process is operating under an MMSE controller, the output Y_t equals a moving average of order $k - 1$. Thus we can investigate the MSE performance of an existing process by checking if there is any significant autocorrelation beyond lag $k - 1$. Evidently, this is only useful for proving that a controlled process is *not* operating under MMSE. For proving that a process is operating under MMSE control, we would have to know the process description. If a process is suspected to be operating under MMSE control but the MSE is still too large, then (Harris, 1989):

1. Reduce the inherent variability (σ_ε^2).
2. Reduce the process delay (k).
3. If possible, introduce feedforward control.
4. Find another controllable factor and try multivariate control (see Chapter 9).

Unstable B($\mathcal{B}$) **Polynomial**
When using a MMSE control law, if the $B(\mathcal{B})$ polynomial is unstable[7] (i.e., if it has zeros inside the unit circle on the $\mathcal{B}$ plane), the control sequence $\{X_t\}$ defined by

$$X_t = \frac{-G(\mathcal{B})}{B(\mathcal{B})F(\mathcal{B})} Y_t$$

will diverge even though the output will remain on target and will exhibit minimum variance (notice how for such controlled process the output is stable but the process is not BIBO stable). The reason for this is that the closed-loop equation cancels all zeros of $B(\mathcal{B})$, including the unstable ones, whereas the unstable zero is not canceled in the MMSE control law. From the closed-loop equation

$$A Y_t = -\frac{GB}{BF} Y_{t-k} + C\varepsilon_t$$

we get

$$Y_t = \frac{BFC}{ABF + BG\mathcal{B}^k} \varepsilon_t$$

where it is seen that $B(\mathcal{B})$ will cancel regardless of the location of its zeros. Thus the unstable zero is not observed in the output, but it will be observed in the input (X_t) since it appears in the denominator of that equation.

5.1.2 Relation with PID Controllers

Sometimes, but not always, MMSE controllers can be written, after some algebraic manipulation, in the following form:

$$X_t = K_p Y_t + K_I \sum_{j=1}^{t} Y_j + K_D \nabla Y_t$$

which is called a *proportional–integral–derivative* (PID) *controller.* PID controllers are the most common type of industrial controllers in practice, with a long history. These controllers are explained in more detail in Chapter 6. The constants K_p, K_I, and K_D are the tuning parameters of the controller. The continuous-time analogy of the discrete PID equation shown above is

$$X(t) = K_p Y(t) + K_I \int_0^t Y(s)\, ds + K_D \frac{dY(t)}{dt}$$

---

[7]Such a process is called a *non-minimum-phase system* in the control engineering literature.

The three terms correspond to setting the controllable variable in a proportional way with respect to the deviation from target, the integral of the deviations from target, and the derivative of the deviations from target, respectively.

5.1.3 Simplified MMSE Controller

Ljung and Söderström (1987) provide a very simple equation for a MMSE control law that is valid whenever the effect of a change in the input on the output is at least partially felt immediately at the next sampling instance. This rule assumes that the input–output model, written as

$$Y_t = H(\mathcal{B})X_t + G(\mathcal{B})\varepsilon_t$$

where H and G are in general ratios of polynomials in $\mathcal{B}$, must satisfy

$$G(0) = 1$$

$$H(\mathcal{B}) = v_1\mathcal{B} + v_2\mathcal{B}^2 + v_2\mathcal{B}^3 + \cdots$$

with $v_1 \neq 0$. With these requirements, the MMSE control equation is simply

$$X_t = \frac{1 - G(\mathcal{B})}{H(\mathcal{B})}Y_t$$

We now illustrate how to find MMSE controllers both with the Diophantine approach described earlier and with Ljung and Söderström's formula.

5.2 EXAMPLES OF MMSE CONTROLLERS

In this section we illustrate the application of equation (5.18) and the simpler Ljung–Söderström formulation for obtaining MMSE control rules.

Example 5.4: Responsive Process with IMA(1, 1) Noise. Consider a zero-order Box–Jenkins transfer function with additive IMA(1, 1) noise:

$$Y_t = g X_{t-1} + \frac{1 - \theta\mathcal{B}}{1 - \mathcal{B}}\varepsilon_t$$

Thus we have that

$$A(\mathcal{B}) = 1 - \mathcal{B}$$

$$B(\mathcal{B}) = g(1 - \mathcal{B})$$

$$C(\mathcal{B}) = 1 - \theta\mathcal{B}$$

and $n = \max\{n_A, n_B, n_C\} = 1$ and $k = 1$. Therefore, the $F(\mathcal{B})$ polynomial is of order zero $(k - 1 = 0)$ and $G(\mathcal{B})$ is of order zero $(n - 1 = 0)$. From the Diophantine equation;

$$C = AF + \mathcal{B}^k G$$

Equating coefficients multiplying $\mathcal{B}$, we get $-\theta = -1 + g_0$ or $g_0 = 1 - \theta$. Then, from equation (5.18), the MMSE controller is

$$X_t = -\frac{1 - \theta}{g(1 - \mathcal{B})} Y_t = -\frac{1 - \theta}{g} \sum_{j=1}^{t} Y_j$$

which constitutes a pure integral controller. That is, the level of the controllable factor is proportional to the cumulative sum of the deviations from target. The proportionality constant is $K_I = (\theta - 1)/g$. This is called the *absolute or level form* of the controller. To implement this control law, it is better to write it in *incremental form*:

$$X_t = X_{t-1} - \frac{1 - \theta}{g} Y_t$$

Ljung and Söderström's formulas can be applied in this example, too. In this case we have

$$H(\mathcal{B}) = g\mathcal{B}$$

$$G(\mathcal{B}) = \frac{1 - \theta\mathcal{B}}{1 - \mathcal{B}}.$$

Thus the MMSE controller is

$$X_t = \frac{1 - G}{H} Y_t = \frac{1 - (1 - \theta\mathcal{B})/(1 - \mathcal{B})}{g\mathcal{B}} Y_t = -\frac{1 - \theta}{g(1 - \mathcal{B})} Y_t$$

the same as before. Note that the presence of a unit root in the denominator (i.e., the integrator) of the control rule implies that $\{X_t\}$ will follow a nonstationary process (the adjustments $\{\nabla X_t\}$ will be stationary). This could be too expensive in some processes.   □

Example 5.5: Process with Second-Order Dynamics. Consider a second-order Bo–Jenkins transfer function with IMA(1, 1) noise:

$$Y_t = \frac{g}{1 - \delta_1\mathcal{B} - \delta_2\mathcal{B}^2} X_{t-1} + \frac{1 - \theta\mathcal{B}}{1 - \mathcal{B}} \varepsilon_t$$

Applying the Ljung–Söderström formula gives us

$$X_t = \frac{1 - (1 - \theta\mathcal{B})/(1 - \mathcal{B})}{g\mathcal{B}/(1 - \delta_1\mathcal{B} - \delta_2\mathcal{B}^2)}Y_t = \frac{(\theta - 1)(1 - \delta_1\mathcal{B} - \delta_2\mathcal{B}^2)}{g(1 - \mathcal{B})}Y_t$$

Since the difference operator $(1 - \mathcal{B})$ appears in the denominator, this controller also contains integral action. □

Example 5.6: Longer Delays. Consider the process defined by

$$Y_t = \frac{X_{t-2}}{1 + a\mathcal{B}} + \frac{1 + c\mathcal{B}}{1 + a\mathcal{B}}\varepsilon_t$$

an $(r, s, k) = (1, 0, 2)$ Box–Jenkins transfer function with ARMA$(1, 1)$ noise. Since $v_1 = 0$ in $H(\mathcal{B}) = v_1\mathcal{B} + v_2\mathcal{B}^2 \cdots$, Ljung and Söderström's simplified formulation cannot be applied. Thus we need to use the Diophantine equation. We have

$$A(\mathcal{B}) = 1 + a\mathcal{B}$$

$$B(\mathcal{B}) = 1$$

$$C(\mathcal{B}) = 1 + c\mathcal{B}$$

We have $k = 2$, so F is of order $k - 1 = 1$ and G is of order $n - 1 = 0$. The Diophantine identity (5.15) gives

$$1 + c\mathcal{B} = (1 + a\mathcal{B})(1 + f_1\mathcal{B}) + \mathcal{B}^2 g_0$$

Equating coefficients of like powers of $\mathcal{B}$, we have

$$\mathcal{B}: \quad c = a + f_1 \Rightarrow f_1 = c - a$$

$$\mathcal{B}^2: \quad 0 = af_1 + g_0 \Rightarrow g_0 = -af_1 = a(a - c)$$

Thus we have that

$$X_t = -\frac{G}{BF}Y_t = \frac{-a(a - c)}{1 + (c - a)\mathcal{B}}Y_t$$

Note that this controller does not include integral action, mainly because the disturbance is stationary. A simple way to ensure that we end up with integral action is to assume that the disturbance is (homogeneous) nonstationary. □

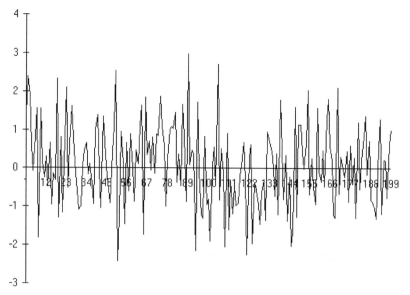

Figure 5.8 Deviations from target of a system controlled with an MMSE controller applied to a transfer function with unstable $B(\mathcal{B})$ polynomial.

Example 5.7: Unstable B Polynomial (Non-minimum-phase Process). Consider the process

$$(1 - 0.7\mathcal{B})Y_t = \frac{0.9 + \mathcal{B}}{1 - \mathcal{B}}X_{t-1} + \frac{1 - 0.7\mathcal{B}}{1 - \mathcal{B}}\varepsilon_t$$

The MMSE controller is

$$X_t = -\frac{1 - 0.7\mathcal{B}}{0.9 + \mathcal{B}}Y_t$$

and it is easy to see that the denominator has a zero at $\mathcal{B} = -0.9 < 1$, thus X_t will diverge. Figures 5.8 and 5.9 illustrate this situation with $\varepsilon_t \sim N(0, 1)$. Notice the extremely large oscillations in X_t. In practice, the controllable factor will hit bounds that the process must obey. From that point on, the output Y_t will show nonstationary behavior. $\qquad\square$

5.3 GENERALIZED MINIMUM VARIANCE CONTROLLER

The large variability that a controllable factor must undergo when manipulated under MMSE control may not always be practical or feasible. This is because the MMSE controller transfers the variability from the quality

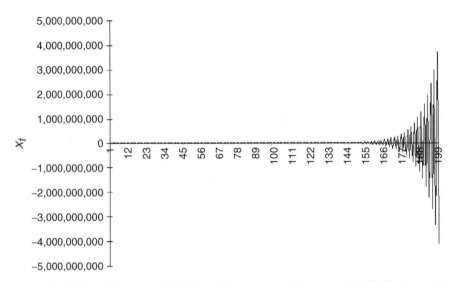

Figure 5.9 Values of the controllable factor for a process with an unstable $B(\mathcal{B})$ polynomial, MMSE controller applied.

characteristics to the controllable factor. Furthermore, non-minimum-phase processes, such as the one encountered in Example 5.7, cannot be regulated in practice by applying the MMSE control formulation given in Section 5.2, and some other control rule must be tried. To illustrate some of these ideas, consider the process defined by

$$\left(1 - \phi\mathcal{B}\right)Y_t = gX_{t-1} + \left(1 - \theta\mathcal{B}\right)\varepsilon_t$$

The MMSE controller is found to be

$$X_t = \frac{\theta - \phi}{g}Y_t$$

which is a proportional controller. Figure 5.10 shows a simulation of this process and controller for the case $\phi = 0.5$, $\theta = -0.4$, $g = 1$, and $\varepsilon_t \sim U(-1, 1)$. In the first case, an *open − loop* simulation was carried out with the input varying as $X_t \sim U(-0.5, 0.5)$. In the second simulation, the closed-loop MMSE performance is achieved by using the controller $X_t = -0.9\ Y_t$. As can be seen, compared to the open-loop performance, an MMSE controller transfers the variability from the output (the response) to the input (the controllable factor[8]).

---

[8]A third case, not shown in the figure, is when we do not control and let X_t equal to some constant. In this case, the process output will follow the stochastic process defined by the disturbance.

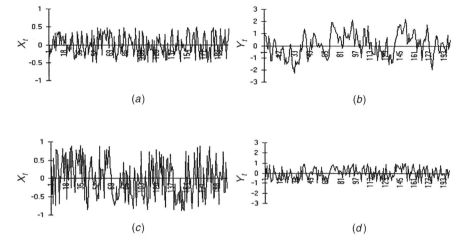

Figure 5.10 Effect on input and output of using an open-loop operation (a, b) as opposed to an MMSE controller (c, d). (a) White noise input; (b) output for white noise input; (c) MMSE control input; (d) output when x_t is an MMSE controller.

The problem in practice is that excessive manipulation of the controllable factor (i.e., large variability of the input) is not desirable in many cases. For example, there might be a cost proportional to the magnitude of the control actions, $|X_t|$. Or, there might be limitations in the *range of values* that the level of the controllable factor can take; for example, X_t could be limited (due to scaling) to a range $(-1, 1)$ as commonly done in the area of design of experiments. One indirect way[9] of achieving this goal is to limit $\text{Var}(X_t)$. There is a need for a controller that minimizes MSE subject to a constraint in the variability of the controllable factor. Although there has been considerable work in this area, the resulting feedback laws are quite complicated [see, e.g., Box et al. (1994) constrained MMSE schemes]. A relatively simple technique that solves this problem and in addition, works for non-minimum-phase systems (such as that of Example 5.7) was proposed by Clarke and Gawthrop (1971) and is described next.

MMSE controllers concentrate on the criterion

$$J_{MV} = E\left[Y_t^2\right]$$

which is minimized by selecting X_t such that $\hat{Y}_{t+k|t} = 0$ for all time periods t. A natural extension of this criterion is to minimize instead

$$J_1 = E\left[Y_{t+k}^2 + \lambda X_t^2\right]$$

where the variable Y_{t+k} is included since it is the one affected by X_t. The

[9]Constrained feedback controllers subject to bounds on X_t constitute a direct way of solving the problem.

constant λ can be thought of as the Lagrange multiplier of

$$E[X_t^2] \leq c$$

which is a constraint in the MSE of the controllable factor. If X_t is scaled such that it is centered at a zero value, and if the uncontrolled system is not drifting, $\text{MSE}(X_t) = \text{Var}(X_t)$.

To minimize J_1, Clarke and Gawthrop noticed that we can always write

$$Y_{t+k} = \hat{Y}_{t+k|} + \tilde{Y}_{t+k|t}$$

where the last term is the forecast error, which is independent of X_t since it is a function only of future errors at time t. Thus

$$E[Y_{t+k}^2 + \lambda X_t^2] = E\left[\hat{Y}_{t+k|t}^2 + 2\hat{Y}_{t+k|t}\tilde{Y}_{t+k|t} + \tilde{Y}_{t+k|t}^2 + \lambda X_t^2\right]$$

which reduces to

$$E[Y_{t+k}^2 + \lambda X_t^2] = \text{Var}\left(\tilde{Y}_{t+k|t}\right) + E\left[\hat{Y}_{t+k|t}^2 + \lambda X_t^2\right]$$

Since $\text{Var}(\tilde{Y}_{t+k|t})$ is not a function of X_t, to minimize J_1 it is sufficient to minimize

$$J_2 = E\left[\hat{Y}_{t+k|t}^2 + \lambda X_t^2\right] = \int_{-\infty}^{\infty} \left(\hat{Y}_{t+k|t}^2 + \lambda X_t^2\right) f(X_t) \, dX_t$$

where $f(X_t)$ is the probability density function of X_t. Thus this expression acknowledges the fact that the control factor is a random variable owing to its dependence on Y_t which in turn depends on the random errors ε_t. It can be seen that there is a difficulty: The density of X_t depends on the feedback control law we wish to find in the first place. If the errors are normally–distributed, this problem is called a *linear quadratic gaussian* (LQG) stochastic control problem in the control engineering literature. It is solved by recourse to dynamic programming techniques (Åström, 1970, Lewis, 1986). Instead, Clarke and Gawthrop propose to minimize the *conditional expectation*:

$$J_3 = E\left[Y_{t+k}^2 + \lambda X_t^2 \big| Y_t, Y_{t-1}, \ldots\right] = E_t\left[Y_{t+k}^2 + \lambda X_t^2\right] = \hat{Y}_{t+k|t}^2 + \lambda X_t^2$$

where the last equality follows from having taken expectation conditioned on current and past observations (MacGregor and Tidwell, 1977). If $\{Y_i\}_{i \leq t}$ are known and given, X_t is not a random variable and the conditional expected value of Y_{t+k}^2 can be computed using the MMSE criterion as

$$\hat{Y}_{t+k|t} = E_t[Y_{t+k}] = \frac{G(\mathcal{B})}{C(\mathcal{B})}Y_t + \frac{B(\mathcal{B})F(\mathcal{B})}{C(\mathcal{B})}X_t$$

Thus, omitting backshift notation, we minimize

$$J_3 = \left(\frac{G}{C}Y_t + \frac{BF}{C}X_t\right)^2 + \lambda X_t^2$$

which can be minimized by making $(dJ_3)/(dX_t) = 0$. Now

$$\frac{d\hat{Y}_{t+k|t}}{dX_t} = \frac{B(0)F(0)}{C(0)} = b_0$$

recognizing that only the first elements of the $B(\mathcal{B})$, $F(\mathcal{B})$, and $C(\mathcal{B})$ polynomials affect the controllable factor at time t. Thus

$$\frac{dJ_3}{dX_t} = 2b_0\left(\frac{G}{C}Y_t + \frac{BF}{C}X_t\right) + 2\lambda X_t = 0$$

from which we obtain

$$X_t = -\frac{G(\mathcal{B})}{B(\mathcal{B})F(\mathcal{B}) + \lambda/b_0 C(\mathcal{B})}Y_t \tag{5.19}$$

which constitutes Clarke and Gawthrop's generalized minimum variance controller. Note that if $\lambda = 0$, we get the MV (MMSE) control formula (5.18).

Example 5.8: Clarke–Gawthrop Controller. Consider a Box–Jenkins first-order transfer function $(r, s, k) = (1, 0, 1)$ with IMA(1, 1) noise:

$$Y_t = \frac{g}{1 - \delta\mathcal{B}}X_{t-1} + \frac{1 - \theta\mathcal{B}}{1 - \mathcal{B}}\varepsilon_t$$

so in this case

$$A(\mathcal{B}) = (1 - \delta\mathcal{B})(1 - \mathcal{B})$$

$$B(\mathcal{B}) = g(1 - \mathcal{B})$$

$$C(\mathcal{B}) = (1 - \theta\mathcal{B})(1 - \delta\mathcal{B})$$

Here we have $k = 1$, so F is of order $k - 1 = 0$ and G is of order $n - 1 = 1$. The Diophantine identity (5.15) in this case gives:

$$1 - (\theta + \delta)\mathcal{B} + \theta\delta\mathcal{B}^2 = \left[1 - (\delta + 1)\mathcal{B} + \delta\mathcal{B}^2\right](1) + \mathcal{B}(g_0 + g_1\mathcal{B})$$

Equating coefficients of like powers of $\mathcal{B}$, we find that

$$\mathcal{B}: \quad -(\theta + \delta) = -\delta - 1 + g_0 \Rightarrow g_0 = 1 - \theta$$

$$\mathcal{B}^2: \quad \theta\delta = \delta + g_1 \Rightarrow g_1 = \delta(\theta - 1)$$

Thus the controller is

$$X_t = -\frac{1 - \theta + \delta(\theta - 1)\mathcal{B}}{g(1 - \mathcal{B}) + (\lambda/g)(1 - \delta\mathcal{B})(1 - \theta\mathcal{B})}Y_t$$

or

$$X_t = \frac{(1 - \theta)(1 - \delta\mathcal{B})}{g(1 - \mathcal{B}) + (\lambda/g)(1 - \delta\mathcal{B})(1 - \theta\mathcal{B})}Y_t$$

The effect of increasing λ is typically of the form depicted in Figure 5.11.[10] As λ increases large reductions in $\mathrm{Var}(X_t)$ are obtained for only small increments in $\mathrm{Var}(Y_t)$.

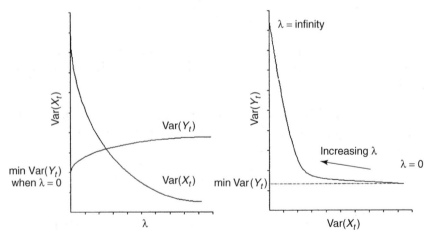

Figure 5.11 Typical behavior of Clarke and Gawthrop's constrained controller.

[10] But not always. The exceptions occur under uncommon conditions (see Modén and Söderström, 1982). It should be pointed out that the stability of the closed-loop process depends on the value of λ, so certain values of λ may indeed lead to unstable behavior.

Example 5.9: Clarke–Gawthrop Controller, Unstable B Polynomial. Consider the case described earlier where $B(\mathcal{B})$ is an unstable polynomial:

$$(1 - 0.7\mathcal{B})Y_t = \frac{0.9 + \mathcal{B}}{1 - \mathcal{B}}X_{t-1} + \frac{1 - 0.7\mathcal{B}}{1 - \mathcal{B}}\varepsilon_t$$

where $\varepsilon_t \sim N(0, 1)$. Using the Diophantine identity as before, we get the generalized minimum variance controller

$$X_t = -\frac{1 - 0.7\,\mathcal{B}}{0.9 + \mathcal{B} + (\lambda/0.9)(1 - 0.7\mathcal{B})}Y_t$$

Figure 5.12 illustrates the behavior of this controller and process when $\lambda = 0.3$. As can be seen, the controllable factor does not "blow up." To illustrate the performance as λ increases, Table 5.1 gives estimated values of the input and output variances as λ increases. (Estimates are simply the average of 10 simulations but illustrate the general behavior.)

The table agrees with previous results, where it was shown that the minimum variance controller ($\lambda = 0$) results in a controllable factor that diverges as time increases. If $\lambda > 0$, the unstable zero does not affect the

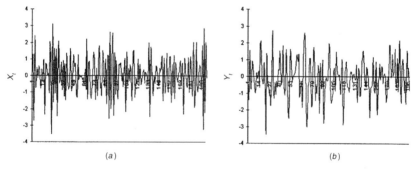

(a)   (b)

Figure 5.12 Simulated behavior of (*a*) input and (*b*) output variables when using Clarke and Gawthrop's GMV controller when $\lambda = 0.3$ ($\sigma_\varepsilon^2 = 1.0$), Example 5.9.

Table 5.1 Input and Output Variances for the Process in Example 5.9

λ	$\widehat{\mathrm{Var}(Y_t)}$	$\widehat{\mathrm{Var}(X_t)}$
0	1.01	Diverges
0.1	1.062	2.41
0.3	1.086	1.17
0.5	1.149	0.86
1.0	1.236	0.57

input. Notice that by doing this, the resulting controller no longer has integral action. □

Var(X_t) **versus** Var(∇X_t)

Depending on the process to be controlled, it might be convenient to constrain the variability of the adjustments (∇X_t) rather than the variability of the level of the controllable factor (X_t) as done so far in this section. To achieve this using the Clarke and Gawthrop approach, we would minimize instead

$$J_{\nabla X} = E_t\left[\hat{Y}_{t+k\,|\,t} + \lambda w_t^2\right] = \hat{Y}_{t+k\,|\,t}^2 + \lambda w_t^2$$

where $w_t \equiv \nabla X_t$. A transfer function model that relates Y_t with ∇X_{t-k} (as opposed to X_{t-k}) is needed. This can be identified and fit directly from $\{Y_t, w_t\}$ data or obtained from a transfer function that relates Y_t and X_t. In the latter case, suppose that we have available the ARMAX model

$$A(\mathcal{B})Y_t = B(\mathcal{B})X_{t-k} + C(\mathcal{B})\varepsilon_t$$

This is equivalent to

$$(1 - \mathcal{B})A(\mathcal{B})Y_t = B(\mathcal{B})\nabla X_{t-k} + (1 - \mathcal{B})C(\mathcal{B})\varepsilon_t \qquad (5.20)$$

Defining $A'(\mathcal{B}) = (1 - \mathcal{B})A(\mathcal{B})$, $C'(\mathcal{B}) = (1 - \mathcal{B})C(\mathcal{B})$ and $w_t = \nabla X_t$, we obtain the generalized minimum variance controller for the adjustments:

$$w_t = \nabla X_t = -\frac{G}{BF + (\lambda/b_0)C'}Y_t$$

where the difference with (5.19) lies on the Diophantine identity, which in this case is

$$C' = \nabla C = FA' + B^k G = F\nabla A + B^k G$$

where F is still of order $k - 1$ but G is of order $\max\{n_{A'}, n_B, n_{C'}\} = \max\{n_A + 1, n_B, n_C + 1\}$.

Example 5.10: Constraining the Variance of the Adjustments. Capilla et al. (1999) illustrate the use of a Clarke–Gawthrop controller applied to a polymerization process. In this application, the quality characteristic is a quantity described as the *melt index*, with a target equal to 0.8, and the controllable factor is the average reactor temperature. From input–out data, the authors identified the process

$$\nabla Y_t = 0.185\nabla X_{t-1} + 0.274\nabla X_{t-2} + \varepsilon_t$$

where Y_t denotes the deviations from target of the melt index. Rearranging terms, we have that

$$\nabla Y_t = (0.185 + 0.274\mathcal{B})\nabla X_{t-1} + \varepsilon_t$$

Following equation (5.20), we have that $A'(\mathcal{B}) = 1 - \mathcal{B}$, $B(\mathcal{B}) = 0.185 + 0.274\mathcal{B}$, and $C'(\mathcal{B}) = 1$. Note that the B polynomial has a zero inside the unit circle, and we cannot use a standard MMSE controller. In this case the F polynomial is of order $k - 1 = 0$ and the G polynomial is of order $n - 1 = 1$. From the Diophantine identity, we have, equating coefficients that multiply $\mathcal{B}^1$, that $g_0 = 1$. Therefore, the Clarke–Gawthrop controller for this process is

$$\nabla X_t = -\frac{1}{0.185 + 0.274\mathcal{B} + \lambda/0.185}Y_t \qquad \square$$

5.4 FIXED ADJUSTMENT COSTS AND DEADBAND ADJUSTMENT POLICIES

Sometimes, particularly in discrete-part manufacturing processes, adjusting a production process involves changing the settings of the process, a task in which there will be a cost incurred independent of the magnitude of the adjustment. The effect of these fixed costs on the optimal adjustment policy were studied by Box and Jenkins (1963). They considered the process

$$Y_t = gX_{t-1} + \frac{1 - \theta\mathcal{B}}{1 - \mathcal{B}}\varepsilon_t \qquad (5.21)$$

where Y_t denoted the deviation from target and $\varepsilon_t \sim N(0, \sigma_\varepsilon^2)$. They found the optimal adjustment policy that would minimize

$$J = E\left[\sum_{t=1}^{\infty} aY_t^2 + C_A\delta(\nabla X_{t-1})\right]$$

where $\delta(z) = 1$ if $z \neq 0$ and equals zero otherwise. For this machine tool problem, as named by Box and Jenkins, there is a quadratic off-target cost and a fixed adjustment cost independent of the magnitude of the adjustment (∇X_t). Figure 5.13 shows typical graphs of these two cost components. Define the standardized off-target cost $C_T = a\sigma_\varepsilon^2$. Then the objective is to minimize

$$J = E\left[\sum_{t=1}^{\infty} C_T\left(\frac{Y_t}{\sigma_\varepsilon}\right)^2 + C_A\delta(\nabla X_{t-1})\right].$$

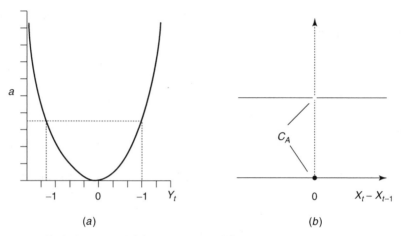

Figure 5.13 Typical graphs of (a) off-target and (b) adjustment costs in Box and Jenkins' machine tool problem.

Box and Jenkins (1963) showed that the optimal adjustment policy is to use a *deadband policy* (called later by Box a *bounded adjustment chart*) that resembles the operation of a Shewhart chart. In the optimal policy, as long as the one-step-ahead forecast of the process, assuming that we do not adjust, falls inside certain deadband, there is no adjustment. Otherwise, an adjustment takes place. Thus the limits of the deadband are determined based on cost considerations, not based on statistical considerations as in a Shewhart chart. An SPC monitoring chart should not be used for adjustment purposes.

As described in Chapter 3, the MMSE one-step-ahead forecast of the process, assuming that $\nabla X_t = 0$, is given by the EWMA equation:

$$\hat{Y}_{t+1|t} = \lambda Y_t + (1 - \lambda)\hat{Y}_{t|t-1}$$

with $\lambda = 1 - \theta$. The deadband policy is as follows. If $\hat{Y}_{t+1|t} \notin (-L, +L)$, adjust by

$$\nabla X_t = -\frac{1}{g}\hat{Y}_{t+1|t}.$$

Whenever an adjustment is made, the EWMA forecasting equation is reset to zero. As the cost of adjustment decreases, the limit L tends to zero, and eventually we get an MMSE control scheme in which an adjustment is made at each time instant.

Two performance measures of deadband policies are the average adjustment interval (AAI) and the increment in standard deviation (ISD) of the output with respect to the minimum possible variance that can be achieved, as provided by an MMSE scheme that always adjusts. The AAI is the average

Table 5.2 Average Adjustment Interval (AAI) and Percentage Increase in Standard Deviation (ISD) for Machine Tool Problems Assuming Model (5.21)

λ	$L/\lambda\sigma_\varepsilon$	AAI	ISD	C_A/C_T
0.5	1.00	2.8	2.4	0.3
	2.00	6.9	10.0	1.9
	3.00	13.1	21.0	7.2
	4.00	21.3	34.0	19.0
0.4	1.25	3.6	2.5	0.3
	2.50	9.8	10.0	2.5
	3.75	19.0	21.0	10.0
	5.00	31.4	34.0	27.0
0.3	1.67	5.3	2.6	0.4
	2.50	9.8	5.8	1.4
	3.33	15.6	10.0	3.6
	5.00	31.4	20.0	15.0
0.2	2.50	9.8	2.6	0.6
	3.75	19.0	5.6	2.4
	5.00	31.4	9.0	6.6
	6.25	47.0	14.0	15.0
0.1	2.50	9.8	0.7	0.2
	5.00	31.4	2.4	1.6
	7.50	66.0	5.2	7.1
	10.00	112.0	9.0	21.0

Source: Box and Luceño (1997).

number of time periods between successive adjustments; as C_A increases, so does the AAI. Box and Luceño (1997) provided graphs and tables from which, given C_T, C_A, θ, and σ_ε, the user can read the values of the deadband limit L and the resulting values of AAI and ISD (see Table 5.2).

Example 5.11: Deadband Adjustment Policy. Suppose that in a process such as (5.21) it is known that $\theta = 0.8$ (so $\lambda = 0.2$), $\sigma_\varepsilon = 11.6$, and the costs are $a = 1$ and $C_A = 28$. Then

$$C_T = a\sigma_\varepsilon = 11.6, \qquad \frac{C_A}{C_T} = 2.4$$

and from Table 5.2 for $\lambda = 0.2$ and $C_A/C_T = 2.4$, we get that

$$\frac{L}{\lambda\sigma_\varepsilon} = 3.75$$

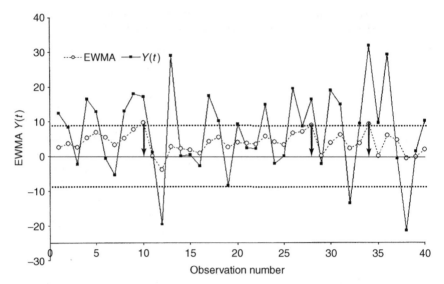

Figure 5.14 Operation of the deadband policy of Example 5.11. Vertical arrows represent the adjustments made. Deadband limits are at ± 8.7 around the value of zero.

or $L = 8.70$. From the table, this corresponds to approximate AAI and ISD values of 19 and 5.6%. Thus, adjusting on the average every 19 time periods produces a very modest increase in the output standard deviation. ☐

In practice, estimating the a and C_A costs is difficult. Instead, empirically using a deadband with limits width L that is varied until an acceptable compromise between AAI and σ_Y is obtained is preferred. Figure 5.14 shows a time series graph illustrating the operation of the resulting deadband policy in Example 5.11.

Box and Kramer (1992) discuss adjustment costs further. Crowder (1992) solves the machine tool problem under the assumption of a finite run length (i.e., the sum in the performance index J is finite). He shows that the optimal deadband limits funnel out as the end of the production run approaches. That is, it is less attractive to adjust if the end of the run is close (at which time it is assumed that a complete renewal of the process starts). Jensen and Vardeman (1993) studied the same cost function as in Crowder but introduce an adjustment error (i.e., the adjustment ∇X_t is not precise). The effect of this is a deadband adjustment policy even if there are no fixed adjustment costs. That is, if adjusting is imprecise, it pays not to make small adjustments. This is in agreement with Figure 1.20.

5.5 SETUP ADJUSTMENT PROBLEMS: GRUBBS'S HARMONIC RULE

In some manufacturing processes, an incorrect setup operation can result in drastic consequences in the quality of the parts produced thereafter, thus there is a need to correct—or adjust— a process upset that occurs due to an incorrect setup operation. In particular, for a machining process, the set point of the machine (i.e., the nominal dimension aimed at prior to perform a particular cutting operation) can easily be modified from part to part in modern CNC (computer numerically controlled) machines.

Grubbs (1954) proposed a method for the adjustment of a machine to bring the process back to target (T) if at startup was off target by d units. Since the target is constant through time, we can model without loss of generality the deviations from target. For the manufactured dimension of the first part, the mean deviation from target is assumed to be equal to

$$\mu_1 = d \tag{5.22}$$

where d is an unknown quantity, not necessarily equal to zero. Let the controllable factor U_t denote the deviation from target aimed at before processing part $t + 1$. Thus, in equation (5.22), we have $\mu_1 = d + U_0 = d$ or $U_0 = 0$. This implies that initially we aim at the target dimension, unaware of any offset d. If d is known, then we would set $U_t = -d$ for all t and completely eliminate the offset; of course, this is seldom possible.

The deviation from target of the generated dimension once the operation takes place is equal to

$$Y_1 = \mu_1 + v_1 \tag{5.23}$$

where $v \sim N(0, \sigma_v^2)$ models both the part-to-part variability and the measurement error.

In contrast to previous sections, where the adjustment affected the observed deviations from target, in a machining process the controllable factor is the set point, which can be assumed to have an effect on the mean of the process (i.e., on the mean deviation from target). That is, the adjustment $\nabla U_1 = U_1 - U_0$ in the set point will result in a new process mean of

$$\mu_2 = \mu_1 + \nabla U_1 = \mu_1 - k_1 Y_1 \tag{5.24}$$

where k_1 is a constant that needs to be determined. From equations (5.23 and 5.24) the new process mean is

$$\mu_2 = (1 - k_1)d - k_1 v_1 \tag{5.25}$$

and the observed deviation from the target will be

$$Y_2 = \mu_2 + \upsilon_2. \tag{5.26}$$

The next adjustment is $\nabla U_2 = -k_2 Y_2$, where k_2 is a constant we need to determine. The mean deviation from target for the third part is

$$\mu_3 = \mu_2 + \nabla U_2 \tag{5.27}$$

and so on.

The adjustment policy is completely determined if we specify the constants k_i. Grubbs proposed to find these constants by solving

$$\begin{aligned} &\min \text{Var}(\mu_{n+1}) \\ &\text{subject to:} \quad E[\mu_{n+1}] = 0. \end{aligned} \tag{5.28}$$

This means that it is desired to have a process that, on average, is on target after n parts have been processed, with a minimum variance around the target. Notice that μ_t becomes a random variable since it is a function of adjustments based on random observations. In an elegant derivation, Grubbs (1954) shows that the constants k_i must satisfy

$$k_t = \frac{1}{t} \tag{5.29}$$

so the adjustments are simply

$$\nabla U_t = -\frac{Y_t}{t}. \tag{5.30}$$

The adjustment rule implies that after producing the first part, the machine set point is adjusted by the full observed deviation. After the second part is produced, the set point is adjusted by half the observed deviation from target, and so on. The adjustments follow the harmonic series $\{1, \frac{1}{2}, \frac{1}{3}, \dots\}$. We emphasize the assumptions behind Grubbs's method: (1) the process is stable, with no autocorrelation or drift in the process mean; (2) adjustments modify the process mean; and (3) adjustments are exact and implemented on every part. Assumption (1) is discussed at length in del Castillo (1998). Modifications to assumption 3 are discussed in Trietsch (1998). We discuss the setup adjustment problem again in Appendix 8A when we look at Kalman filters.

PROBLEMS

5.1. Find the MMSE adjustment rule for the process

$$Y_t = \frac{W}{(1 - \delta B)(1 - B)} X_{t-1} + \frac{1 - \theta B}{1 - B} \varepsilon_t$$

5.2. Find the MMSE adjustment rule for the process in Example 5.5 using the Diophantine equation approach as opposed to Ljung and Söderström's formula.

5.3. Find Clarke and Gawthrop's generalized minimum variance control scheme for the process

$$(1 - \mathcal{B})Y_t = gX_{t-1} + (1 - \theta\mathcal{B})\varepsilon_t$$

5.4. Find the Clarke and Gawthrop control scheme for the process

$$(1 - \mathcal{B})Y_t = \frac{w}{1 - \delta_1\mathcal{B} + \delta_2\mathcal{B}^2}X_{t-1} + (1 - \theta\mathcal{B})\varepsilon_t$$

5.5. Simulate the process in Example 5.9 for increasing values of λ. What is the largest value of λ that will provide a stable output?

5.6. Find the generalized minimum variance controller for the *adjustments* (∇X_t) of the process

$$(1 - 0.7\,\mathcal{B})Y_t = \frac{0.9 + \mathcal{B}}{1 - \mathcal{B}}X_{t-1} + \frac{1 - 0.7\mathcal{B}}{1 - \mathcal{B}}\varepsilon_t$$

Simulate for increasing values of λ and estimate $\mathrm{Var}(Y_t)$ and $\mathrm{Var}(\nabla X_t)$. (*Hint:* Use a spreadsheet.)

5.7. Find the *closed-loop description* of the controlled process $\{Y_t\}$, obtained by substituting the controller equation into the process equation, for the process in Example 5.5.

In problems 5.8 to 5.11 it is suggested that spreadsheet software be used.

5.8. Simulate the controlled process in Example 5.4 if $\varepsilon_t \sim N(0, 5)$, $g = 1$, and $\theta = 0.4$. Plot a single realization of the input and output.

5.9. Simulate the controlled process in Example 5.5 if $\delta_1 = 0.4$, $\delta_2 = 0.4$, $\theta = 0.4$, and $\varepsilon_t \sim N(0, 5)$. Plot a single realization of the input and the output.

5.10. Simulate the controlled process in Example 5.6 if $a = -0.4$, $c = 0.4$, and $\varepsilon_t \sim N(0, 5)$. Plot a single realization of the input and output.

5.11. Simulate the controlled process in Example 5.5 if $\delta_1 = 0.4$, $\delta_2 = 0.4$, $\theta = 0.4$, but the errors ε_t instead follow a uniform distribution between -15 and 15. Plot a single realization of the input and the output.

5.12. (Continuation of Problem 5.8). From a simulation run of at least 100 observations, estimate and plot the autocorrelation and partial autocorrelation function of the controlled process output. What type of process do the estimated autocorrelations indicate? Was this expected?

5.13. (Continuation of Problem 5.10). From a simulation run of at least 100 observations, estimate and plot the autocorrelation and partial autocorrelation function of the controlled process output. What type of process do the estimated autocorrelations indicate? Was this expected?

5.14. Find conditions on λ (Lagrange multiplier) that are necessary for the stability of the process

$$(1 - \mathcal{B})Y_t = \frac{w}{1 - \delta\mathcal{B}}X_{t-1} + \frac{1 - \theta\mathcal{B}}{1 - \mathcal{B}}\varepsilon_t$$

assuming that this process is adjusted using Clarke and Gawthrop's generalized minimum variance controller.

5.15. Consider the process in Example 5.4, where $\sigma_\varepsilon = 1$, $\theta = 0.7$, the off-target cost a equals 1, and the adjustment cost C_A equals 15 monetary units. Find the action limit of a deadband adjustment policy. What is the approximate AAI and ISD?

5.16. Repeat Problem 5.15 with $\theta = 0.6$, $\sigma_\varepsilon = 10$, $a = 2$, and $C_A = 25$.

5.17. Repeat Problem 5.15 with $\theta = 0.8$, $\sigma_\varepsilon = 5$, $a = 3$, and $C_A = 5$.

5.18. Using a spreadsheet, simulate the controlled process in Problem 5.15. Assume that $g = 1$.

5.19. How would one use Grubbs's adjustment rule for adjusting for process upsets that occur during manufacturing, as opposed to at startup, as originally discussed by Grubbs?

5.20. Show that the constants $k_t = 1/t$ provide the solution to problem (5.28) (Grubbs, 1954).

BIBLIOGRAPHY AND COMMENTS

Section 5.1 follows in general terms the classic presentation by Åström (1970) and Box and Jenkins (1976). These authors derived the expressions for minimum MSE control at about the same time in the mid-1960s. Palmor and

Shinnar (1979) provide an extensive analysis of the robustness of this type of controllers for a variety of processes. For more information on linear quadratic Gaussian (LQG) control, see Åström (1970). Deadband policies were first shown to be optimal for fixed setup costs by Box and Jenkins (1963) in a seminal paper. Readers interested in seeking more details on this type of procedures should consult the book by Box and Luceño (1997). The setup adjustment problem proposed by Grubbs (1954) evidently is of major importance in machining operations, and perhaps this is why it has not been discussed much in the process adjustment literature which has been heavily influenced by chemical and electrical engineering applications and problems. The paper, however, was reprinted in the *Journal of Quality Technology* in 1983. Del Castillo and Pan (2001) show the equivalence of Grubbs's solution, and Robbins and Monro (1951) celebrated stochastic approximation technique. Connections between Grubbs's problem and the Kalman filter are also made in that paper, and we discuss them in Appendix 8A.

CHAPTER 6

Discrete-Time PID Controllers

The basic proportional–integral–derivative (PID) controller was introduced at the end of Section 5.1. By far the most common type of industrial feedback controller, it is not based on any optimality criterion. However, several authors have noticed the robustness properties of PID controllers with respect to different disturbances and process dynamics. These are analyzed in this chapter. We first describe different ways to parameterize a PID controller. Then properties of the three terms that make up the controller (P, I, and D, respectively) are discussed in detail. The design and robustness of PI controllers (i.e., no derivative action) are discussed next. The focus of this chapter is on *discrete-time* PID controllers.

6.1 PARAMETERIZATION OF PID CONTROLLERS

There are several ways in which PID controllers have been parameterized in the literature. The most standard form of a PID controller used by many authors is the *positional* or *level form*:

$$X_t = K\left(Y_t + \frac{1}{T_I} \sum_{j=1}^{t} Y_j + T_D \nabla Y_t\right)$$

where K is the proportional (also called controller) gain, T_I the integral time, and T_D is called the derivative time. These are tuning constants that determine the performance of the controller. The practical value of a PID controller rests in good part on the ability that a process engineer has to manipulate these tuning parameters to achieve a desired performance. Thus PID controllers provide three "knobs" that a process engineer can "tweak."

218

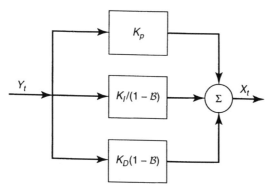

Figure 6.1 Parallel form of a discrete PID controller for the level form of the controllable factor.

The level form can also be written in what is called the *parallel form* of a PID controller:

$$X_t = \underbrace{K_p\, Y_t}_{P} + \underbrace{K_I \sum_{j=1}^{t} Y_j}_{I} + \underbrace{K_D \nabla Y_t}_{D}$$

where K_p, K_I, and K_D are tuning parameters. Here the three terms (P, I, and D) do not interact as in the standard form (Figure 6.1). Clearly, the relations between the standard and parallel forms are $K_p = K$, $K_I = K/T_I$, and $K_D = KT_D$.

A more convenient way of writing a PID controller for quality control applications is the *incremental* form, in which the adjustments ∇X_t are specified instead:

$$\nabla X_t = \underbrace{K_p\, \nabla Y_t}_{P} + \underbrace{K_I Y_t}_{I} + \underbrace{K_D\, \nabla^2 Y_t}_{D}$$

Notice that in incremental form, the term $K_I Y_t$ is *not* what authors usually refer to as the proportional (P) term but refers instead to the integral term. Similarly, the term $K_p \nabla Y_t$ is the proportional, not the derivative term. In other words, the P, I, and D terms usually relate to the positional or level form of the controller.

6.2 DISCRETE-TIME PROPORTIONAL CONTROLLERS

Consider first an off–on controller, also called a *bang-bang controller* in the control engineering literature (an example of this is a thermostat):

$$X_t = \begin{cases} X_{\max} & \text{if } Y_t > 0 \quad (T > Q_t) \\ X_{\min} & \text{if } Y_t \le 0 \quad (T \le Q_t) \end{cases}$$

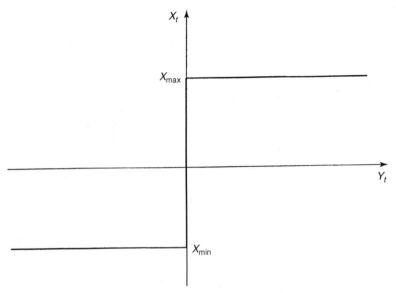

Figure 6.2 Bang-bang or off–on controller.

Figure 6.2 shows the input–output relation for such a controller. In this type of control rule, the level of the adjustable variable is always at one of its two extreme values, and in such a case it is said that the input is *saturated*.

Bang-bang control is easy to implement since there is no need to define tuning parameters. However, it usually results in excessive oscillation.[1] Because of this, a better alternative in general is to use a controller that manipulates X_t in a proportional way with respect to Y_t, as in Figure 6.3.

As long as the level of the controllable factor is in the linear region, the controller is

$$X_t = K_p Y_t + X_0$$

where X_0 is the value required by the controllable factor to keep the response on target (i.e., keep $Y_t = 0$). Usually, the origin is translated, so the value of X_t denotes deviation from what we call in the figure X_0. With this, a proportional controller is simply

$$X_t = K_p Y_t$$

It is readily seen that a bang-bang controller is a rather extreme case of a proportional controller where $K_p \to \infty$.

---

[1]A controller rule of the bang-bang form can be shown to be optimal for certain types of control problems in which the control action is constrained and the system obeys a deterministic linear difference equation (see Lewis, 1986, p. 254).

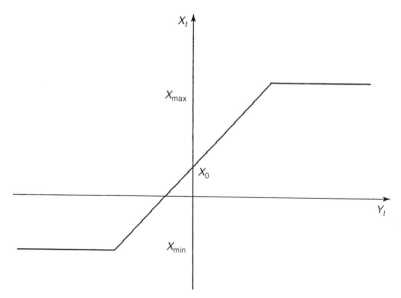

Figure 6.3　Proportional controller.

Static Analysis of a Proportional Controller

It is very instructive to study the behavior of a proportional (P) controller when there are no dynamics and no random variables in a process. Define, as before,

$$Y = T - Q$$

where we are dropping the time subscripts because this is a static process. Consider Figure 6.4 from which we have the following equations:

$$s = g(X + l) \tag{6.1}$$

$$Q = s + n = g(X + l) + n$$

$$Y = T - Q$$

$$X = K_p Y = K_p(T - Q) = K_p(T - s - n) \tag{6.2}$$

Substituting (6.2) into (6.1), we get

$$s = g\big(K_p(T - n - s) + l\big)$$

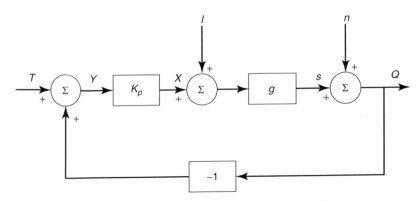

Figure 6.4 Static process controlled with a P controller.

Solving for s, we obtain

$$s = \frac{gK_p}{1 + gK_p}(T - n) + \frac{g}{1 + gK_p}l$$

where gK_p is called the *loop gain*. Suppose first that $n = 0$. Then if we can make $gK_p \to \infty$, we will get $s \to T$ regardless of the disturbance l, and then we will have $Q = s = T$, that is, an on-target process. This is achieved if we set or tune K_p to a very large number, which according to this static, deterministic analysis results in a "robust" control strategy. This explains in part why bang-bang controllers work sometimes in practice, since in these controllers, $K_p = \infty$.

Now suppose that $n(\neq 0)$ is a constant or load disturbance at the output. Then if $g K_p \to \infty$, we have that $s \to T - n$, but $Q = s + n \to T$, since the load disturbance n cancels. However, if n were a random disturbance, this cancellation will not occur (n_t and n_{t-1} do not cancel). Furthermore, when there are process dynamics, it can be seen that as K_p is increased, oscillations will occur up to a point when the stability of the process is lost. Figure 6.5 shows the trajectory of the deterministic process,

$$Q_t = \alpha + \frac{g}{1 - g\mathcal{B}}X_{t-1}$$

under the actions of the controller $X_t = K_p Y_t = K_p(T - Q_t)$. The values $T = 10$, $g = 0.5$, and $\alpha = 2.0$ were used. The value α is the offset that will result if no control action is exercised. Increasing values of K_p were tried. As K_p increases, the offset decreases, but before obtaining the target value of 10, oscillations appear and eventually make the system unstable. Thus *proportional control alone results in an offset (deviation from target) when subject to shifts or loads*. The offset can be seen as a consequence of our inability to

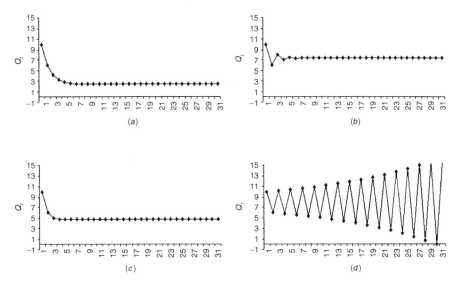

Figure 6.5 Effect of increasing K_p in a P controller applied to a deterministic first-order dynamical system, $T = 10$: (a) $K_p = 0.05$; (b) $K_p = 2.0$; (c) $K_p = 0.5$; (d) $K_p = 3.1$.

make K_p as large as we want it to be. For example, in the static analysis, if both n and l are zero but K_p is not very large, then

$$s = \frac{gK_p}{1 + gK_p} T < T$$

and therefore there will be an offset from the target for small gK_p.

6.3   DISCRETE INTEGRAL CONTROL

To eliminate the offset not compensated for by a pure proportional controller, integral action is added to the control equation, and we get a proportional–integral (PI) controller. A controller with integral action has the term

$$\text{integral action} = I_t = K_I \sum_{i=1}^{t} Y_i$$

added to the equation defining the level of the control factor, X_t. Thus a pure integral controller is

$$X_t = I_t = K_I \sum_{i=0}^{t} Y_i$$

or

$$\nabla X_t = K_I Y_t$$

Thus, as long as $Y_t > 0$ $(T > Q_t)$, X_t will increase. As long as $Y_t < 0$ $(T < Q_t)$, X_t will decrease. The long-run result of integral action is to make $Y_t \to 0$ (i.e., an elimination of the offset).

That an I controller always eliminates the steady-state offset[2] can be proved by contradiction (Åström and Häglund, 1995). Consider an I controller and suppose that there is a *constant*, perhaps zero, steady-state control variable level X_∞ (so $X_t = X_\infty$ for all t) that has resulted in a constant steady-state error $Y_\infty \neq 0$. If an integral controller was used, we have that

$$X_t = K_I \sum_{i=1}^{t} Y_\infty = \underbrace{t K_I Y_\infty}_{\text{not constant}}$$

Thus X_t is not constant. But if X_t is not constant in time, Y_∞ will not be constant in time, contrary to our assumption. Therefore, the only way out of this contradiction is that either the control factor was not really constant or that our assumption of a non-zero error steady-state must be false. Since we know the input is constant (because we can fix it), the steady-state error must be zero. Therefore, an integral controller eventually eliminates an offset or shift.

For SPC purposes, the important corollary of this is that *a controller with integral action will compensate and eliminate sudden shifts in the quality characteristic*.[3] Sudden shifts are perhaps the most common types of effects due to assignable causes of variation in industrial processes and are by far the most extensively studied in the process monitoring literature.

The effect of increasing the integral constant K_I in a PI controller can be illustrated numerically. The first-order dynamical system

$$Q_t = \alpha + \frac{g}{1 - g\mathcal{B}} X_{t-1}$$

is controlled by $X_t = K_p Y_t + K_I \sum_{i=1}^{t} Y_i$. The same values of $T = 10, \alpha = 2$, $g = 0.5$ were used as before and K_p was fixed at a value of 2.0. As K_I increases from a value of zero (pure P controller), we observe in Figure 6.6 how the steady-state offset is reduced more rapidly, although eventually, for large values of K_I, the stability of the process is lost, due to excessive oscillation.

---

[2] This is true for any controller with integral action, not only for pure I controllers.
[3] More precisely, as long as the controllable factor X_t is unconstrained, a controller with I action will compensate eventually against shifts of *any* magnitude, bringing the process back to target after a transient period that depends on the magnitude of the disturbance and on the value of K_I.

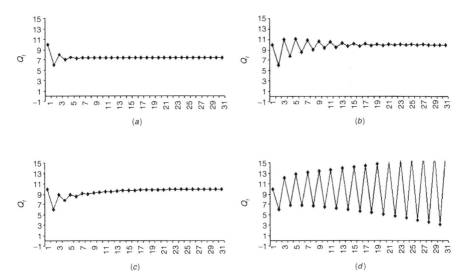

Figure 6.6 Effect of increasing K_I in a PI controller applied to a deterministic first-order dynamical system ($K_p = 2$), $T = 10$: (a) $K_I = 0.0$; (b) $K_I = 1.5$; (c) $K_I = 0.5$; (d) $K_I = 2.0$.

6.4   DISCRETE DERIVATIVE CONTROL

As shown in Section 6.3, a PI controller can bring the quality characteristic closer to target than a pure proportional controller can. However, in some applications even tighter control may be necessary. This is the rationale behind the addition of derivative control to a PI controller, forming what is called a PID controller. It should be pointed out, however, that in industrial PID controllers, derivative action is frequently not used (Åström and Häglund, 1995).

Derivative action can be thought of as a control action proportional to a predicted deviation from target. The idea is to predict where the response will be and anticipate it, since given the input–output delay, a PI controller may be late in correcting the process.

In Figure 6.7 it can be seen that $Y_{t+T_D} \approx Y_t + T_D \nabla Y_t$, where recall that $\nabla Y_t = Y_t - Y_{t-1}$ and where from the figure, T_D is evidently the distance along the time axis from t to $t + T_D$. Thus a controller of the form

$$X_t = K(Y_t + T_D \nabla Y_t)$$

has a *control action* proportional to a T_D-period-ahead prediction of Y (this is, in fact, a PD controller). This similar in spirit to what an MMSE controller does, but note that the prediction is rather crude, as can be seen from the figure. Intuitively:

- T_D too large $\Rightarrow$ prediction poor $\Rightarrow$ poor control.
- $T_D \rightarrow 0 \Rightarrow$ get P (pure proportional) control.

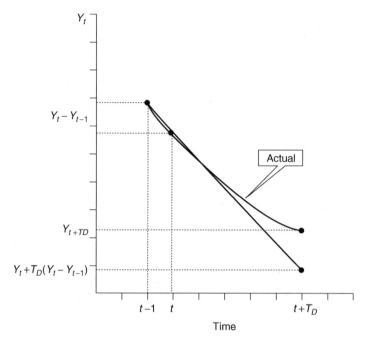

Figure 6.7   Derivative action as a prediction.

Derivative action is sometimes used for second-order dynamical systems that require very tight control.

6.5   PROPORTIONAL–INTEGRAL CONTROLLERS

We now look in more detail at PI controllers of the form

$$X_t = K_p Y_t + K_I \sum_{i=1}^{t} Y_i \quad \text{(position form)}$$

or

$$\nabla X_t = K_p \nabla Y_t + K_I Y_t \quad \text{(incremental form)}.$$

Therefore, in a PI controller the adjustment is a function only of the last two observed errors (deviations from target). Box and Luceño (1997) parametrized a PI controller as

$$\nabla X_t = -G(Y_t + P \nabla Y_t)$$

where the negative sign is due to their convention of using a negative sign (due to *negative* feedback) in the control equation, whereas we have assumed

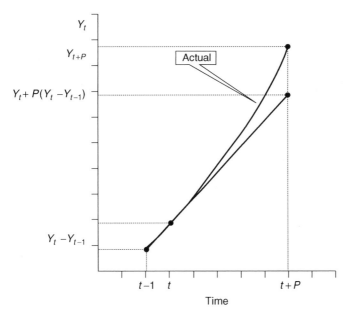

Figure 6.8 PI control action as a prediction.

so far that the minus sign was on the feedback loop (see Section 5.1). The relationship between the last two formulations is

$$G = -K_I \qquad P = \frac{K_p}{K_I}$$

The Box–Luceño formulation can be understood better from the graph in Figure 6.8. Thus we see that in a PI controller, the *adjustment* ∇X_t is proportional to a P-period-ahead prediction of the response (extrapolation beyond Y_t if $P > 0$). If $P < 0$, the adjustment is proportional to an interpolation between Y_t and Y_{t-1}. In both cases, $-G$ is the proportionality constant. This interpretation of PI controllers led Box and Luceño to propose their *Feedback adjustment charts.*

Example 6.1: Feedback Adjustment Charts. Consider the task of controlling the polishing of silicon wafers, as in semiconductor manufacturing. Here X is the back pressure applied to the wafers, Q the removal rate of silicon oxide with a target $T = 1.6$, and Y the deviation from the target. Consider applying the pure integral controller

$$\nabla X_t = -0.3 Y_t$$

Such a pure integral controller can be implemented in a graphical manner by setting a chart with two scales, a scale for the adjustments on the left and a

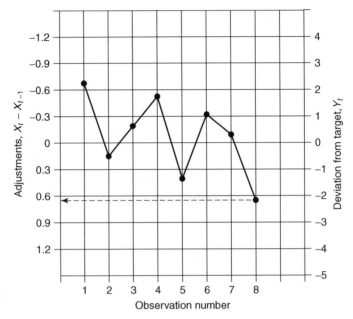

Figure 6.9 I feedback adjustment chart, Example 6.1. At the eighth observation, the deviation from target is -2.1. This corresponds to an adjustment of 0.63.

scale for the quality characteristic on the right. The scales in these feedback adjustment charts can be determined from the controller equation. In this example, to a value of $Y_t = 1$ corresponds a value of $\nabla X_t = -0.3$, and so on. An operator plots the values of Y_t on the chart and reads on the left-hand side the adjustment suggested by the I scheme (see Figure 6.9). If, instead, a PI controller such as

$$\nabla X_t = -0.3\,(Y_t + 0.25\nabla Y_t)$$

is used, this implies that we want to extrapolate the last two values of the deviations observed from the target 0.25 $(= P)$ time units ahead. To implement a feedback adjustment chart for a PI controller, it is necessary to add vertical dashed lines at each time instant plus the value of P; in this case a dashed line has been added every $\frac{1}{4}$ time unit ahead of each sampling time. To obtain the adjustment, the last two values are linearly extrapolated until they reach the vertical dashed lines. Then the adjustment is read on the left-hand side (see Figure 6.10). The charts in Figures 6.9 and 6.10 were called I and PI feedback adjustment charts by Box and Luceño (1997). These authors also propose rounded feedback adjustment charts that decrease the number of adjustments by rounding the adjustment to some easy-to-implement multiple of a basic unit. □

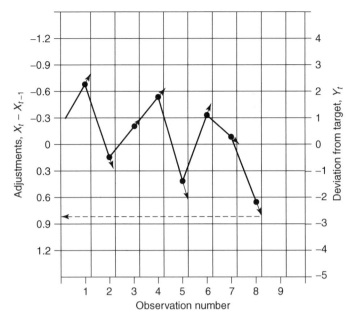

Figure 6.10 A PI feedback adjustment chart, Example 6.1. Since $Y_8 = -2.1$ and $Y_7 = 0.3$, we have that $\nabla X_8 = -0.3[-2.1 + 0.25(-2.1 - 0.3)] = 0.81$.

6.6 CONSTRAINED-INPUT PI CONTROLLERS

Using a PI controller, we can balance the input and output variances of a process using a method similar to that used by Clarke and Gawthrop's controller (Section 5.3). To achieve this goal, Box and Luceño (1997) considered a PI controller

$$\nabla X_t = - G(Y_t + P\nabla Y_{t-1}) = c_1 Y_t + c_2 Y_{t-1}$$

that controls a first-order system with IMA(1, 1) noise:

$$Y_t = \frac{g(1 - \delta)}{1 - \delta \mathcal{B}} X_{t-k} + \frac{1 - \theta B}{1 - \mathcal{B}} \varepsilon_t$$

If the delay k is one period, the closed-loop equation is

$$\left\{ 1 + [\delta + 1 + g(1 - \delta)c_1]\mathcal{B} + [\delta - g(1 - \delta)c_2]\mathcal{B}^2 \right\} Y_t$$
$$= (1 - \theta \mathcal{B})(1 - \delta \mathcal{B}) \varepsilon_t$$

which is an ARMA(2, 2) process with variance easy but messy to compute (See section 3.3). The variance of the adjustments is given by

$$\mathrm{Var}(\nabla X_t) = \left(c_1^2 + c_2^2 \right) \mathrm{Var}(Y_t) + 2c_1 c_2 \, \mathrm{Cov}(Y_t, Y_{t-1})$$

where the last term is a function of the lag 1 autocovariance of Y_t. Box and Luceño (1997) provided tables for choosing the values of G and P by numerically minimizing

$$\min_{G, P} \left\{ \frac{\mathrm{Var}(Y_t)}{\sigma_\varepsilon^2} + \alpha \frac{\mathrm{Var}(\nabla X_t)}{\sigma_\varepsilon^2} \right\} \qquad (6.3)$$

for several values of δ, $\lambda = 1 - \theta$, α, and for $k = 1$ and $k = 2$. From a table of possible PI controller designs, a user needs to choose the values of G and P that give a trade-off between $\mathrm{Var}(\nabla X_t)/\sigma_\varepsilon^2$ and $\mathrm{Var}(Y_t)/\sigma_\varepsilon^2$ that is considered acceptable.

Instead of using the tables in Box and Luceño (1997), a user could solve model (6.3) directly for different values of α. Del Castillo (2001b) provides a spreadsheet optimization model (see the file *PIOptimization.xls*) that will solve (6.3) for a simpler pure delay transfer function model ($\delta = 0$) but for the more complicated disturbance

$$N_t = \omega + N_{t-1} - \theta \varepsilon_{t-1} + \varepsilon_t \qquad |\theta| \leq 1$$

instead of the IMA(1, 1) noise model assumed by Box and Luceño. Here ω is a drift parameter that allows us to model unidirectional drift, an important consideration in manufacturing systems that wear off with time. In Chapter 7 we take a closer look at this type of disturbance. If in the transfer function we have that $\delta \neq 0$, the tables in Box and Luceño (1997) should be consulted instead.

Example 6.2. Suppose that we have fitted a first-order transfer function system with IMA(1, 1) noise and the parameters are $\theta = 0.4$ (so $\lambda = 1 - \theta = 0.6$), $k = 1$, $g = 1$, and $\delta = 0.0$. Using the *PIOptimization.xls* spreadsheet, the controller designs in Table 6.1 can be obtained by varying the relative weight α. Thus we see that the familiar trade-off curve between adjustment and output variance shown in Figure 5.11 is produced. Large reductions in the variance of the adjustments result in small increments in the variance of the quality characteristic.   □

Table 6.1 Constrained PI Controller Designs for Different Values of the Relative Weight α

α	G	P	$\mathrm{Var}(Y_t)/\sigma_\varepsilon^2$	$\mathrm{Var}(\nabla X_t)/\sigma_\varepsilon^2$
0	0.60	0.00	1.0	0.36
0.5	0.48	−0.10	1.03	0.20
5	0.28	−0.23	1.25	0.08
10	0.22	−0.25	1.41	0.05

Box and Luceño (1997) show that generic choices of $G = 0.3$ and either $P = -0.25$ or $P = 0$ provide a large reduction in Var(Y_t) for a moderate increase in Var(∇X_t). The case $P = 0$ gives a pure integral controller,

$$\nabla X_t = -GY_t$$

Note that in this case we have that

$$\text{Var}(\nabla X_t) = G^2 \text{Var}(Y_t)$$

or

$$\frac{\sigma_{\nabla X}}{\sigma_Y} = G$$

Thus the recommendation $G = 0.3$, $P = 0$, implies that the ratio of standard deviations must be kept at around 30%.

A self-tuning version of a discrete-time PI controller that accounts for a constraint in the variability of the adjustments is described by del Castillo (2000). Self-tuning controllers are described in Chapter 8.

6.7 INFLATION OF OUTPUT VARIANCE FOR A PROCESS THAT IS IN A STATE OF STATISTICAL CONTROL

Suppose that the process obeys Shewhart's model if left uncontrolled:

$$Y_t = \varepsilon_t.$$

Thus there is no offset and no autocorrelation or drift. Let us assume that we nevertheless apply a pure integral controller to the process by manipulating some controllable factor X:

$$\nabla X_t = -GY_t$$

Assume, further, that the system delay is one period and that this is a pure delay process (what Box and Luceño call a *responsive process*). Then the closed-loop output equation is

$$\nabla Y_t = -GY_t + \nabla \varepsilon_t$$

or

$$\left[1 - (G - 1)\mathcal{B}\right]Y_t = (1 - \mathcal{B})\varepsilon_t$$

which is an ARMA(1, 1) process with parameters $\phi = 1 - G$ and $\theta = 1$. From this we can compute the variance of the closed-loop Y_t:

$$\frac{\text{Var}(Y_t)}{\sigma_\varepsilon^2} = \frac{1 + \theta^2 - 2\phi\theta}{1 - \phi^2} = \frac{2}{2 - G}$$

So if, for example, $G = 0.3$ is used, this implies an inflation in standard deviation over the minimum possible we can achieve of

$$\frac{\sigma_Y}{\sigma_\varepsilon} = \sqrt{\frac{2}{2 - 0.3}} = 1.084.$$

Thus there is only a 8.4% increase in the output standard deviation. This is a small price to pay for the assurance that if the process does not obey Shewhart's model but instead has offsets or shifts in the mean of the quality characteristic, or experiences process dynamics and therefore autocorrelation, the I controller in use will compensate for them. This has been discussed by Box and Luceño (1997) and by del Castillo (2001a).

6.8   RELATION BETWEEN INTEGRAL CONTROL AND THE EWMA STATISTIC

A pure integral feedback controller is closely related to the exponentially weighted moving average (EWMA). Assume that the process under control can be described by

$$Y_t = gX_{t-1} + N_t$$

where N_t is any stochastic disturbance or noise. The corresponding discrete-integral controller for this process is

$$\nabla X_t = -GY_t = -G(N_t + gX_{t-1})$$

where the last equality follows immediately from the process description. This is equivalent to

$$X_t - (1 - Gg)X_{t-1} = -GN_t$$

or

$$X_t - \tau X_{t-1} = \frac{\tau - 1}{g}N_t$$

where $\tau = 1 - Gg$. Therefore,

$$X_t = \tau X_{t-1} + \frac{\tau - 1}{g}N_t$$

which is a difference equation with solution

$$X_t = -\frac{1}{g}\underbrace{(1-\tau)(N_t + \tau N_{t-1} + \tau^2 N_{t-2} + \cdots)}_{\text{EWMA of the disturbance}}$$
$$= -\frac{1}{g}\hat{N}_{t+1|t}$$

where we can think of the EWMA as providing a one-step-ahead forecast of the disturbance:

$$\hat{N}_{t+1|t} = (1-\tau)N_t + \tau\hat{N}_{t|t-1}$$

Thus a pure I controller is equivalent to setting the level of the controllable factor to cancel a one-step-ahead forecast of the disturbance, a forecast that is made with an EWMA statistic. Note (see Problem 6.11) that if the disturbance is IMA(1, 1), then setting $G = (1 - \theta)/g$ gives MMSE control.

PROBLEMS

In Problems 6.1 to 6.3 it is suggested that a spreadsheet software be used.

6.1. Simulate a P controller applied to a deterministic first-order dynamical process:

$$Q_t = \alpha + \frac{g}{1 - g\mathcal{B}}X_{t-1}$$

Extend your simulation so that the quality characteristic Q_t can only be observed with additive IMA(1, 1) noise with parameter $\theta = 0.4$. Using increasingly larger values of K_P, compare the trajectories of the deterministic and the stochastic first-order models for the case $g = 0.5, \alpha = 2$, $T = 10$ if the errors of the IMA(1, 1) are $N(0, 1)$ random variables.

6.2. Simulate a PI controller applied to the same deterministic and stochastic first-order process as in Problem 6.1. Keeping a constant value for K_P, use increasing values of K_I and compare the deterministic and stochastic models.

6.3. Repeat Problem 6.2 for the case of a PID controller. Keep K_P and K_I constant, but vary the K_D parameter and compare the deterministic behavior with that of the stochastic behavior.

6.4. Using the spreadsheet program *PIOptimization.xls*, find a table of PI controller designs for a responsive model with IMA(1, 1) noise as in

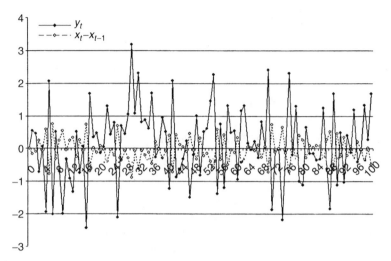

Figure 6.11 Simulated IMA(1, 1) process controlled with a PI adjustment scheme. The process experiences a sudden shift in the mean at time $t = 30$ (see Problem 6.7).

Example 6.2 with $\theta = 0.3$ and $g = 1$. Suppose that the drift $\delta = \sigma_\varepsilon r$, with $r = 1$. Vary the relative weight α from 0 to large values.

6.5. Using the spreadsheet program *PIOptimization.xls*, find a table of PI controller designs for a responsive model with IMA(1, 1) noise as in Example 6.2 with $\theta = 0.5$ and $g = 1.2$. Suppose that the drift $\delta = \sigma_\varepsilon r$, with $r = 0.5$. Vary the relative weight α from 0 to large values.

6.6. Repeat Problem 6.5 if there is no drift ($r = 0$) and if $\theta = 0.6$ and $g = 2$ instead.

6.7. *Monitoring a PI-controlled process*. Figure 6.11 indicates the observed deviations from target of a simulated process controlled with a PI controller with parameters $G = 0.3$ and $P = -0.25$. A sudden shift occurred at time $t = 30$. Notice how the controller brings the process back to target, so for SPC purposes, this makes detection of the shift harder, since there is only an exponentially decreasing signal that shows in the data. Apply a CUSUM chart for the data (given in the file *prob6-7.txt*) to determine if the process was in control in the Shewhart sense. Is the chart able to detect the shift?

6.8. Find the inflation in variance that results by applying an I controller to a process that is in a state of statistical control if the controller parameter G equals 0.1, 0.2, 0.3, 0.4, 0.5, and 0.9.

6.9. Consider the following 10 deviations from target $\{Y_t\}$ observed in an uncontrolled process: -4.9, -10.8, -6.1, 2.5, -18.3, -36.4, 14.4, 0.16, -8.3, and -8.2. Draw an I feedback adjustment chart applied to these deviations if the integral controller parameter is $G = 0.3$.

6.10. Repeat Problem 6.9 constructing instead a PI feedback adjustment chart for a controller with parameters $G = 0.3$ and $P = 0.25$.

6.11. Consider a pure gain process with unit delay and IMA(1, 1) noise. Show that a pure integral controller with $G = (1 - \theta)/g$ gives MMSE performance.

6.12. Consider a first-order dynamical system with noise:

$$Y_t = \frac{g(1 - \delta)}{1 - \delta \mathcal{B}} X_{t-1} + N_t$$

where the noise N_t is an IMA(1, 1) process. Show that a PI controller with $G = (1 - \theta)/g$ and $P = \delta/(1 - \delta)$ gives MMSE performance.

6.13. Derive explicit formulas for $\mathrm{Var}(Y_t)$ and $\mathrm{Cov}(Y_t, Y_{t-1})$ for the process in Section 6.6. Use these equations to find an expression for $\mathrm{Var}(\nabla X_t)$.

6.14. Show that if $w \neq 0$ in the disturbance $N_t = w + N_{t-1} - \theta \varepsilon_{t-1} + \varepsilon_t$ and a PI controller is applied, a nonzero bias control results (i.e., $E[Y_t] \neq 0$).

BIBLIOGRAPHY AND COMMENTS

The static analysis of Section 6.2 is taken from Åström and Hägglund (1995), a book with emphasis on tuning PID controllers based on physically observing the step response of a process, including the well-known Ziegler–Nichols tuning rules. Although used formally for the first time in ship steering control systems in the 1920s, the basic principles behind PID controllers date back much longer ago. See Lewis (1992) for a historical account. Feedback adjustment charts are treated in detail by Box and Luceño (1997) (and the references therein), which we follow closely in Sections 6.5 to 6.8. Tsung et al. (1998) provide further analysis on the robustness of PID controllers.

CHAPTER 7

EWMA Feedback Controllers and the Run-to-Run Control Problem

As noted in Section 6.8, there is a close relation between pure integral controllers and the exponentially weighted moving average (EWMA) statistic. Integral action eliminates offsets or shifts and provides robustness. For this reason, most control engineers will find integral action a desirable feature of almost any adjustment scheme. An I controller can be thought of implicitly providing EWMA forecasts of the disturbance affecting the process. In this chapter, feedback adjustment methods that make *explicit* use of the EWMA statistic are presented. EWMA controllers, developed for the semiconductor manufacturing industry, provide a simple, yet robust scheme against a variety of realistic process disturbances. However, EWMA controllers have application not only in semiconductor manufacturing but in any batch-oriented process where adjustments are necessary with every batch and where there is considerably drift in the quality characteristics (e.g., due to tool wear). Perhaps not surprisingly, it will be shown in this chapter that EWMA controllers *are* pure integral controllers or at least contain integral action.

In the particular area of semiconductor manufacturing, EWMA feedback controllers are used for compensating against disturbances that affect the batch-to-batch [or run-to-run (R2R)] variability in the quality characteristics of silicon wafers at a given fabrication step (Butler and Stefani, 1994; Sachs et al., 1995; del Castillo and Hurwitz, 1997). Figure 7.1 illustrates a simple run-to-run controller as applied to a semiconductor manufacturing step. The basic idea is that while the batch of wafers is processed, the manufacturing equipment regulates the quality characteristics, thanks to various built-in feedback controllers. However, due to aging effects and wear-out conditions, there is a need to retune the operating conditions (or *recipe*) under which the machine is operating from run to run. A problem in R2R environments is that in some processes, measuring the quality characteristics while the processing takes place (called *in-situ measurements*) is very difficult, if not impossible, under current technology. Thus each completed batch of wafers

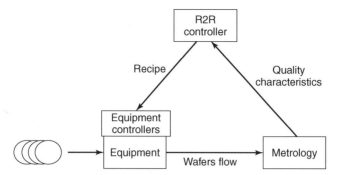

Figure 7.1   Basic structure of a run-to-run controller as applied in semiconductor manufacturing.

is taken to a metrology station for *ex-situ measurement*.[1] Then the measured quality characteristics Y_t will be entered into the R2R controller, which will suggest a new recipe X_t for use in the next batch or run. Currently, R2R controllers usually operate in laptop computers not connected with either metrology or equipment, but the technological trend is to have electronic communications for automatic downloading of the recipe to the equipment and automatic uploading of metrology data from metrology to the R2R controller (Moyne et al., 2000). In most cases, both the recipe and the quality measurements are vectors. In this chapter, however, we discuss single-input, single-output R2R controllers. Multivariate generalizations of EWMA controllers are discussed in Chapter 9.

Note that this manufacturing environment is very similar to that found in computer numerically controlled (CNC) machines in which the set points for a new machining operation eventually need to be modified due to wear-out of the cutting tool. Similar to some semiconductor manufacturing processes, in-situ measuring of machined dimensions in a CNC machine is quite difficult; usually, the part needs to be dismounted, cleaned, and measured in a separate machine (e.g., in a coordinate measurement machine).

Although EWMA-based controllers have been in use in the semiconductor industry since the early 1990s, analysis of this type of controllers is relatively recent. The main problem addressed in this chapter is how to select or *tune* the parameters of this type of controller to achieve adequate performance. As before performance is defined by the mean square error of the quality characteristic and by the variability of the adjustments. We will also be concerned with the stability conditions of the process controlled. Controllers based on a single EWMA equation are discussed in Section 7.1; controllers based on two EWMA equations used in tandem are discussed in Section 7.2.

[1]This may imply an input–output delay for control purposes. Such delays due to taking measurements in the product are common in chemical processes.

7.1 SINGLE-EWMA CONTROLLER

Let us assume that the input–output relation of a manufacturing process can be modeled by

$$Y_t = \alpha + \beta X_{t-1} + \varepsilon_t \tag{7.1}$$

where Y_t denotes deviation from target and $\{\varepsilon_t\}$ is a white noise process. This is a simple linear regression model where the parameters α (offset or intercept) and β (gain or slope) need to be estimated. The rationale behind such linear regression model is that within a small region in the space of the controllable factor (X_t), it can provide a linear approximation to the true underlying transfer function. Equation (7.1) implies that there are no process dynamics and that the effect of making a change in the controllable factor is fully observed in the next measurement of the quality characteristic. Such responsive processes are common in discrete-part manufacturing.

If the process parameters were known, a very simple MMSE controller will be equal to

$$X_t = -\frac{\alpha}{\beta}$$

which is essentially a feedforward controller of the type discussed in Example 5.1. In practice, the parameters are unknown and the noise is not white. Common practice in semiconductor manufacturing is to estimate the gains off-line using design of experiments techniques and fitting the linear regression model (7.1). This provides $\hat{\beta} = b$, an estimate of the gain and can provide an estimate of the intercept α. However, to protect against unmodeled drift and other process dynamics, EWMA controllers reestimate the offset term from run to run.[2] The differences

$$Y_t - bX_{t-1}$$

$$Y_{t-1} - bX_{t-2}$$

$$\vdots$$

provide estimates of α at each sampling point t. We can average these to compute a better estimate, or, even better, we can use an exponentially weighted average to give more weight to the more recent observations:

$$a_t = \lambda(Y_t - bX_{t-1}) + (1 - \lambda)a_{t-1}$$

$$= \lambda\left[Y_t - bX_{t-1} + (1 - \lambda)(Y_{t-1} - bX_{t-2}) + (1 - \lambda)^2(Y_{t-2} - bX_{t-3}) + \cdots\right] \tag{7.2}$$

---

[2] In the semiconductor manufacturing literature, EWMA controllers are sometimes called bias tuning controllers since α is referred to as the bias term.

Table 7.1   Single-EWMA Controller Computations, Example 7.1

t	Y_t	a_t	X_t	∇X_t
0	—	0.00	0.00	—
1	3.25	0.32	-0.22	-0.22
2	3.16	0.64	-0.43	-0.21
3	4.06	1.05	-0.70	-0.27
4	0.77	1.12	-0.75	-0.05
5	3.13	1.44	-0.96	-0.21
6	0.91	1.53	-1.02	-0.06
7	0.06	1.53	-1.02	0.00
8	0.92	1.62	-1.08	-0.06
9	0.09	1.63	-1.09	-0.01
10	0.17	1.65	-1.10	-0.01
$\vdots$	$\vdots$	$\vdots$	$\vdots$	$\vdots$

Thus the single-EWMA controller is given by equation (7.2) used in conjunction with

$$X_t = -\frac{a_t}{b} \tag{7.3}$$

Example 7.1. Suppose it has been found that $\hat{\beta} = b = 1.5$, $\alpha = 2$, and the value $\lambda = 0.1$ is going to be used to control a process. Table 7.1 illustrates the computations required to utilize a single-EWMA controller for these parameters for a set of hypothetical data. As can be seen from the table, the recursive EWMA equation was started from a value of 0.0, although other values could be used. To use this feedback adjustment method, the quality characteristic Y_t needs to be measured, then the EWMA statistic a_t is updated, and finally, the formula $X_t = -a_t/b$ gives the suggested value for the controllable factor in the next run or time instant t. If the adjustment is judged adequate by a process engineer, the control factor is adjusted by ∇X_t units.   □

The single-EWMA controller is equivalent to a pure I controller. To show this, consider the control equation

$$X_t = -\frac{a_t}{b}$$

with

$$a_t = \lambda(Y_t - bX_{t-1}) + (1 - \lambda)a_{t-1}$$
$$= \lambda[(Y_t + a_{t-1})] + (1 - \lambda)a_{t-1} = \lambda Y_t + a_{t-1}$$
$$= \lambda Y_t + \lambda Y_{t-1} + a_{t-2}$$
$$\vdots$$
$$a_t = \lambda \sum_{i=1}^{t} Y_i + a_0$$

where a_0 is the initial value of the recursive EWMA equation. Thus

$$X_t = X_0 - \frac{\lambda}{b} \sum_{i-1}^{t} Y_i$$

where $X_0 = -a_0/b$; that is, we obtain a pure integral controller with integration constant $K_I = -\lambda/b$. From Section 6.3 we know that as λ (and the integral constant K_I) increases, the offsets will be eliminated more rapidly, but there will be more oscillations. The only difference between single-EWMA controllers and the I controllers described in Chapter 6 is the effect produced by the initial value of the EWMA recursion (a_0) and the initial control factor level (X_0).

If X_0 is *not* set equal to $-a_0/b$ in both an EWMA controller and an integral controller (i.e, a controller such that $\nabla X_t = K_I Y_t$ with $K_I = -\lambda/b$), the transient behavior of the two controllers will differ although their asymptotic behavior will still be identical (see Problem 7.1).

7.1.1 Long-Run Mean Square Error: Deterministic Drift

Suppose that the deviations from target of the process obey the model

$$Y_t = \alpha + \beta X_{t-1} + \delta t + \varepsilon_t \tag{7.4}$$

where δ is the average drift per run if no control is exercised and the other parameters are as in equation (7.1). Suppose that the adjustment variance is *not* a concern, so we simply want to minimize the asymptotic mean square error[3]:

$$\text{AMSE}(Y_t) = \lim_{t \to \infty} E(Y_t^2)$$

The word *asymptotic*, which was not used before, recognizes the fact that for this type of controller there are transient effects due to initialization of the EWMA equations that need to be accounted separately from the steady-state behavior. Assuming that we are dealing with a process described by (7.4), the closed-loop equation is

$$Y_t = \alpha - \xi a_{t-1} + \delta t + \varepsilon_t \tag{7.5}$$

where we define $\xi = \beta/b$. Substituting this back into the EWMA equation,

---

[3]As noted before, this should be called more precisely the (asymptotic) mean square *deviation* from target, but we follow the terminology used in the time-series literature.

we get

$$a_t = \lambda(\alpha + \delta t + \varepsilon_t) + (1 - \lambda\xi)a_{t-1} \qquad (7.6)$$

To solve difference equations of the type involved in this controller, a state-space formulation (see Appendix 4A) is useful.[4] Define the state vector

$$\mathbf{z}'_t = (a_{t-1}, t)$$

With this definition we can represent equations (7.5) and (7.6) in state-space form as follows:

$$\mathbf{z}_{t+1} = A\mathbf{z}_t + \boldsymbol{\omega}_t$$
$$y_t = \mathbf{C}'\mathbf{z}_t + R_t$$

where

$$A = \begin{pmatrix} 1 - \lambda\xi & \lambda\delta \\ 0 & 1 \end{pmatrix} \qquad \boldsymbol{\omega}_t = \begin{pmatrix} \lambda(\alpha + \varepsilon_t) \\ 1 \end{pmatrix}$$

$\mathbf{C}' = (-\xi, \delta)$, and $R_t = \overset{\circ}{\alpha} + \varepsilon_t$. As described in Appendix 4A, solution of the state equation is given by (see also Åström and Wittenmark, 1997)

$$\mathbf{z}_t = A^t = \mathbf{z}_0 + \sum_{j=0}^{t-1} A^{t-j-1} \mathbf{w}_j \qquad (7.7)$$

The matrix A^t is computed by diagonalization, namely, $A^t = P\Gamma^t P^{-1}$, where Γ is a diagonal matrix with the eigenvalues of A, and P is a matrix with the corresponding eigenvectors. As it turns out, it is not too difficult to show that in our case we have

$$\Gamma = \begin{pmatrix} 1 - \lambda\xi & 0 \\ 0 & 1 \end{pmatrix}$$

Therefore, from the results in state-space theory (Appendix 4A), *the quality characteristic will be stable if and only if* $|1 - \lambda\xi| < 1$. If $\mathbf{z}'_0 = (0,0)$ is assumed, then from (7.7) we obtain

$$Y_t = -\xi \sum_{j=0}^{t-1} \left\{ \lambda(\alpha + \varepsilon_t)(1 - \lambda\xi)^{t-j-1} + \frac{\delta}{\xi}[1 - 1(1 - \lambda\xi)^{t-j-1}] \right\} + \delta t + \alpha + \varepsilon_t$$

---

[4]The advantages of using a state-space formulation will become clearer in the next section.

Assuming that the process is stable,

$$\lim_{t \to \infty} \sum_{j=0}^{t-1} (1 - \lambda\xi)^{t-j-1} = \frac{1}{\lambda\xi}$$

so we have that

$$Y_t = \frac{\delta}{\lambda\xi} - \xi\lambda \sum_{j=0}^{t-1} (1 - \lambda\xi)^{t-j-1} \varepsilon_j + \varepsilon_t$$

From this we obtain

$$\text{AMSE}(Y_t) = \lim_{t \to \infty} E[Y_t^2] = \sigma_\varepsilon^2 + \underbrace{\frac{\lambda\xi\sigma_\varepsilon^2}{2 - \lambda\xi}}_{\text{asymptotic Var}} + \underbrace{\left(\frac{\delta}{\lambda\xi}\right)^2}_{\text{(asymptotic bias or offset)}^2}$$

Thus, as can be seen, there is a trade-off when selecting λ. The asymptotic variance is minimized when $\lambda = 0$ (which implies no control), but this results in a diverging offset and a diverging quality characteristic that will trend with a slope equal to δ. Conversely, the asymptotic offset or bias is minimized by making $\lambda = 1$, but this can result in very large variances, particularly when ξ approaches 2 (in general, $\lambda = 2/\xi$ is a very poor choice from a variance and stability points of view). Thus, for $0 < \lambda < 1$, it is clear from the AMSE formula that the EWMA controller does not provide minimum variance (or MMSE) control. Note that even when there is no system-model mismatch ($\xi = 1$), there will be an offset as long as $\delta \neq 0$. We also point out that to minimize AMSE(Y_t), we need to know ξ, so this limits the practical applicability of this formula. However, it is possible to study how sensitive the AMSE is with respect to misidentification of β and different choices of λ.

To study the sensitivity of the AMSE formula, consider first what would happen if the process is in a state of statistical control; that is, suppose that $\delta = 0$ but we still apply the EWMA controller for adjustment purposes. Then the inflation in AMSE (or AVAR[5]) over the minimum possible variance σ_ε^2 is given by the ratio

$$\frac{\sigma_\varepsilon^2 + \lambda\xi\sigma_\varepsilon^2/(2 - \lambda\xi)}{\sigma_\varepsilon^2} = \frac{2}{2 - \lambda\xi}$$

which can be recognized as a generalization of the formula derived in Section 6.7 for I controllers for cases when $b \neq \beta$. Figure 7.2 illustrates the inflation

---

[5]Recall that for this case, if $\delta = 0$, then AMSE = AVAR.

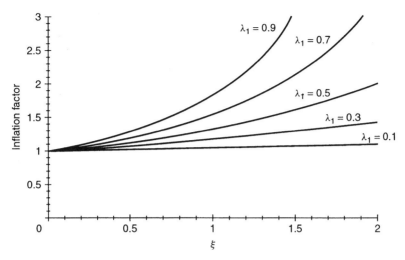

Figure 7.2 Inflation in AMSE versus the process-model mismatch parameter ξ ($= \beta/b$) when compensating a process in statistical control using a single EWMA controller.

factors for various values of ξ and λ. From the graph it seems that small values of λ should be preferred. The graph has the same corollary as in Section 6.7: The price to pay for "overadjusting" an in-control process using an EWMA chart (with a small λ weight) is small for the assurance that if there are disturbances other than white noise, such as shifts and drift, the controller will compensate for them. The graph provides stronger evidence for such a recommendation, as it indicates that the penalty is small for a wide range of estimation errors in β, the process gain. Interestingly, it seems that if there were no drift, overestimating the gain ($\xi = \beta/b < 1$) is better than underestimating the gain ($\xi = \beta/b > 1$).

As a second step in analyzing the sensitivity of the AMSE function, consider the graphs in Figure 7.3, where the AMSE is plotted against ξ for the case when there is a drift ($\delta = 0.1$) and $\sigma_\varepsilon = 1.0$. From the figures it can be seen that the effect of the drift drastically alters the AMSE function over the no-drift case of Figure 7.2. In none of these two cases, a value of $\xi = 1$ minimizes the AMSE(Y_t). More important, when there is drift, the AMSE function increases rapidly for extreme values of ξ.

7.1.2 Design Procedure for EWMA Controllers: Balancing the Adjustment and Output Variances

It is of interest to study the performance of any adjustment scheme in the presence of a variety of disturbances. For this end, let us consider the more

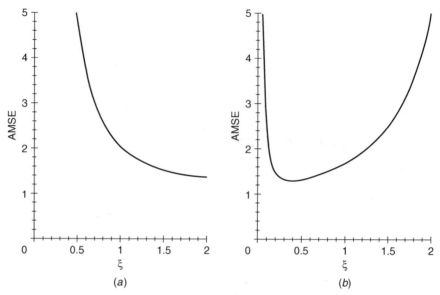

Figure 7.3   AMSE versus $\xi (= \beta/b)$ for $\delta = 0.1$, $\sigma_\varepsilon^2 = 1$: (a) $\lambda = 0.1$; (b) $\lambda = 0.8$.

general disturbance:

$$N_t = \delta + N_{t-1} - \theta \varepsilon_{t-1} + \varepsilon_t \qquad |\theta| \le 1 \tag{7.8}$$

where θ and δ are parameters. Model (7.8) was described in Chapter 6; it contains the following particular cases:

- Random walk (RW) process if $\theta = 0$ and $\delta = 0$
- Random walk with drift (RWD) if $\theta = 0$ and $\delta \ne 0$
- IMA(1, 1) if $\theta \ne 0$ ($|\theta| < 1$) and $\delta = 0$
- IMA(1, 1) with drift if $\theta \ne 0$; ($|\theta| < 1$) and $\delta \ne 0$
- Deterministic trend (DT) model if $\theta = 1$ and $\delta \ne 0$
- White noise if $\theta = 1$ and $\delta = 0$

If the process model is described by expression (7.1), it can be shown (see Problem 7.3) that

$$\text{AMSE}(Y_t) = \frac{1 + \theta^2 - 2(1 - \lambda \xi)\theta}{\lambda \xi (2 - \lambda \xi)} \sigma_\varepsilon^2 + \left(\frac{\delta}{\lambda \xi} \right)^2 \tag{7.9}$$

Similarly, it can be shown that the variance of the adjustments is

$$\text{Var}(\nabla X_t) = \frac{\lambda\left[1 + \theta^2 - 2(1 - \lambda\xi)\theta\right]}{b^2\xi(2 - \lambda\xi)}\sigma_\varepsilon^2 \tag{7.10}$$

With equations (7.9) and (7.10), the following optimization model can be solved for trading off the variability of the adjustments with the output mean square error:

$$\min_\lambda J = \frac{\text{AMSE}(Y_t)}{\sigma_\varepsilon^2} + \rho\frac{\text{Var}(\nabla X_t)}{\sigma_\varepsilon^2}$$

$$\text{subject to:} \quad |1 - \lambda\xi| < 1 \tag{7.11}$$

where ρ is a relative cost that needs to be defined by the process engineer. The spreadsheet *EWMAOptimization.xls* solves this model for given b, δ, θ, and ρ assuming that our gain estimate is perfect (i.e., assuming that $\xi = 1$ or that $\beta = b$).

Example 7.2. Suppose that a process is as described by model (7.1) under the disturbance given by (7.8). Suppose that the estimated process gain is $b = 1.2$, the estimated drift is $\delta = 1.5\sigma_\varepsilon$, and the IMA(1, 1) parameter is $\theta = 0.4$. To find EWMA designs by varying the relative weight ρ from 0 to 100, we use the *EWMAOptimization.xls* spreadsheet and assume that the gain estimate is correct. Table 7.2 shows the corresponding designs. As can be seen, a design that neglects the variability of the adjustments will use large values of λ. As the relative weight given to the variance of the adjustments increases, the EWMA weight λ decreases. □

Table 7.2   EWMA Designs, Example 7.2

ρ	λ^*	AMSE(Y_t)	Var(∇X_t)
0	1.0	3.41	0.8055
5	0.835	4.28	0.5121
10	0.729	5.24	0.3764
20	0.630	6.66	0.2761
50	0.512	9.58	0.1840
100	0.434	12.96	0.1362

7.2 DOUBLE-EWMA CONTROLLER[6]

Given that the single-EWMA controller (and by extension, I controllers) can exhibit considerable offset if there is a severe drift in the quality characteristic, researchers at Texas Instruments (see Butler and Stefani, 1994) extended the single-EWMA controller with a second EWMA equation that would compensate for the offset. Let the process be described by equation (7.4) (i.e., assume that the disturbance is a deterministic trend). The double EWMA controller is given by

$$X_t = \frac{-a_t - D_t}{b} \tag{7.12}$$

where

$$a_t = \lambda_1(Y_t - bX_{t-1}) + (1 - \lambda_1)a_{t-1} \qquad 0 < \lambda_1 \le 1 \tag{7.13}$$

$$D_t = \lambda_2(Y_t - bX_{t-1} - a_{t-1}) + (1 - \lambda_2)D_{t-1} \qquad 0 < \lambda_2 \le 1 \tag{7.14}$$

Example 7.3. Suppose that a double-EWMA controller with parameters $\lambda_1 = 0.3$ and $\lambda_2 = 0.05$ is applied to a process using a historical gain estimate of $b = 3$. Table 7.3 shows a worksheet of the first few computations required to obtain the level of the controllable factor (X_t) and the adjustment $(\nabla X_t = X_t - X_{t-1})$ required to control this process. □

The idea of a double-EWMA controller is that successive values of $Y_t - bX_{t-1} - a_{t-1}$ provide estimates of the amount the process has drifted (δt), which we can average using an EWMA. If $\xi = \beta/b = 1$ (i.e., if the gain β is known), it can be shown (del Castillo, 1999) that the quantity $a_t + D_t$ provides an asymptotically unbiased estimate (forecast) computed at run t of $\alpha + \delta(t + 1)$, the one-step-ahead level of the process if no control actions were exercised.[7]

By substituting equation (7.12) into equation (7.4), we obtain the closed-loop description of the output given by

$$Y_t = \alpha - \xi a_{t-1} - \xi D_{t-1} + \delta t + \varepsilon_t \tag{7.15}$$

Substituting this into equations (7.13) and (7.14), we get, respectively,

$$a_t = \lambda_1(\alpha \delta t + \varepsilon_t) + (1 - \lambda_1 \xi)a_{t-1} + \lambda_1(1 - \xi)D_{t-1} \tag{7.16}$$

$$D_t = \lambda_2(\alpha + \delta t + \varepsilon_t) - \lambda_2 \xi a_{t-1} + (1 - \lambda_2 \xi)D_{t-1} \tag{7.17}$$

---

[6]Section 7.2 contains relatively more advanced material and may be skipped on first reading.
[7]The reader familiar with forecasting methods will notice that equations (7.13) and (7.14) require computations similar to those in a double exponential smoothing forecasting method.

Table 7.3   Illustrative Computations Required for Application of the Double-EWMA Controller, Example 7.3

t	Y_t	a_t	D_t	X_t	∇X_t
0	—	0	0	0	—
1	3.0160	$0.3(3.0160 - 3(0)) +$ $0.7(0) = 0.9048$	$0.05(3.0160 - 3(0) - 0) +$ $0.95(0) = 0.1508$	$(0 - 0.9048 - 0.1508)/3$ $= -0.3518$	-0.3518
2	1.0227	$0.3(1.0227 - 3(-0.3518)) +$ $0.7(0.9048) = 1.2568$	$0.05(1.0227 - 3(-0.3518) - 0.9048) +$ $0.95(0.1508) = 0.2019$	$(0 - 1.2568 - 0.2019)/3$ $= -0.4863$	-0.1344
3	0.7861	1.5532	0.2412	-0.5982	-0.1119
4	1.3010	2.0159	0.3063	-0.7741	-0.1759
5	-0.2606	2.0296	0.2933	-0.7743	-0.0002
...	...	...	...	...	...

247

Equations (7.15) to (7.17) represent a system of three coupled difference equations. As in the single-EWMA case, to analyze the stability conditions of the system, define the state vector $z'_t = (a_{t-1}, D_{t-1}, t)$, where the apostrophe means transpose. With this setting we have the state-space representation

$$z_{t+1} = Az_t + w_t \qquad (7.18)$$
$$Y_t = C'z_t + R_t$$

where

$$A = \begin{pmatrix} 1 - \lambda_1 \xi \lambda_1 & (1 - \xi) & \lambda_1 \delta \\ -\lambda_2 \xi & 1 - \lambda_2 \xi & \lambda_2 \delta \\ 0 & 0 & 1 \end{pmatrix} \qquad w_t = \begin{pmatrix} \lambda_1(\alpha + \varepsilon_{t+1}) \\ \lambda_2(\alpha + \varepsilon_{t+1}) \\ 1 \end{pmatrix}$$

$C' = (-\xi, -\xi, 0)$, and $R_t = \alpha + \delta t + \varepsilon_t$. To solve the state equation with

$$z_t = A^{t-k_0} z_{k_0} + \sum_{j=k_0}^{t-1} A^{t-j-1} w_j \qquad (7.19)$$

[where k_0 is the point in time where initial conditions (z_{k_0}) are known] we need to compute the powers of the transition matrix A. From $A = P\Gamma P^{-1}$ we get

$$\Gamma = \begin{pmatrix} 1 - 0.5\xi(\lambda_1 + \lambda_2) + 0.5z & 0 & 0 \\ 0 & 1 - 0.5\xi(\lambda_1 + \lambda_2) - 0.5z & 0 \\ 0 & 0 & 1 \end{pmatrix} \quad (7.20)$$

where

$$z = \sqrt{\xi\left[\xi(\lambda_1 + \lambda_2)^2 - 4\lambda_1\lambda_2\right]} = e_1 - e_2 \qquad (7.21)$$

and e_1 and e_2 are the first two eigenvalues of A. Thus the process will be asymptotically stable if and only if

$$\left|1 - 0.5\xi(\lambda_1 + \lambda_2) + 0.5z\right| < 1$$
$$\left|1 - 0.5\xi(\lambda_1 + \lambda_2) - 0.5z\right| < 1$$

Note that if $\lambda_2 = 0$, the stability conditions reduce to $|1 - \lambda_1 \xi| < 1$, the condition for stability in a single-EWMA controller.

The stability conditions describe the equation of a circle on the (λ_1, λ_2) plane. The shaded region in Figure 7.4 correspond to weight values that

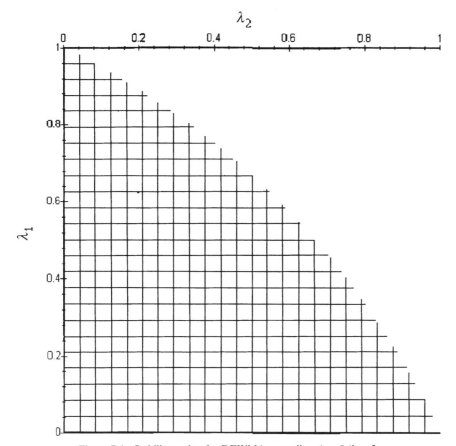

Figure 7.4 Stability region for DEWMA controller, $\xi = \beta/b = 2$.

satisfy the stability condition for $\xi = 2.0$. As ξ increases, the smaller the weights (λ_1, λ_2) should be to achieve stability. These stability conditions are also valid for a more general ARMA$(1, q)$ with drift disturbance (del Castillo, 2001a).

7.2.1   Design Procedure for DEWMA Controllers: Balancing Transient and Steady-State Performance

If the variability of the adjustments $[\text{Var}(\nabla X_t)]$ can be neglected, the following approach for tuning DEWMA controllers was suggested by del Castillo (1999, 2001a). The idea is that when choosing the weights (λ_1, λ_2) there is a trade-off between the magnitude of the transient effect and the long-run (asymptotic) variance. In the long run, the double-EWMA controller eliminates the offset and the process will on average be on target (i.e., for this controller, AMSE = AVAR). A measure of the severeness of the *expected* transient up to a specified run number m is given by the average mean square deviation:

$$\overline{\text{MSD}} = \frac{1}{m} \sum_{t=1}^{m} E[Q_t - T]^2 = \frac{1}{m} \sum_{t=1}^{m} E[Y_t]^2$$

Small values of the weights (close, but not equal to zero) make $\overline{\text{MSD}}$ very large, but they tend to minimize AVAR. Conversely, large values of the weights (close to 1) make AVAR large but minimize the transient effect. For the particular case of a deterministic trend drift, expressions for $\overline{\text{MSD}}$ and AVAR were provided in del Castillo (1999). With such expressions it is then possible to solve

$$\min_{\lambda_1, \lambda_2} w_1 \text{AVAR}(Y_t) + w_2 \overline{\text{MSD}} \tag{7.22}$$

$$\text{subject to:} \qquad 0 < \lambda_1 \le 1 \tag{7.23}$$

$$0 < \lambda_2 \le 1 \tag{7.24}$$

$$-1 < e_1 < 1 \tag{7.25}$$

$$-1 < e_2 < 1 \tag{7.26}$$

where e_1 and e_2 are the first two eigenvalues of the transition matrix A. The last two constraints guarantee a stable solution. The parameters (w_1, w_2) are set by the process engineer according to the nature of the process. If $w_1 = 0$ and $w_2 = 1$, an *all-bias solution* is obtained, that is, a solution that gives all weight to the transient bias objective. If, on the contrary, we set $w_1 = 1, w_2 = 0$, we will get an *all-variance solution* (we do not call this a *minimum variance* solution since it does not achieve the minimum possible variance). It was found by del Castillo (1999) that setting $w_1 = w_2 = 1$ avoids introducing additional parameters into the model and provides adequate performance in a variety of cases. Such a solution is called a *trade-off solution*.

The spreadsheet model *DEWMAOptimization.xls* solves the optimization problem above for the case when one can assume that our gain estimate b is indeed very good ($\xi = 1$). If $\xi = 1$, the expressions for AVAR and $\overline{\text{MSD}}$ (for a DT disturbance) are relatively simpler than if $\xi \ne 1$. The expressions are

$$\text{AVAR}(Y_t) = \frac{\sigma^2}{(\lambda_1 - \lambda_2)^2} \left(\frac{\lambda_1^2 \lambda_2 + \lambda_2(\lambda_1 - \lambda_2)^2}{2 - \lambda_2} \right.$$

$$\left. + \frac{\lambda_1(\lambda_1 - \lambda_2)^2 + \lambda_1 \lambda_2^2}{2 - \lambda_1} \right) + \sigma^2 \tag{7.27}$$

(since the asymptotic bias is zero, we have that AMSE = AVAR) and

$$\overline{MSD}_m(Y_t) = \frac{1}{m}\sum_{t=1}^{m} E[Y_t]^2$$

$$= \frac{(\delta - \alpha\lambda_2)^2\left[1 - (1 - \lambda_2)^{2(m+1)}\right]}{m(\lambda_1 - \lambda_2)^2\left[1 - (1 - \lambda_2)^2\right]}$$

$$+ \frac{2(\delta - \alpha\lambda_2)(\alpha\lambda_1 - \delta)\left[1 - (1 - \lambda_2)^{m+1}(1 - \lambda_1)^{m+1}\right]}{m(\lambda_1 - \lambda_2)^2\left[1 - (1 - \lambda_1)(1 - \lambda_2)\right]}$$

$$+ \frac{(\alpha\lambda_1 - \delta)^2\left[1 - (1 - \lambda_1)^{2(m+1)}\right]}{m(\lambda_1 - \lambda_2)^2\left[1 - (1 - \lambda_1)^2\right]}$$

Example 7.4: Semiconductor Manufacturing Process. Consider a chemical–mechanical planarization (CMP) process, a critical polishing process in the fabrications of semiconductors. A CMP machine consists of a rotating platen that contains a polishing pad lubricated with a slurry. Silicon wafers are mounted on vertical holders that push the wafers onto the platen. The combined effect of the pressure/rotation (mechanical effect) and the slurry (chemical effect) polishes the wafers to make them as flat as possible. Two quality characteristics are the nonuniformity (a standard deviation measure of the thickness of the wafer taken across several locations on the wafer) and the removal rate of silicon oxide. Evidently, what manufacturers like in this process is rapid polishing (maximize removal rate) while planarizing the wafers. Controllable factors include rotation speed and downforce among others. In this example we apply the DEWMA controller for a simulated CMP process where the response is removal rate (in Å/min) and the controllable factor is the back pressure on the wafer. From equipment data, the parameters $\alpha = 2$, $\delta = 0.1$, and $\sigma_\varepsilon^2 = 1$ were obtained.

Designing as if $\xi = 1$ (i.e., assuming that the gain estimate is "perfect"), the all-variance solution obtained with the *DEWMAOptimization.xls* spreadsheet model yields $\lambda_1 = 0.1265$ and $\lambda_2 = 0.001$. Figure 7.5 shows a simulation of the processing of 200 wafers in a CMP machine controlled by a DEWMA scheme with these parameters. As can be seen, there is a significant transient that would make such strategy not very useful in practice (the estimated MSE is 1.55).

Giving all weight to the transient bias term in the optimization model, a solution with $\lambda_1 = 0.05$ and $\lambda_2 = 1.00$ is obtained instead if the transient during the first $m = 200$ observations is considered.[8] While the transient is

_____

[8]The values $\lambda_1 = 1$ and $\lambda_2 = 0.05$ are also optimal.

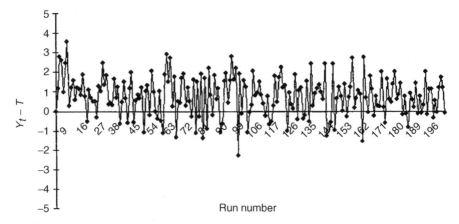

Figure 7.5   DEWMA controller simulation, CMP example, all-variance solution.

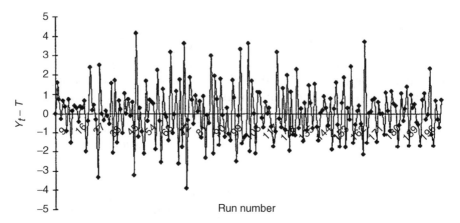

Figure 7.6   DEWMA controller simulation, CMP example, all-bias solution.

practically eliminated (see Figure 7.6), the resulting estimated variance is about twice the minimum possible variance (the estimated MSE is 1.95 in this simulation). This happens because the long-run variability is neglected with this solution.

Finally, consider the trade-off solution, in which equal weights are given to the two objectives in the optimization problem. The solution $\lambda_1 = 0.2982$, $\lambda_2 = 0.0036$ is obtained. This seems to indicate that *unless the drift rate is very strong, a single-EWMA controller is enough*. Figure 7.7 shows a 200-wafer simulation. The transient is rather moderate, and the overall variance is relatively close to the minimum possible variance (the estimated MSE is 1.1836).   □

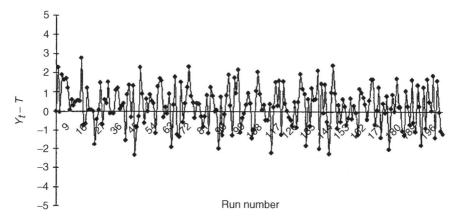

Figure 7.7 DEWMA controller simulation, CMP example, trade-off solution.

7.2.2 EWMA Controllers as Internal Model Controllers

The double-EWMA controller can be displayed in *internal model control* (IMC) form (Garcia and Morari, 1985). In fact (see Appendix 7A), any model-based controller can be written in IMC form. This form of feedback controller is particularly useful to investigate the stability conditions and the sensitivity of the controller to model misidentification. Figure 7.8 gives the DEWMA controller in IMC form. If $\lambda_2 = 0$, a single-EWMA controller in IMC form results.

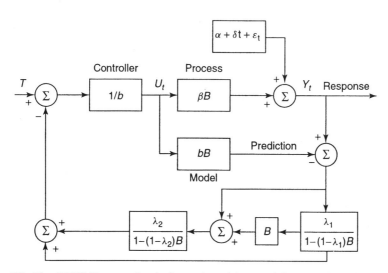

Figure 7.8 The DEWMA controller in internal model control form. A deterministic trend disturbance is assumed.

PROBLEMS

7.1. Use spreadsheet software to simulate a single-EWMA controller and an I controller applied to the same process with pure gain transfer function and IMA(1, 1) noise. Corroborate that if $X_0 = -a_0/b$ is used in both controllers, the behavior is identical; however, any other value of X_0 results in different transient behavior in each controller.

7.2. Show that if $\theta = 1$ and $\delta \neq 0$, disturbance (7.8) results in a deterministic drift disturbance.

7.3. Derive equation (7.9).

7.4. Derive equation (7.10).

7.5. *Designing EWMA controllers.* Using the *EWMAOptimization.xls* spreadsheet model, obtain the optimal EWMA controller solution for the following cases:
  (a) $\rho = 0$, $\delta = 0$, $\theta = 1$ and any value of b
  (b) $\rho = 0$, $\theta = 0.5$, $\delta = 0$ and any value of b
  (c) $\rho = 1$, $\theta = 0.5$, $\delta = 0$, $b = 1$
  (d) $\rho = 1$, $\theta = 0.5$, $\delta = 1$, $b = 0.5$

In each case, explain the solutions you get.

7.6. *Designing double-EWMA controllers.* Suppose in the deterministic trend process considered in Section 7.2 we have that $\alpha = 2$, the drift rate is $\delta = 0.5$, and $\sigma^2 = 1$. Using the *DEWMAOptimization.xls* spreadsheet, find the all-bias, all-variance, and trade-off controller designs for **(a)** $m = 100$; **(b)** $m = 50$; **(c)** $m = 20$.

7.7. Show that the inflation in variance for applying an EWMA controller to a process in a state of statistical control is given by $2/(2 - \lambda\xi)$, which agrees with the discussion in Section 6.7.

7.8. Suppose that an EWMA controller acts on an IMA(1, 1) process with no drift ($\delta = 0$) in equation (7.8). Show that the optimal weight from the AMSE point of view is given by $\lambda^* = (1 - \theta)/\xi$.

7.9. Show that $\lambda^* = (4\delta^2 - \sigma^2 - \sigma\sqrt{8\delta^2 + \sigma^2})/2(\delta^2 - \sigma^2)\xi$ is the AMSE-optimal weight if the disturbance is a random walk with drift δ.

7.10. Show that if the real root of $\sigma^2\xi^3\lambda^3 - \delta^2\xi^2\lambda^2 + 4\delta^2\xi\lambda - 4\delta^2 = 0$ in an EWMA controller satisfies the stability condition $|1 - \lambda\xi| < 1$, it is the

AMSE optimal weight for a deterministic trend disturbance (Smith and Boning, 1997).

7.11. *Using Matlab/Simulink to simulate a process from its block diagram.* Matlab and Simulink are two powerful software packages that allow us to simulate a process from its block diagram structure.

(a) Duplicate the DEWMA block diagram of Figure 7.8 and simulate the process in this environment.

(b) Modify your Simulink program so that it can also simulate an IMA(1, 1) with drift disturbance.

7.12. Write a MMSE controller applied to a generic Box–Jenkins transfer function with ARIMA noise in internal model control form. What is the equivalent filter used by the MMSE control law?

7.13. (Continuation of Problem 7.12). Show that if the disturbance is an IMA(1, 1), the filter in the internal model control form of the corresponding MMSE controller is an EWMA filter.

7.14. Write Clarke and Gawthrop's generalized minimum variance controller (see Chapter 5) in internal model control form. What is the equivalent filter used by this control law?

BIBLIOGRAPHY AND COMMENTS

For an overview on run-to-run or run-by-run control methods with an introduction to the complexities of semiconductor manufacturing, see Moyne et al. (2000). This book contains considerable discussion of the implementation of control methods in such sophisticated manufacturing processes. Edgar et al. (2000) describe control problems in semiconductor manufacturing. A paper by Sachs et al. (1995) originated much of the current work in the run-to-run control area. The internal model control principle is very popular among chemical engineers and was first proposed in a series of papers in the 1980s by Garcia and Morari (see, e.g., Garcia and Morari, 1985). An elementary introduction to these methods appears in MacGregor (1988) and Seborg et al. (1989). An advanced treatment of this subject, with emphasis on robust control methods, is Morari and Zafiriou (1989).

APPENDIX 7A: INTERNAL MODEL CONTROL

The internal model control (IMC) principle is very useful for designing model-based controllers and analyzing their robustness properties with respect to errors in the model parameters. In particular, obtaining the closed-

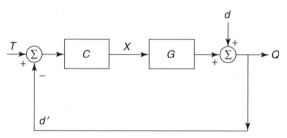

Figure 7.9   Conventional feedback controller.

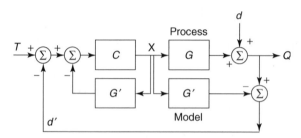

Figure 7.10   Adding a model (and its prediction) to a conventional feedback controller.

loop stability conditions of the system is relatively easy once a controller is set up in IMC form. Consider a conventional feedback controller as shown in Figure 7.9. Here all blocks are assumed to be functions (polynomials or ratios of polynomials) of the backshift operator $\mathcal{B}$. Suppose that there is a model G' of the process. To transform a conventional controller to internal model control form, add and subtract from the loop $\hat{Q} = XG'$, the prediction given by the model (the net effect of this will not alter the feedback loop), as shown in Figure 7.10. From Figure 7.10 we have that

$$X = C(T - d' - XG')$$

so

$$X = \frac{C(T - d')}{1 + CG'} \equiv G_C(T - d')$$

so to transform to IMC form we make the IMC controller gain, G_C, equal to

$$G_C = \frac{C}{1 + CG'}$$

Simplifying the block diagram, we get a structure such as that in Figure 7.11.

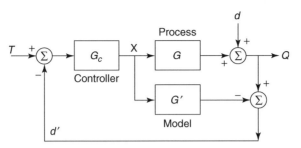

Figure 7.11 Resulting IMC controller after simplification.

From the IMC block diagram, we have

$$d' = XG + d - XG' \tag{7.28}$$

$$X = G_c(T - d') \tag{7.29}$$

Substituting (7.29) in (7.28), we get

$$d = G_c(T - d')(G - G') + d \tag{7.30}$$

Clearly, if the model were perfect and there were no disturbance, $d' = 0$, but this implies *no need for feedback* (i.e., the controller would become the feedforward law,

$$X = G_c T.$$

In the general case, solving (7.30), we get

$$d' = \frac{G_c T(G - G') + d}{1 + G_c(G - G')}$$

Plugging this into (7.29), after some algebra we obtain the closed-loop input equation

$$X = \frac{G_c}{1 + G_c(G - G')}(T - d)$$

Finally, since $Q = XG + d$, we obtain the closed-loop output equation

$$Q = \frac{G_c G}{1 + G_c(G - G')}(T - d) + d$$

Therefore, the stability of the system will be given by the roots of the characteristic equation:

$$1 + G_c(G - G') = 0$$

In internal model controllers, the controller G_c is selected as

$$G_c = \frac{1}{G'_s}F$$

where F is a filter (usually, but not always, an EWMA filter) and G'_s is the stable part of the model G' without the delay. In other words, the controller equals a filter times the inverse of that part of the model that can be inverted (this is done to avoid processes with noninvertible polynomials, also called non-minimum-phase processes, as discussed at the end of Section 5.1).

Example 7.5.. Consider a single-EWMA controller applied to a pure gain (no dynamics) system with unit delay. Here, $G' = b\mathcal{B}$, $F = \lambda/(1 - (1 - \lambda)\mathcal{B})$, and we make

$$G_c = \frac{\lambda/b}{1 - (1 - \lambda)\mathcal{B}}$$

since $G'_s = 1/b$, which is the model inverse (clearly invertible), excluding the delay. Therefore, the characteristic equation is given by

$$1 + \frac{\mathcal{B}\lambda(\xi - 1)}{1 - (1 - \lambda)\mathcal{B}} = 0$$

or, after simplifying,

$$1 + \mathcal{B}(\lambda\xi - 1) = 0$$

which has a root at

$$\frac{1}{1 - \lambda\xi}$$

Therefore, for stability we require that

$$\left| \frac{1}{1 - \lambda\xi} \right| > 1$$

which implies that $|1 - \lambda\xi| < 1$, the same condition as derived in Section 7.1.   □

CHAPTER 8

Recursive Estimation and Adaptive Control

In some practical applications it is necessary to estimate the parameters of a model on-line using a recursive, or sequential, estimation algorithm. Consider, for example, the case described in Chapter 7, semiconductor manufacturing. If one has to wait until at least 75 to 100 observations are collected in order to fit a transfer function model and use it for process control purposes, this might be too many observations, probably more than the total number of parts produced. This will be the case in expensive-part, short-run manufacturing environments.

To illustrate the concept of recursive estimation we wish to address in this chapter, suppose that we want to compute recursively the average of t observations, $y_1, y_2, \ldots, y_t$. As we know, the offline formula is:

$$\bar{y}_t = \frac{1}{t} \sum_{i=1}^{t} y_i$$

Now, consider the following algebraic manipulation:

$$\bar{y}_t = \frac{1}{t} \sum_{i=1}^{t} y_i = \frac{1}{t} \left(\sum_{i=1}^{t-1} y_i + y_t \right)$$

Since $\sum_{i=1}^{t-1} y_i$ equals $(t-1)\bar{y}_{t-1}$, we have that

$$\bar{y}_t = \frac{t-1}{t} \bar{y}_{t-1} + \frac{1}{t} y_t$$

which is our online version of the average. This is a recursive equation that gives equal weight to all past data. An exponentially weighted moving average (EWMA) equation will also provide a recursive formula for $\bar{y}_t$, but with

259

weights that decrease with the age of the data points, as discussed in Chapter 1. Provided we have some initial value for the recursion $(\bar{y}_0)$, we can compute $\bar{y}_1, \bar{y}_2, \ldots$, with such recursive formulas. Thus a recursive estimator is a function only of the previous estimate and the last value observed in the sample.

In this chapter we present first some general methods for recursive estimation. In Section 8.1 we present a recursive version of the least squares estimation principle; in Section 8.2 we provide some extensions to the basic recursive least squares algorithm. In this chapter we focus mainly on the recursive version of the least squares estimation algorithm. Appendix 8A reviews the closely related Kalman filter, which is a recursive algorithm for estimation of the state in a state-space model.

The idea of coupling a recursive estimation scheme with a feedback adjustment mechanism dates back to work by Kalman (1958). Asymptotic properties were studied by Åström and his collaborators in the early 1970s[1] from where successful applications began to emerge. The resulting self-tuning controllers are a particular type of adaptive controllers relevant for quality control because they allow control early in the process. In a manufacturing environment, large identification experiments (usually conducted in open loop) are too costly to undertake. There is a need for "learning as doing"; that is, use the parameter estimates obtained *online* via recursive least squares for control purposes. Self-tuning controllers are useful in processes with unknown but constant parameters. In the last sections of this chapter we provide a very succinct introduction to the topic of self-tuning control. In Section 8.3 we provide examples of the two main types of self-tuning controllers (direct and indirect). In Section 8.4 we look in more detail at when it is possible to estimate the parameters of the models, called the *identifiability problem* in the control engineering literature. Finally, in Section 8.5 we discuss the problem of closed-loop estimation, a problem that naturally arises in self-tuning and adaptive control.

8.1   RECURSIVE LEAST SQUARES ESTIMATION

In this section our goal is to find a recursive version of the least squares estimator of θ in the linear statistical model:

$$\mathbf{Y}_t = Z_t \theta_t + \boldsymbol{\varepsilon}_t$$

where $\mathbf{Y}_t$ is a $(t \times 1)$ vector containing the last t observations, θ_t is a $(n \times 1)$ vector of parameters, Z_t is a $(t \times n)$ matrix of regressors, and $\boldsymbol{\varepsilon}_t$ is $(t \times 1)$

---

[1]Adaptive control originated in the late 1950s from work by Kalman and Bellman, among others. See the "Bibliography and Comments" section for more specific references.

vector of errors. The recursive version of the least squares estimator seems to have been first proposed by Plackett (1950).

In the discussion below, it will be useful to use the following notation for matrix Z_t:

$$
Z_t = \begin{pmatrix} \mathbf{z}'_1 \\ \mathbf{z}'_2 \\ \vdots \\ \mathbf{z}'_t \end{pmatrix} = \begin{pmatrix} z_{11} & z_{12} & \cdots & z_{1n} \\ z_{21} & z_{22} & \cdots & z_{2n} \\ \vdots & \vdots & \vdots & \vdots \\ z_{t1} & z_{t2} & \cdots & z_{tn} \end{pmatrix}
$$

where each vector $\mathbf{z}_i$ contains the values of the regressors used when obtaining observation i.

The least squares criterion consists of minimizing the sum of squared errors, that is,

$$
\min_{\boldsymbol{\theta}_t} SS(\boldsymbol{\theta}_t) = \boldsymbol{\varepsilon}'_t \boldsymbol{\varepsilon}_t = (\mathbf{Y}_t - Z_t \boldsymbol{\theta})'(\mathbf{Y}_t - Z_t \boldsymbol{\theta})
$$

$$
= \frac{1}{t} \sum_{i=1}^{t} (y_i - \boldsymbol{\theta}'_i \mathbf{z}_i)^2
$$

As is well known (e.g., Myers, 1990), the minimizer of this function is the (offline) least squares estimator:

$$
\hat{\boldsymbol{\theta}}_t = (Z'_t Z_t)^{-1} Z'_t \mathbf{Y}_t
$$

$$
= \left[\sum_{i=1}^{t} \mathbf{z}_i \mathbf{z}'_i \right]^{-1} \sum_{i=1}^{t} \mathbf{z}_i y_i
$$

where y_i is the ith observation. We now derive a recursive version of this formula.

Define the $(n \times n)$ matrix

$$
P_t = \left(\sum_{i=1}^{t} \mathbf{z}_i \mathbf{z}'_i \right)^{-1} = (Z'_t Z_t)^{-1}
$$

The matrix P_t is called the precision matrix and equals $\text{Var}(\hat{\boldsymbol{\theta}}_t)/\sigma_\varepsilon^2$; thus it equals the variance–covariance matrix of the parameter estimates scaled by the variance of the errors. Then

$$
\hat{\boldsymbol{\theta}}_t = P_t \sum_{i=1}^{t} \mathbf{z}_i y_i = P_t \left(\sum_{i=1}^{t-1} \mathbf{z}_i y_i + \mathbf{z}_t y_t \right)
$$

following exactly the same steps as earlier when developing the recursive version of the sample mean. Now since

$$P_{t-1}^{-1} = \sum_{i=1}^{t-1} \mathbf{z}_i \mathbf{z}_i' = \sum_{i=1}^{t} \mathbf{z}_i \mathbf{z}_i' - \mathbf{z}_t \mathbf{z}_t' = P_t^{-1} - \mathbf{z}_t \mathbf{z}_t'$$

we get an updating formula for the inverse of the precision matrix:

$$P_t^{-1} = P_{t-1}^{-1} + \mathbf{z}_t \mathbf{z}_t'$$

We also have that

$$
\begin{aligned}
\hat{\boldsymbol{\theta}}_t &= P_t\Big(\hat{\boldsymbol{\theta}}_{t-1}\big(P_t^{-1} - \mathbf{z}_t \mathbf{z}_t'\big) + \mathbf{z}_t y_t\Big) \\
&= P_t\Big(\hat{\boldsymbol{\theta}}_{t-1} P_t^{-1} + \mathbf{z}_t\big(y_t - \mathbf{z}_t'\hat{\boldsymbol{\theta}}_{t-1}\big)\Big) \\
&= \hat{\boldsymbol{\theta}}_{t-1} + P_t \mathbf{z}_t\big(y_t - \mathbf{z}_t'\hat{\boldsymbol{\theta}}_{t-1}\big) \qquad\qquad (8.3)
\end{aligned}
$$

Equations (8.2) and (8.3) give the recursive least squares (RLS) algorithm. At each t, we observe scalar y_t and vector $\mathbf{z}_t$, update P_t^{-1} using (8.2), compute $P_t = (P_t^{-1})^{-1}$, and update $\hat{\boldsymbol{\theta}}_t$ using (8.3).

If the number of parameters n is large, inverting a matrix at each iteration might be too time consuming for some applications, particularly in electrical engineering, and this was a major concern in the early days of adaptive control. With modern computing power, inverting P_t^{-1} at each iteration is not very time consuming. For quality control applications, the dimension n will be rather small, and in fact, recomputing the offline estimate (8.1) at every sampling instance t is a possibility. However, this alternative is not easy to implement manually if several time periods are considered and for large number of observations will require considerably redundant computational effort. If we were to have a very short time between samples available to make the computations, it would be desirable to avoid the matrix inversions required in either (8.1) or (8.3). To avoid the matrix inversion, we can make use of the *matrix inversion lemma* (Ljung and Söderström, 1983).

Lemma 1: Matrix Inversion Lemma. Let A, B, C, and D be matrices of conformable dimensions. Then

$$[A + BCD]^{-1} = A^{-1} - A^{-1}B(DA^{-1}B + C^{-1})^{-1}DA^{-1}$$

Proof. We use the basic fact that if $X^{-1} = W$, then $XW = I$, and the converse also holds. Thus

$$(A + BCD)\left[A^{-1} - A^{-1}B(DA^{-1}B + C^{-1})^{-1}DA^{-1}\right]$$

$$= I + BCDA^{-1} - B(C^{-1} + DA^{-1}B)^{-1}DA^{-1}$$

$$- BCDA^{-1}B(C^{-1} + DA^{-1}B)^{-1}DA^{-1}$$

$$= I + BCDA^{-1} - (B + BCDA^{-1}B)(C^{-1} + DA^{-1}B)^{-1}DA^{-1}$$

$$= I + BCDA^{-1} - BC(C^{-1} + DA^{-1}B)(C^{-1} + DA^{-1}B)^{-1}DA^{-1}$$

$$= I + BCDA^{-1} - BCDA^{-1} = I \qquad \square$$

The significance of the lemma is that if we want to compute the inverse of the matrix on the left of the equality sign and have available the inverse of A, if we can rearrange the B, C, and D matrices in such a way that the term $D A^{-1}B + C^{-1}$ becomes a scalar, we can avoid in this way computing any further inverses.

Applying this result to P_t yields

$$P_t = (Z_t'Z_t)^{-1} = \left(\underbrace{Z_{t-1}'Z_{t-1}}_{A} + \underbrace{z_t}_{B} \ \underbrace{z_t'}_{D}\right)^{-1}$$

where we have defined the A, B, and D matrices as above and let $C = 1$ (a scalar). Thus using the lemma, we have that

$$P_t = (Z_{t-1}'Z_{t-1})^{-1}$$

$$- (Z_{t-1}'Z_{t-1})^{-1}z_t\underbrace{\left[z_t'(Z_{t-1}'Z_{t-1})^{-1}z_t + 1\right]}_{\text{a scalar}}^{-1}z_t'(Z_{t-1}'Z_{t-1})^{-1}$$

which can be written as

$$P_t = P_{t-1} - \frac{P_{t-1}z_tz_t'P_{t-1}}{1 + z_t'P_{t-1}z_t}$$

A simpler formulation is obtained if we introduce the $n \times 1$ *Kalman gain vector*, defined as

$$\mathbf{K}_t \equiv P_t \mathbf{z}_t = P_{t-1} \mathbf{z}_t - \frac{P_{t-1} \mathbf{z}_t \mathbf{z}_t' P_{t-1} \mathbf{z}_t}{1 + \mathbf{z}_t' P_{t-1} \mathbf{z}_t}$$

$$= \frac{P_{t-1} \mathbf{z}_t}{1 + \mathbf{z}_t' P_{t-1} \mathbf{z}_t}.$$

With this, we can write the RLS algorithm as follows. For $t = 1, 2, \ldots$, compute:

$$1.\ \mathbf{K}_t = \frac{P_{t-1} \mathbf{z}_t}{1 + \mathbf{z}_t' P_{t-1} \mathbf{z}_t}$$

$$2.\ P_t = (I - \mathbf{K}_t \mathbf{z}_t') P_{t-1}$$

$$3.\ \hat{\boldsymbol{\theta}}_t = \hat{\boldsymbol{\theta}}_{t-1} + \mathbf{K}_t \left(y_t - \mathbf{z}_t' \hat{\boldsymbol{\theta}}_{t-1} \right)$$

Remarks

- P_t is invertible after time $t_0 = \dim(\boldsymbol{\theta}) = n$ (i.e., until when we have at least equal number of observations than parameters to estimate). Thus we could start the recursions at $t_0 = n$ with

$$P_n = \left(\sum_{i=1}^{n} \mathbf{z}_i \mathbf{z}_i' \right)^{-1}$$

$$\hat{\boldsymbol{\theta}}_n = P_n \sum_{i=1}^{n} \mathbf{z}_i y_i$$

Alternatively, we can give a Bayesian interpretation to $\hat{\boldsymbol{\theta}}_0$ and P_0. In this interpretation, $\hat{\boldsymbol{\theta}}_0$ are the a priori parameter estimates and P_0 is their associated variance–covariance matrix (times a constant). In the absence of any prior data, it is a common recommendation to set all elements of $\hat{\boldsymbol{\theta}}_0$ equal to zero and set $P_0 = \rho I$, where ρ is on the order of 100 or 1000. This constitutes a "diffuse" prior in Bayesian terminology. If more confidence exists a priori on some other initial parameter estimates, ρ should be decreased accordingly.[2]

- A well-known statistical result (see Myers and Milton, 1991) is that if $\varepsilon_t \sim (0, \sigma_\varepsilon^2)$ and the sequence of errors are uncorrelated, the ordinary least squares estimate (8.1) is BLUE (best linear unbiased estimator).

---

[2] This interpretation of RLS is closely related to Kalman filters, see Appendix 8A.

This means that among all estimators that are *linear* functions of the observations and that provide unbiased estimates, the least squares estimator has minimum variance (note that here "best" means minimum variance, not minimum MSE). If, in addition, the errors are normally distributed, then the least squares estimator (8.1) is UMVUE (uniformly minimum variance unbiased estimator). This stronger result indicates that among the class of *all* estimators that yield unbiased estimates, the LS estimator provides minimum variance. This *does not* imply that the least squares estimates, even under normality, are best in the mean square error (MSE) sense. As frequently happens in statistics, sacrificing the unbiasedness property can result in a lower MSE.

Example 8.1. Consider an AR(2) time-series process:

$$Y_t = \phi_1 Y_{t-1} + \phi_2 Y_{t-2} + \varepsilon_t$$

where $\phi_1 = 0.5$, $\phi_2 = -0.5$, and $\varepsilon_t \sim U(-2,2)$. In this case, the regressor's vector is

$$\mathbf{z}'_t = (Y_{t-1}, Y_{t-2})$$

and the parameter vector is

$$\mathbf{\theta}' = (\phi_1, \phi_2)$$

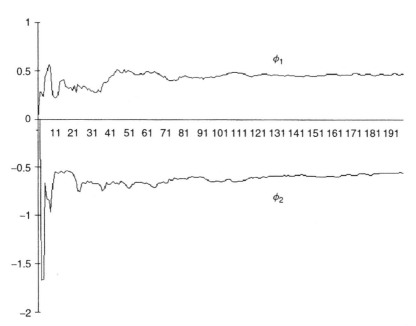

Figure 8.1 RLS algorithm applied to a simulated AR(2) process ($\phi_1 = 0.5$, $\phi_2 = -0.5$). $\hat{\mathbf{\theta}}'_0 = (0, 0)$ and $P_0 = 100I$ were used.

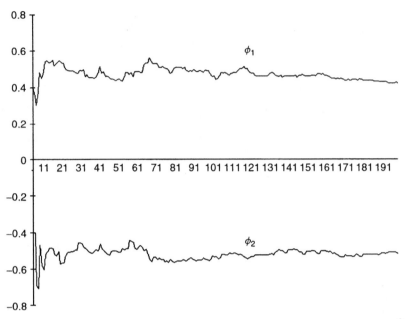

Figure 8.2 RLS algorithm applied to a simulated AR(2) process ($\phi_1 = 0.5$, $\phi_2 = -0.5$). $\hat{\boldsymbol{\theta}}'_0 = (0.4, -0.4)$ and $P_0 = 0.1I$ were used.

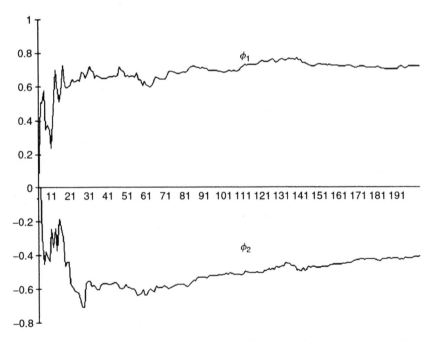

Figure 8.3 RLS algorithm applied to a simulated AR(2) process ($\phi_1 = 0.5$, $\phi_2 = -0.5$) when the error terms are not white. $\hat{\boldsymbol{\theta}}'_0 = (0,0)$ and $P_0 = 100I$ were used.

so the model can be written as $Y_t = z_t'\theta + \varepsilon_t$. Figure 8.1 shows the parameter estimates obtained by applying the RLS algorithm to a simulation of this process. The initial values $\hat{\theta}_0' = (0,0)$ and $P_0 = 100\ I$ were used. As can be seen, after a short transient, the parameter estimates quickly converge to their true values. Figure 8.2 shows the trajectory of the parameter estimates of a simulation of the same process as above when the RLS algorithm is started from the relatively better initial values $\hat{\theta}_0' = (0.4, -0.4)$ and $P_0 = 0.1\ I$. The transient behavior is much more moderate given the better initial estimates and the smaller precision matrix coefficients. Finally, Figure 8.3 shows the trajectories of the parameter estimates for the same process as before when the errors follow a $U(-1, 2)$ distribution (i.e., when the errors do not constitute a white noise sequence). Initial values of $\hat{\theta}_0' = (0,0)$ and $P_0 = 100\ I$ were used. As can be seen, the parameter estimates do not converge to their true values, there is bias in both estimates.   □

8.2 EXTENSIONS

As shown in Example 8.1, the recursive least squares algorithm provides biased estimates when the errors v_t in vector $\mathbf{v}_t$ in

$$Y_t = z_t'\theta_t + \mathbf{v}_t$$

are not white. An important instance when this can happen is in time-series models where there is a moving average (MA) polynomial term. For example, consider the ARMAX model

$$A(\mathcal{B})Y_t = B(\mathcal{B})X_t + C(\mathcal{B})\varepsilon_t$$

where $\{\varepsilon_t\}$ is a white noise sequence. This can be written as

$$Y_t = -\big[A(\mathcal{B}) - 1\big]Y_t + B(\mathcal{B})X_t + \underbrace{C(\mathcal{B})\varepsilon_t}_{v_t}$$

where evidently, v_t is not white noise when $C(\mathcal{B}) \neq 1$. As mentioned in Section 3.7, ARIMA (or ARMAX) models with an MA term are nonlinear in the parameters, and therefore the ordinary least squares formula (8.1) cannot be applied. Thus recourse of direct minimization of the sum of squares or maximum likelihood function is necessary. In recursive estimation, a simple way around this dilemma is to use the *extended least squares* (ELS) *algorithm*. We explain this method for an ARMAX model.

Write the ARMAX model in difference equation form:

$$Y_t + a_1 Y_{t-1} + \cdots + a_{n_a} Y_{t-n_a}$$
$$= b_1 X_{t-1} + b_2 X_{t-2} + \cdots + b_{n_b} X_{t-n_b} + \varepsilon_t + c_1 \varepsilon_{t-1} + \cdots + c_{n_c} \varepsilon_{t-n_c}$$

and define

$$\boldsymbol{\theta}' = \left(a_1, a_2, \ldots, a_{n_a}; b_1, \ldots, b_{n_b}; c_1, \ldots, c_{n_c} \right)$$
$$\mathbf{z}_t' = \left(-Y_{t-1}, -Y_{t-2}, \ldots, -Y_{t-n_a}; X_{t-1}, \ldots, X_{t-n_b}; \varepsilon_{t-1}, \ldots, \varepsilon_{t-n_c} \right)$$

The model can then be written as

$$Y_t = \mathbf{z}_t' \boldsymbol{\theta} + \varepsilon_t$$

which appears to be a linear statistical model, as required for RLS. However, this is not so since the error terms in the regressors vector are not directly observable. A simple remedy to this is to estimate the errors from the residuals as

$$\hat{\varepsilon}_t = e_t = Y_t - \hat{\boldsymbol{\theta}}_t' \mathbf{z}_t$$

and use the RLS equations with the ε_t's substituted by the residuals.

Example 8.2. Consider a first-order transfer function with ARMA(1, 1) noise:

$$\left(1 - \phi \mathcal{B} \right) Y_t = g X_{t-1} + \left(1 - \theta \mathcal{B} \right) \varepsilon_t$$

which can be written as

$$Y_t = \phi Y_{t-1} + g X_{t-1} - \theta \varepsilon_{t-1} + \varepsilon_t$$

Define

$$\mathbf{z}_t' = \left(Y_{t-1}, X_{t-1}, \hat{\varepsilon}_{t-1} \right)$$
$$\hat{\boldsymbol{\theta}}_t' = \left(\hat{\phi}_t, \hat{g}_t, \hat{\theta}_t \right)$$

where using $\hat{\varepsilon}_{t-1} = e_{t-1} = Y_{t-1} - \hat{\boldsymbol{\theta}}_{t-1}' \mathbf{z}_{t-1}$, we can apply RLS. Figure 8.4 shows the trajectories of the parameter estimates produced by RLS applied to a simulation to this process with $\phi = 0.5$, $g = 1.0$, $\theta = -0.4$, and $\varepsilon_t \sim U(-1,1)$. Initial values $\hat{\boldsymbol{\theta}}_0 = (0,0)$ and $P_0 = 00I$ were used. In this case, the input X_t was a white noise process with each $X_t \sim U(-0.5, 0.5)$ (i.e., the process was simulated in open loop without the action of any control rule).

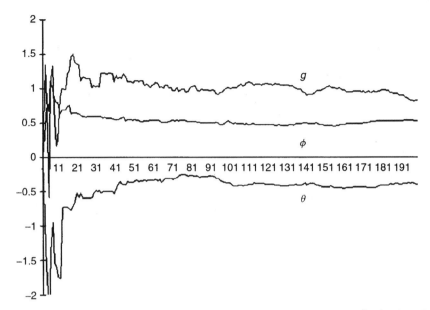

Figure 8.4 RLS algorithm applied to a first-order transfer function with ARMA(1, 1) noise. The input (X_t) is a white noise sequence. Here $\phi = 0.5$, $g = 1$, and $\theta = -0.4..$ $\hat{\theta}_0 = (0, 0, 0)$ and $P_0 = 100I$ were used.

Consider what would happen if a MMSE is applied while the RLS algorithm estimates the parameters of the model. The MMSE controller for this process is given by

$$X_t = -\frac{\phi - \theta}{g} Y_t$$

or, using the true values of the parameters, $X_t = -0.9Y_t$, a proportional (P) controller. From Figure 8.5 it can be seen that the parameter estimates do not converge to their true values; that is, biased estimates are obtained. In general, a constant feedback rule such as the MMSE rule used here introduces a linear dependency among the columns of the Z_t matrix, which makes the estimates unreliable in the recursive version of the LS algorithm. More details about closed-loop estimation are given in Section 8.5.

Consider next the case where the same process as before is run in open loop (no control law) and when X_t equals to a constant. Figures 8.6 and 8.7 show the trajectories of the parameter estimates resulting from applying RLS to the process when $X_t = 1$ and $X_t = 0$, respectively. As can be seen, all estimates converge to their true values for $X_t = 1$, although in general such an input signal is not recommended. Had the dynamics been more compli-

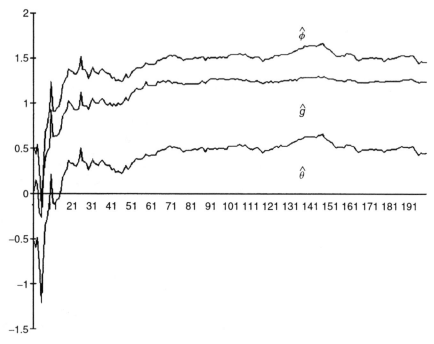

Figure 8.5   RLS algorithm applied to a first-order transfer function with ARMA(1, 1) noise. The input (X_t) is a MMSE controller $(X_t = -0.9Y_t)$. Here, $\phi = 0.5$, $g = 1$, and $\theta = -0.4$. $\hat{\theta}'_0 = (0, 0, 0)$ and $P_0 = 100I$ were used.

cated, "identifiability" of the parameters could be lost. As will be seen in Section 8.4, the regressor vectors $\{\mathbf{z}_t\}$ must vary enough to "excite" the process in such a way that we can gather information about its parameters. The case $X_t = 0$ illustrates this point. Here no information can be obtained from the parameters of the transfer function (the estimate of g is always zero), although unbiased parameter estimates are obtained for the ARMA(1, 1) parameters.   □

Lack of Excitation

A necessary condition for the convergence of recursive estimators is that the vector of regressors must vary or *excite* the process enough so that we obtain enough information about it Åström and Wittenmark, 1989). For example, if $X_t = 0$ for all t, no information about the transfer function parameters is gathered (see Example 8.2). A useful definition to study this phenomenon is the following: The sequence of $n \times 1$ vectors $\{\mathbf{z}_t\}$ is said to be *persistently*

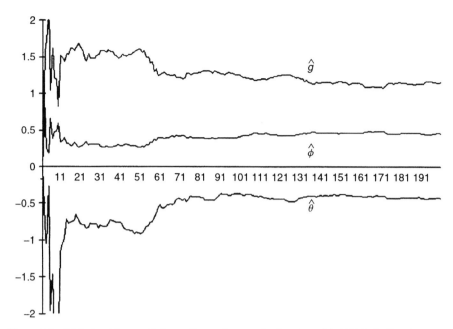

Figure 8.6 RLS algorithm applied to a first-order transfer function with ARMA(1, 1) noise. The input (X_t) was always set to 1. Here $\phi = 0.5$, $g = 1$, and $\theta = -0.4$. $\hat{\boldsymbol{\theta}}'_0 = (0, 0, 0)$ and $P_0 = 100I$ were used.

exciting of order n if the $n \times n$ matrix of long-run regressor's covariances

$$\lim_{N \to \infty} \frac{1}{N} \sum_{t=1}^{N} \mathbf{z}_t \mathbf{z}'_t = \lim_{N \to \infty} \frac{1}{N} P_N^{-1}$$

is positive definite and hence invertible.

Recall that if the model is linear in the parameters, then

$$\text{var}(\hat{\boldsymbol{\theta}}) = \sigma_\varepsilon^2 (Z'_t Z_t)^{-1} = \sigma_\varepsilon^2 P_t$$

Thus, in the long run, the variance of the parameter estimates is bounded if $(Z'_t Z_t)^{-1}$ exists as $t \to \infty$. Thus a necessary condition for *consistent* estimates is persistent excitation of order n.

Lack of excitation can occur if the elements of the regressor vector do not vary enough. If this occurs, the precision matrix P_t will eventually become ill-conditioned. This means that the point estimates will be increasingly unreliable, and eventually this causes a burst in the parameter estimates known in the adaptive control literature as *estimation windup* (see Shah and Cluett, 1991). The condition of the precision matrix can be monitored by

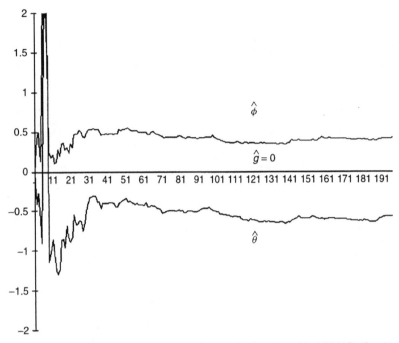

Figure 8.7 RLS algorithm applied to a first-order transfer function with ARMA(1, 1) noise. The input (X_t) was always set to zero. As before, $\phi = 0.5$, $g = 1$, and $\theta = -0.4$. $\hat{\boldsymbol{\theta}}_0' = (0, 0, 0)$ and $P_0 = 100I$ were used.

checking $\mathrm{tr}(P_t)$, the trace of P_t. Lack of excitation occurs when there is little new information in $\mathbf{z}_t$ or when estimation is carried on in closed-loop operation under the actions of a constant linear controller. Lack of excitation problems are discussed further in Section 8.5 after we show first the essentials of adaptive controllers, which are based on recursive least squares.

8.3 SELF-TUNING AND ADAPTIVE SCHEMES FOR QUALITY CONTROL

The central idea of a self-tuning controller (see Figure 8.8) is that a recursive estimator provides parameter estimates to a (given) controller that uses the parameters as if they were the true parameters. That is, the uncertainty associated with the estimates is neglected. This is called the *certainty equivalence principle* and the resulting controllers are sometimes called *certainty*

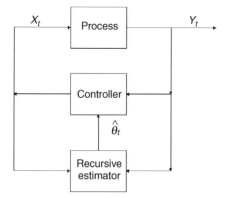

Figure 8.8 Basic structure of a self-tuning controller.

equivalence controllers.[3] Adaptive controllers assume, in general, that the parameters of a process change with time so that the controller must vary (adapt) its own parameters accordingly. A self-tuning controller is a particular instance of an adaptive controller in which the process parameters are constant but unknown.

A controller is said to have the *self-tuning property* if it converges to the one that could be designed had the process model be known a priori. This implies that if a controller is self-tuning, its long-run performance coincides with the performance that would be obtained if the process parameters were known perfectly.

8.3.1 Direct and Indirect Controllers

Indirect self-tuning (ST) *controllers* (called *explicit ST controllers* by some authors) have a recursive estimator that estimates the parameters of the *process*. Then these parameters are used by the controller. In contrast, in a *direct ST controller* (also called an *implicit ST controllers*), the parameters of the *controller* are estimated directly by the recursive scheme. The key is to find an equivalent model for the process so that its parameters are directly in correspondence with the controller's parameters. Reparametrization of this type usually reduces the number of parameters to estimate. A couple of examples will illustrate the differences between direct and indirect ST controllers.

Example 8.3: Indirect Self-Tuning Controller. Consider again a first-order transfer function with ARMA(1, 1) noise:

$$(1 - \phi \mathcal{B})Y_t = gX_{t-1} + (1 - \theta \mathcal{B})\varepsilon_t$$

---

[3]This does not imply that the certainty equivalence strategy is always optimal. In fact, this principle holds only in a reduced number of cases, as discussed in Appendix 8B.

or

$$Y_t = \phi Y_{t-1} + gX_{t-1} + (1 - \theta \mathcal{B})\varepsilon_t$$

We use Extended Least Squares with $\mathbf{z}'_t = (Y_{t-1}, X_{t-1}, -\hat{\varepsilon}_{t-1})$ and $\hat{\boldsymbol{\theta}}'_t = (\hat{\phi}_t, \hat{g}_t, \hat{\theta}_t)$, where $\hat{\varepsilon}_{t-1} = e_{t-1} = Y_{t-1} - \hat{\boldsymbol{\theta}}'_{t-1}\mathbf{z}_{t-1}$. Suppose that we want an MMSE controller. The MMSE controller is

$$X_t = -\frac{\phi - \theta}{g}Y_t$$

and the corresponding MMSE ST controller is simply

$$X_t = -\frac{\hat{\phi}_t - \hat{\theta}_t}{\hat{g}_t}Y_t$$

Figures 8.9 to 8.11 show the parameter estimates, outputs, and inputs, respectively, of a simulation of this self-tuning controller and process for the case $g = 1$, $\phi = 0.5$, and $\theta = -0.4$. Two aspects are worth pointing out. First, note that the estimates of g and ϕ converge rapidly to their true values. However, the estimate of θ is clearly varying much more than the others. This happens because θ is parameter associated with an estimated,

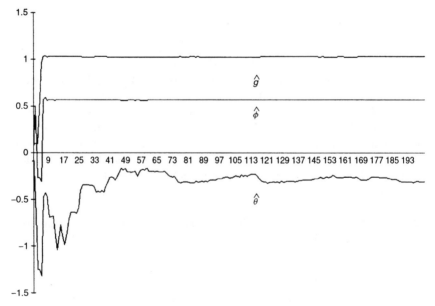

Figure 8.9 Parameter estimates used by the indirect ST controller. The true values are $\phi = 0.5$, $g = 1$, and $\theta = -0.4$. $\hat{\boldsymbol{\theta}}'_0 = (0, 0.1, 0)$ and $P_0 = 100I$ were used.

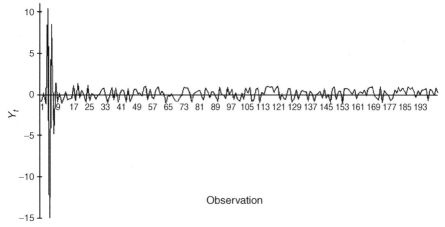

Figure 8.10 Resulting values of the quality characteristic while using the indirect ST controller in Example 8.3.

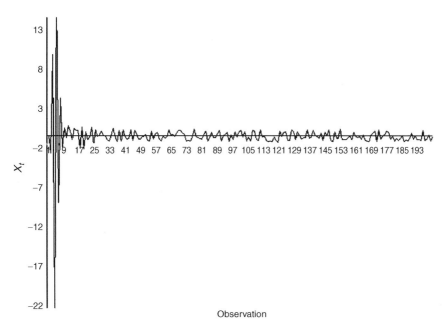

Figure 8.11 Values of the controllable factor suggested by the indirect ST controller in Example 8.3.

not with a directly observable, regressor ($\hat{\varepsilon}_{t-1}$). Second, note that there is a short but rather severe transient in all three graphs (parameter estimates, X_t and Y_t). This occurs because initially, the precision matrix indicates large uncertainties. To reduce the uncertainties, the controller first needs to probe the process by making drastic changes in the controllable factor. This quickly results in more precise parameter estimates and better control. The two features of the behavior of a ST controller, namely, probing and control, are usually referred to in the control literature as the *dual control effect*, (see Appendix 8B).   □

Example 8.4: Direct Self-Tuning Controller. Consider the same process as in Example 8.3, and suppose again that we seek MMSE control. The MMSE controller

$$X_t = -\frac{\phi - \theta}{g}Y_t = -LY_t$$

has only one parameter, the proportional gain. To estimate such gain, consider the process

$$Y_t = X_{t-1} + LY_{t-1} + \varepsilon_t$$

It is easy to see that this process also has as its MMSE controller the equation

$$X_t = -LY_t$$

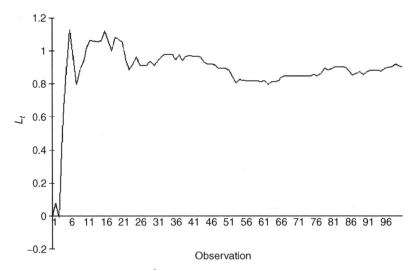

Figure 8.12 Parameter estimates ($\hat{L}_t$) used by the direct ST controller. The true value of L is 0.9. $\hat{\theta}'_0 = (1, 0)$ and $P_0 = 100I$ were used.

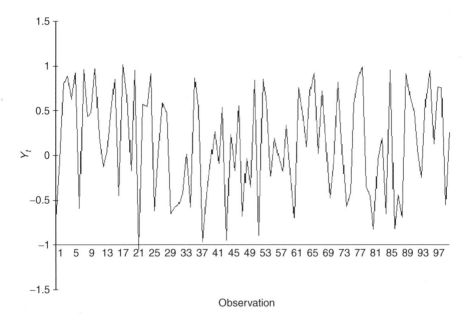

Figure 8.13 Resulting values of the quality characteristic when using the direct ST controller of Example 8.4.

Therefore, we can estimate L using RLS with

$$\hat{\boldsymbol{\theta}}'_t = \left(1, \hat{L}_t\right)$$

$$\mathbf{z}'_t = \left(X_{t-1}, Y_{t-1}\right)$$

so that the process model can be written as $Y_t = \mathbf{z}'_t \boldsymbol{\theta} + \varepsilon_t$. Figure 8.12 shows the trajectories of the parameter estimates for the same values of ϕ, θ, and white noise as before for a simulation of the controlled process. Figures 8.13 and 8.14 give the quality characteristics and values of the controllable factors, respectively, for the same simulation. As can be seen from this example, it is evident that in the direct ST controller the transient behavior is much more moderate than in the indirect self-tuning controller, a consequence of only one parameter being estimated.   □

Note from our earlier discussion that if a controller has the self-tuning property, this does not say anything about the transient behavior of the process controlled.[4] As shown in the examples above, the transient behavior is typically investigated through simulation. For the simple type of processes

---

[4]In particular, a minimum variance self-tuning controller is not optimal for any finite horizon performance criterion, such as $J_1 = E[Y_t^2]$ (for finite t) or $J_2 = E[1/n\sum_{t=1}^{N}Y_t^2]$ (see Åström and Wittenmark, 1973).

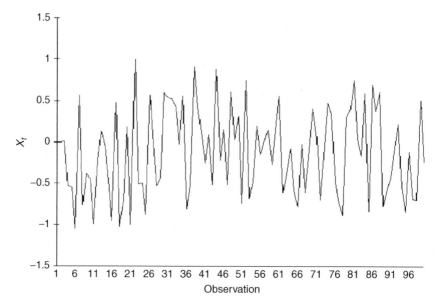

Figure 8.13 Values of the controllable factor suggested by the direct ST controller of Example 8.4.

with which we are concerned, the transient is usually not very prolonged, although the initial fluctuations can be very large if the P_0 matrix has large entries. This is particularly true of indirect self-tuning controllers.

Self-tuning controllers other than minimum variance are possible, of course. For example, the Clarke–Gawthrop controller discussed in Chapter 5 has a straightforward self-tuning implementation (Clarke and Gawthrop, 1975). Self-tuning PID controllers have also been proposed. See the "Bibliography and Comments" section for some references on these topics.

8.3.2 Stability and Convergence

Results about the stability and convergence of self-tuning and adaptive controllers, in general, are sparse and very technical, so we provide only a few remarks here that may guide interested readers who wish to get a deeper understanding from the literature. Convergence can be observed in the parameter estimates (convergence to their true values), in the performance measure (converging to the performance achieved by a known parameter controller), or in the controller (convergence to the controller we could obtain if the parameters were known). Stability analysis is very difficult because there is a nonlinear relation between the estimator and the controller. The remarks depend on whether we are talking about an indirect or a direct self-tuning controller:

- To ensure convergence of the parameter estimates in an indirect self-tuning controller, it is necessary that the model structure be correct, that the process input ($\{X_t\}$) be persistently exciting, and that the closed-loop system be stable.
- In an MMSE direct self-tuning controller, if the model parameter estimates converge, the corresponding regulator will actually converge to the minimum MMSE controller. This was shown by Åström and Wittenmark (1973).
- Parameter estimates convergence depends mainly on the recursive estimator in use. The ODE (ordinary differential equation) approach developed by Ljung (1977) is a very general method used for proving convergence in recursive algorithms. The idea is to associate a system of deterministic differential equations to the stochastic difference equations that the recursive estimation algorithm defines for θ and P_t. Then the trajectories of these vectors and their convergence points are studied using differential equations theory. This study may involve the simulation of the differential equations to demonstrate the convergence properties.
- Stability analysis of the indirect self-tuning controller is difficult because the controller parameters depend on model parameters.
- Stability can be proved for direct self-tuning controllers in the ideal case that there are no disturbances Åström and Wittenmark, 1989; Narendra and Annaswamy, 1989).

A general practical recommendation before utilizing a self-tuning controller is that the controller be simulated for a wide variety of possible model processes and disturbances before implementation. An instance of such a sensitivity analysis can be found in del Castillo and Hurwitz (1997). Several safety devices may be needed to ensure successful implementation. This is particularly true with respect to abrupt, unpredictable disturbances, which one may wish to detect as soon as they occur. These issues are discussed briefly next.

8.3.3 Monitoring Adaptive Controllers

The bulk of the literature on integration of SPC and EPC methods has centered around the idea of combining control charts for process monitoring and nonadaptive feedback control methods for process adjustment. As described in Chapter 1, the rational is that detecting unpredictable process upsets and generating signals (alarms) allows us to improve the process by correcting the underlying cause of the upsets, even if the feedback mechanism compensates some variables to bring the process back to a more desirable state.

Adaptive controllers are designed to be able to track process changes and modify their internal parameters according to some goal or objective. Åström and Wittenmark (1989) indicate that the parameter changes that these controllers can compensate for take place gradually. In other words, there is still a need to detect abrupt changes in an adaptive controller. Häglund and Åström (2000) mention how in electrical engineering and robotics applications an adaptive controller must be provided with a supervisory shell to take care of operating conditions for which the adaptive controller was not designed. However, the same can be said of any *automatic process controller*. Industrial PID controllers, for instance, are implemented with a variety of additional control logic and safety rules in case the controller encounters abnormal conditions. Some of these safety considerations are described by Åström and Häglund (1995) and resemble SPC charts.

In contrast with complete automatic control, the focus of this book has been on adjustment of the quality characteristics of a process where quality engineers or operators make final adjustment decisions. Self-tuning controllers can be useful in this environment to control a process rapidly after startup. Monitoring devices are also needed to supervise the performance of these control methods, since they would be very helpful for an operator or engineer wishing to detect assignable causes whose removal can lead to improving the process.

8.4 IDENTIFIABILITY ISSUES

Recall that

$$\text{Var}\big(\hat{\boldsymbol{\theta}}_N\big) = \sigma^2 P_N$$

where $P_N = (\sum_{t=1}^{N} \mathbf{z}_t \mathbf{z}_t')^{-1}$. For $\text{Var}(\boldsymbol{\theta}_N)$ to have finite entries, $\sum_{t=1}^{N} \mathbf{z}_t \mathbf{z}_t'$ must be positive definite and therefore invertible. To achieve this, the regressor vector must vary "enough." Recall that $\{\mathbf{z}_t\}$ is *persistently exciting of order n* if

$$\lim_{N \to \infty} \frac{1}{N} \sum_{t=1}^{N} \mathbf{z}_t \mathbf{z}_t' = \lim_{N \to \infty} \frac{1}{N} P_N^{-1}$$

is nonsingular. It follows that a necessary condition for the mean square convergence of the recursive least squares estimator is that the sequence $\{\mathbf{z}_t\}$ is persistently exciting of order n, where n is the number of parameters to estimate.

Example 8.5. Suppose that the true process description is

$$Y_t = aY_{t-1} + bX_{t-1} + \varepsilon_t$$

which we use as a model with $\mathbf{z}'_t = (Y_{t-1}, X_{t-1})$ and $\boldsymbol{\theta}'_t = (a_t, b_t)$. Prior to observing $\{Y_t\}, \{X_t\}$ we have that

$$E[\mathbf{z}_t\, \mathbf{z}'_t] = \begin{pmatrix} E[Y_{t-1}^2] & E[X_{t-1}Y_{t-1}] \\ E[X_{t-1}Y_{t-1}] & E[X_{t-1}^2] \end{pmatrix} \tag{8.4}$$

Case 1. $X_t = 0$ for all periods t. Then we have that

$$E[\mathbf{z}_t\mathbf{z}'_t] = \begin{pmatrix} E[Y_{t-1}^2] & 0 \\ 0 & 0 \end{pmatrix}$$

which implies that $\{\mathbf{z}_t\}$ is persistently exciting of order 1 since $E[Y_{t-1}] > 0$ (a positive definite matrix of dimension 1×1). Thus we can identify a but cannot identify b.

Case 2. $X_t = 1$ for all t. Then

$$E[\mathbf{z}_t\mathbf{z}'_t] = \begin{pmatrix} E[Y_{t-1}^2] & E[Y_{t-1}] \\ E[Y_{t-1}] & 1 \end{pmatrix}$$

Recall that the condition for a 2×2 matrix to be positive definite is that its determinant be greater than zero. Thus $\det E[\mathbf{z}_t\mathbf{z}'_t] = E[Y_{t-1}^2] - (E[Y_{t-1}])^2$ $= \mathrm{MSE}(Y_{t-1})$, which is always greater than zero; thus in this case, we can identify both a and b.

Case 3. X_t is white noise. We have in this case that

$$E[\mathbf{z}_t\mathbf{z}'_t] = \begin{pmatrix} E[Y_{t-1}^2] & 0 \\ 0 & \sigma_X^2 \end{pmatrix}$$

and since $\det E[\mathbf{z}_t\mathbf{z}'_t] > 0$, we can identify both a and b.

Case 4. $X_t = KY_t$ for all t (constant feedback). We have

$$E[\mathbf{z}_t\mathbf{z}'_t] = \begin{pmatrix} E[Y_{t-1}^2] & KE[Y_{t-1}^2] \\ KE[Y_{t-1}^2] & K^2E[Y_{t-1}^2] \end{pmatrix}$$

and since $\det E[\mathbf{z}_t\mathbf{z}'_t] = 0$, neither a nor b is identifiable.

Case 5. $X_t = KY_t + r_t$, where $r_t \sim (0, \sigma_r^2)$ is a white noise "dither" signal. In this case

$$E[\mathbf{z}_t\mathbf{z}'_t] = \begin{pmatrix} E[Y_{t-1}^2] & KE[Y_{t-1}^2] \\ KE[Y_{t-1}^2] & K^2E[Y_{t-1}^2] + \sigma_r^2 \end{pmatrix}$$

Thus we have that $\det E[\mathbf{z}_t \mathbf{z}'_t] = E[Y^2_{t-1}]\sigma^2_r > 0$, and therefore we can identify both a and b.

Case 6. $X_t = KY_{t-1}$ for all t. We have in this final case that

$$E[\mathbf{z}_t \mathbf{z}'_t] = \begin{pmatrix} E[Y^2_{t-1}] & KE[Y_{t-1}Y_{t-2}] \\ E[Y_{t-1}Y_{t-2}] & K^2 E[Y^2_{t-2}] \end{pmatrix}$$

and $\det E[\mathbf{z}_t \mathbf{z}'_t] = K^2 E[Y^2_{t-1}]E[Y^2_{t-2}] - K^2 \operatorname{Cov}(Y_{t-1}, Y_{t-2}) > 0$ (note that the first term is very close to $[\operatorname{Var}(Y_t)]^2$) and we can identify both parameters.

$\square$

8.5 CLOSED-LOOP ESTIMATION

Open-loop identification experiments, as required by the methods discussed in Chapter 4 may be too costly, particularly if the process exhibits severe drift when left uncontrolled. It is desirable to have available an identification technique that selects a transfer function and disturbance models while a controller is operating, because in doing so scrap and cost are reduced even if the controller in use is not optimal in any sense. Knowing the disturbance model affecting a process and having parameter estimates of such model, the controller can be tuned to achieve a more desirable performance.

Box and MacGregor (1974) discussed identification of transfer function models for processes operating under closed-loop control. They show how the cross-correlation between input and output does not provide any information about the process transfer function if there is a linear feedback controller in operation while the data were collected. To alleviate this difficulty, these authors propose to add a dither signal to the input, as in case 5 in the Example 8.5. Then identification is possible by looking at the cross-correlation of the dither signal with the output (assuming that the controller being used is known, which usually is the case).[5] In a subsequent paper, Box and MacGregor (1976) consider estimating the parameters of a transfer function plus noise model assuming a linear feedback controller of the form

$$X_t = -K(\mathcal{B})Y_t \tag{8.5}$$

is operating while the input–output data are collected. Following their 1974 paper, they assume that a dither signal is added to the right-hand side of equation (8.5) to improve the estimation of the parameters in the model.

[5]In the systems identification literature, a two-level random sequence, sometimes called a *pseudo random binary sequence* (PRBS), is frequently used as an experimental design for fitting dynamic models. See Goodwin and Payne (1977) for a discussion of experimental designs and identification.

If no dither signal is utilized in (8.5) while estimating the parameters, this may give *estimation windup problems* due to the presence of a linear relation in the matrix of regressor values. Suppose that the model is of the form

$$A(\mathcal{B})Y_t = B(\mathcal{B})X_t + \varepsilon_t \tag{8.6}$$

or

$$Y_t = a_1 Y_{t-1} + a_2 Y_{t-2} + \cdots + a_n Y_{t-n} + b_1 X_{t-1} + b_2 X_{t-2} + \cdots + b_n X_{t-n} + \varepsilon_t$$

Then the matrix of regressors is

$$
Z_t = \begin{pmatrix} \mathbf{z}_n \\ \vdots \\ \mathbf{z}_{t-1} \\ \mathbf{z}_t \end{pmatrix}
$$

$$
= \begin{pmatrix}
Y_n & Y_{n-1} & \cdots & Y_1 & X_1 & X_{n-1} & \cdots & X_1 \\
\vdots & \vdots & \vdots & \vdots & \vdots & \vdots & \vdots & \vdots \\
Y_{t-2} & Y_{t-3} & \cdots & Y_{t-n-1} & X_{t-2} & X_{t-3} & \cdots & X_{t-n-1} \\
Y_{t-1} & Y_{t-2} & \cdots & Y_{t-n} & X_{t-1} & X_{t-2} & \cdots & X_{t-n}
\end{pmatrix}
$$

and for $\hat{\boldsymbol{\theta}}_t = (Z_t' Z_t)^{-1} Z_t' Y_t$ to exist, $Z_t' Z_t$ must be invertible. But the feedback law (8.5) introduces a linear dependency among the columns of Z_t, causing $Z_t' Z_t$ to be noninvertible. Using linear regression terminology, the feedback law introduces (a rather extreme case of) multicollinearity. This will also occur in the recursive version of least squares estimation, where one way to detect the presence of this problem is to monitor $\operatorname{tr}(P_t)$, the trace of the precision matrix (Shah and Cluett, 1991).

To solve this problem once it is detected, several solutions have been proposed to break the linear dependency created (Åström and Wittenmark, 1989).

1. Use a linear feedback law of sufficiently high order [i.e., the order of the $K(\mathcal{B})$ polynomial in (8.6) should be greater or equal to the order in the forward path]. This occurs in case 6 in Example 8.5.
2. Add a persistently exciting dither signal to X_t. This was done in case 5, Example 8.5.
3. Use a nonlinear control law. In particular, the use of time − varying feedback gains, i.e., coefficients in $K(\mathcal{B})$ that change through time while the model is being estimated is sufficient to break the linear dependencies. Ljung and Söderström (1983) show that this can be

achieved if we alternate between

$$X_t = -K_1(\mathcal{B})Y_t$$

and

$$X_t = -K_2(\mathcal{B})Y_t$$

with $K_1 \neq K_2$. Luceño (1997) used a similar idea by switching between two different PI controllers while doing the estimation in closed-loop mode. Sometimes, if a process has been operating under manual control (i.e., a human operator has adjusted the process) such feedback cannot be modeled with a linear controller such as (8.5). In this case, closed-loop estimation *is* possible. Chapter 9 gives an example of this for a multivariate process. See Akaike and Nakagawa (1988) for a real-life example in a cement plant.

If certain assumptions can be made on the process disturbance and controller, identification from the autocorrelation function of the closed-loop output is possible:

1. The unknown (r, s, k) transfer function describing the process has $r \leq 1$, $s \leq 1$, and $k \leq 2$.
2. A PI (or EWMA) controller with known parameters is utilized while data were collected.
3. The performance of the controller used in assumption 2 is far from minimum MSE control.
4. The disturbance is one from the family given by

$$N_t = \delta + N_{t-1} - \theta\varepsilon_{t-1} + \varepsilon_t \qquad \theta \leq 1 \tag{8.7}$$

   that is, a possibly noninvertible IMA(1, 1) process with drift.

If these assumptions are true, identification of the transfer function and disturbance parameters is possible from the correlation functions of the closed-loop output Y_t. Once this is done, the controller can be retuned to achieve better performance [see del Castillo (2001b) and Pan and del Castillo (2001) for details]. The need for assumption (3) can easily be explained when the input–output delay (k) equals one period. If the controller is giving minimum MSE performance, recall from Chapter 5 that the closed-loop output will be MA$(k - 1)$ = MA(0), a white noise process that is completely uninformative. Evidently, we cannot infer process model parameters from white noise data (other than the properties of the noise, which are not needed for control).

PROBLEMS

8.1. Derive a recursive formula for the sample variance statistic, s^2.

8.2. Write down the formulation necessary to implement the RLS algorithm applied to estimating the parameter θ in an IMA(1, 1) time-series model. Simulate using a spreadsheet.

8.3. The spreadsheet *SelfTune.xls* simulates the indirect self-tuning controller of Example 8.2 (it also simulates the known parameter MMSE controller). Extend this spreadsheet simulation to include a Clarke–Gawthrop controller instead of a MMSE controller for the same process as in Example 8.2. Are the parameter estimates still biased?

8.4. Modify the spreadsheet *SelfTune.xls* (MMSE controller) by adding at each iteration a $(0, \sigma^2)$ random error to each element of the regressor vector. Use $\sigma^2 < \sigma_\varepsilon^2$. Do the RLS parameter estimates seem to improve?

8.5. Consider a $(r, s, k) = (1, 1, 2)$ transfer function with IMA(1, 1) noise. Write down the regressor and the parameter vector necessary to use RLS estimation in this process.

8.6. Modify the direct ST controller in Example 8.4 by defining $w_t = Y_t - X_{t-1}$, $\hat{\theta}_t = \hat{L}_t$, and $z_t = y_{t-1}$. Simulate and compare the performance of the two controllers. Which controller seems to converge faster? (*Suggestion*: Use a spreadsheet to simulate.)

8.7. Find a direct self-tuning controller for:

$$Y_t = gX_{t-1} + \frac{1 - \theta\mathcal{B}}{1 - \mathcal{B}}\varepsilon_t$$

Simulate the corresponding controlled process for $g = 1.5$, $\theta = 0.4$, and $\varepsilon_t \sim N(0, 3)$.

8.8. Find an indirect controller for the process in Problem 8.7. Simulate and compare its behavior against the direct ST controller using the same random streams.

8.9. Derive the cross-covariance in case 5, Example 8.5.

8.10. Suppose an EWMA (or I) controller is applied to a responsive process that experiences the disturbance given by equation (8.7). Show that the closed-loop description of the quality characteristic is given by an ARMA(1, 1) process. How can you use this fact to estimate the parameters of the disturbance?

8.11. Suppose that a PI controller is applied to a $(r, s, k) = (1, 0, 1)$ transfer function process under disturbance (8.7). Show that the closed-loop output is given by an ARMA(2, 2) process.

Problems 8.12 to 8.18 refer to material in the appendices to this chapter.

8.12. Show that the solution of the recursive equation for the variance update in the setup adjustment problem (Appendix 8A) can be solved to yield

$$P_t = \frac{P_0 \sigma_v^2}{\sigma_v^2 + t P_0}$$

where P_0 is the initial prior variance associated with the a priori estimate of the offset.

8.13. Show that if in the setup adjustment problem we apply the control rule $X_t = -\hat{d}_t$ where $\hat{d}_t$ is as in equation (8.19), then

$$\text{Var}(\hat{d}_t) = \frac{\sigma_v^2}{(\sigma_v^2/P_0) + t}$$

8.14. In the setup adjustment problem, show that if $P_0 \to \infty$, the harmonic rule of section 5.6 is obtained.

8.15. What happens in the Kalman filter solution to the setup adjustment problem when $\sigma_v^2 \to \infty$? What interpretation can you give?

8.16. Use a Kalman filter to estimate the state in the steady model $\theta_t = \theta_{t-1} + v_t$; $Y_t = \theta_t + \varepsilon_t$, where v_t and ε_t are two independent, normal white noise sequences. Simulate the process and operation of the filter for different values of the signal-to-noise ratio $\sigma_v^2/\sigma_\varepsilon^2$.

8.17. Using a spreadsheet, simulate the cautious and certainty equivalence controller for case 4 considered in Appendix 8B. Simulate 100 times for $N = 20$ and compute from these simulations an estimate of J for each controller. Use the values $\sigma_v = 1$, $a = 5$, $X_0 = 10$, and $b \sim N(5, 1)$. What happens if $\sigma_b^2 = 5$?

8.18. Consider the state-space process in Appendix 8B, but suppose that $\sigma_v = 0$, $\sigma_\varepsilon > 0$, b is a known constant, but assume that a is a random variable with mean μ_a and variance σ_a^2 that changes only from realization to realization of the process, that is, it stays constant throughout a single simulation or realization of the process. Find the optimal control rule that minimizes the objective $J = E[Z_N^2]$. Does the certainty equivalence principle hold in this case?

BIBLIOGRAPHY AND COMMENTS

Recursive least squares methods are based on remarks made by Plackett (1950). Interestingly, Box and Wilson mentioned the practicality of these results in their seminal paper on response surface methods (Box and Wilson, 1951). A book about recursive estimation with emphasis on the statistical aspects of these procedures applied to time series is Young (1984). Our presentation is based on Ljung and Söderström (1983). Convergence analysis of recursive algorithms was studied by Ljung (1977), who proposed the ODE method. Applied to adaptive control, this method assumes the stability of the closed-loop process (see Åström and Wittenmark, 1989). Section 8.4 is based on Wellstead and Zarrop (1991). Experimental design for identification of dynamic processes is treated in the book by Goodwin and Payne (1977). A history of adaptive control can be found in Åström and Wittenmark (1989). This is a story that includes the fundamental work of Kalman (1958, 1960). Kalman (1960) is the original reference to the celebrated Kalman filters. Readers interested in Kalman filtering applications to SPC should consult the Ph.D. dissertation by Crowder (1986). Self-tuning PID controllers have been proposed by many authors (see e.g., Cameron and Seborg, 1983; Katende and Jutan, 1993; Del Castillo (2000). Appendix 8B follows Söderström (1994). For more details about dual and cautious control, see Åström and Wittenmark (1989).

APPENDIX 8A: INTRODUCTION TO KALMAN FILTERS

An important recursive estimation method applied to state-space models was developed by Kalman[6](1960) and has had very extensive application in engineering and time-series analysis over the years. In this appendix we show the derivation of the celebrated Kalman filter, an estimate of the state vector, using a Bayesian point of view.[7] This was not the point of view that Kalman used in his original work, but is probably easier to explain. [See Meinhold and Singpurwalla (1983), who we follow closely in this appendix.] We then show some of the properties and applications of Kalman filters.

[6]Rudolf Kalman is a professor emeritus at the University of Florida.
[7]This appendix assumes some elementary knowledge of Bayesian statistics.

Derivation

Suppose that we have a process in which the observed measurements obey the model

$$\mathbf{Y}_t = \mathbf{F}_t \mathbf{\theta}_t + \mathbf{\varepsilon}_t \tag{8.8}$$

where $\mathbf{\varepsilon}_t \sim N(\mathbf{0}, \mathbf{V}_t)$. Here $\mathbf{Y}_t$ is a $p \times 1$ vector of variables of interest that we can measure. In addition, suppose that the parameters in $\mathbf{\theta}$ vary randomly in time according to the state equation

$$\mathbf{\theta}_t = \mathbf{G}_t \mathbf{\theta}_{t-1} + \mathbf{w}_t \tag{8.9}$$

where $\mathbf{G}_t$ is a known $n \times n$ transition matrix and $\mathbf{w}_t \sim N(\mathbf{0}, \mathbf{W}_t)$. The random vectors $\mathbf{\varepsilon}_t$ and $\mathbf{w}_t$ are assumed not to depend on each other, and $\mathbf{V}_t$ and $\mathbf{W}_t$ are assumed known.

Our goal is to find out, using Bayes' theorem, the posterior distribution of $\mathbf{\theta}_t$ once we have observed $\mathcal{Y}^t = \{\mathbf{Y}_1, \mathbf{Y}_2, \ldots, \mathbf{Y}_t\}$. We denote such a distribution as $P(\mathbf{\theta}_t | \mathcal{Y}^t)$. To reach our goal[8] we start by looking at the posterior distribution of $\mathbf{\theta}_{t-1}$ given $\mathcal{Y}^{t-1}$:

$$\mathbf{\theta}_{t-1} | \mathcal{Y}^{t-1} \sim N\left(\hat{\mathbf{\theta}}_{t-1}, \mathbf{\Sigma}_{t-1}\right) \tag{8.10}$$

where $\hat{\mathbf{\theta}}_{t-1}$ and $\mathbf{\Sigma}_{t-1}$ are the expectation and variance of $\mathbf{\theta}_{t-1} | \mathcal{Y}^{t-1}$.

Now, *prior to observing* $\mathbf{Y}_t$, we have, from equations (8.9) and (8.10),

$$\mathbf{\theta}_t | \mathcal{Y}^{t-1} \sim N\left(\mathbf{G}_t \hat{\mathbf{\theta}}_{t-1}, \mathbf{G}_t \mathbf{\Sigma}_{t-1} \mathbf{G}_t' + \mathbf{W}_t\right) = P\left(\mathbf{\theta}_t | \mathcal{Y}^{t-1}\right)$$

which constitutes our prior distribution.

After observing $\mathbf{Y}_t$, we want to compute the posterior distribution (i.e., the distribution of $\mathbf{\theta}_t | \mathcal{Y}^t$). For this we also need the likelihood function $P(\mathbf{Y}_t | \mathbf{\theta}_t, \mathcal{Y}^{t-1})$, which is obtained as follows from the prediction errors $\mathbf{e}_t$:

$$\mathbf{e}_t = \mathbf{Y}_t - \hat{\mathbf{Y}}_{t|t-1} = \mathbf{Y}_t - \mathbf{F}_t \hat{\mathbf{\theta}}_t = \mathbf{Y}_t - \mathbf{F}_t \mathbf{G}_t \hat{\mathbf{\theta}}_{t-1} \tag{8.11}$$

Since $\mathbf{F}_t$, $\mathbf{G}_t$ and $\hat{\mathbf{\theta}}_{t-1}$ are all known at the end of time $t-1$, observing $\mathbf{e}_t$ is equivalent to observing $\mathbf{Y}_t$ (so $\mathbf{e}_t$ has all the new information and that is why we call these the innovations). Therefore, we can work with the likelihood of

---

[8] In the literature on time-series analysis based on state-space models, the goal is to estimate the state $\mathbf{\theta}_t$ based on observations $\mathcal{Y}^s$. If $s = t$, the problem is called *filtering*, which is what we address in this appendix. If $s < t$, the estimation problem is a forecasting or prediction problem. Finally, if $s > t$, the problem is called *smoothing* (this terminology was coined by N. Weiner). In each case the goal is to find the estimates and provide a measure of their precision, goals that the Kalman filter fulfills for the filtering and prediction problems.

e_t, $P(e_t|\theta_t, \mathcal{Y}^{t-1})$, instead. We then have from Bayes' theorem that

$$P(\theta_t|\mathcal{Y}^t) = \frac{P(e_t|\theta_t, \mathcal{Y}^{t-1})P(\theta_t|\mathcal{Y}^{t-1})}{\underbrace{\int_{\text{all }\theta_t} P(e_t|\theta_t, \mathcal{Y}^{t-1})P(\theta_t|\mathcal{Y}_{t-1})\,d\theta_t}_{P(e_t,\theta_t|\mathcal{Y}^{t-1})}}$$

From equation (8.11), the distribution of $\theta_t|\mathcal{Y}^t$ is the same as that of $\theta_t|e_t, \mathcal{Y}^{t-1}$; thus

$$\theta_t|e_t, \mathcal{Y}^{t-1} \sim N(\hat{\theta}_t, \hat{\Sigma}_t) \quad \text{and} \quad \hat{\theta}_t|\mathcal{Y}^t \sim N(\hat{\theta}_t, \Sigma_t)$$

and the mean of the posterior of $\theta_t|\mathcal{Y}^t$ is

$$\hat{\theta}_t = G_t\hat{\theta}_{t-1} + R_t F_t'(V_t + F_t R_t F_t')^{-1}e_t = G_t\hat{\theta}_{t-1} + K_t e_t \qquad (8.12)$$

where $R_t = G_t\Sigma_{t-1}G_t' + W_t$, $e_t = Y_t - F_t G_t\hat{\theta}_{t-1}$ and the variance of the posterior is

$$\Sigma_t = R_t - R_t F_t'(V_t + F_t R_t F_t')^{-1} F_t R_t$$
$$= R_t - K_t F_t R_t = (I - K_t F_t)R_t \qquad (8.13)$$

Equations (8.12) and (8.13) constitute the Kalman filter estimate of the state θ, which can be updated recursively as each new observation Y_t is collected, starting from values $\hat{\theta}_0$ and Σ_0.

Relation with Recursive Least Squares
The RLS procedure can be seen as the Kalman filter estimate of the state-space model:

$$\theta_t = \theta_{t-1} \qquad (8.14)$$

$$Y_t = z_t'\theta_t + \varepsilon_t \qquad (8.15)$$

where we use the notation in Section 8.1 and where $\varepsilon \sim (0, 1)$. Since the state is constant, the filtering, prediction, and smoothing problems are all solved by the Kalman filter equations. This constant-state description is the basis for self-tuning controllers, as opposed to more general adaptive controllers that allow for parameter changes.

Properties of the Kalman Filter Estimate
Duncan and Horn (1972) show how if *no* normality assumption is made on either ε_t or w_t, the Kalman filter is a minimum mean square error *linear*

estimator of the state, that is, among all estimators that are linear combinations of the observations, equations (8.12) and (8.13) are best in the MSE sense. If the errors *are* normal, the Kalman filter estimate of the state is minimum MSE in the class of all (and not just linear) estimators. Contrast this with the properties of recursive least squares discussed in Section 8.1. The RLS estimate is a minimum variance unbiased estimator, but this does not guarantee minimum MSE. The difference is in the update of the covariance matrix. In RLS this is

$$P_t = P_{t-1} - \frac{P_{t-1}\mathbf{z}_t\mathbf{z}_t'P_{t-1}}{1 - \mathbf{z}_t'P_{t-1}\mathbf{z}_t}$$

using notation from Section 8.1. The corresponding Kalman filter would substitute the "1" in the denominator for σ_ε^2 since in general $\sigma_\varepsilon^2 \neq 1$.

We point out that the variances of the errors in the state-space formulation, $\mathbf{V}_t$ and $\mathbf{W}_t$, are assumed known. If this is not the case, maximum likelihood methods have been proposed for estimating not only the state but also the covariance matrices (this is called *adaptive filtering* in the control and signal processing literature; see Mehra, 1972). Shumway and Stoffer (2000) give an excellent discussion and illustration of maximum likelihood methods.

Applications of Kalman Filters in Time-Series Analysis and Control Theory

The main application of Kalman filters is in estimation of the state vector in state-space models, which historically was first used in aeronautical applications. The Kalman filter is used in adaptive controllers that utilize the state-space form instead of the polynomial transfer function form of Chapter 4. One main advantage is that the Kalman filter is multivariate, so it can be applied to processes where we have multiple quality characteristics, as described in Chapter 9.

In the time-series literature, Kalman filters have been used to estimate the parameters of univariate ARIMA models (for this the models need to be written in state-space form as indicated in Appendix 4A) and have also been used to estimate multivariate ARIMA models. A particularly useful application is estimation of time-series models when some observations are missing. This turns out to be straightforward using Kalman filters; see Reinsel (1997) and Shumway and Stoffer (2000) for details.

We end this appendix with a practical application of Kalman filters to a very common adjustment problem in the quality control of discrete parts, the setup adjustment problem that we discussed in Section 5.6.

Example 8.6: Application of Kalman Filters to the Setup Adjustment Problem.

The Kalman filter provides a general solution to the setup adjustment problem discussed in Section 5.6. Recall that in this type of problem, a machine experiences setup difficulties that result in an offset in the quality characteristic of an independent, identically distributed (i.i.d.) process. If, as in Section 5.6, Y_t denotes the observed deviation from target and d denotes the unknown offset the machine experiences at startup, we have that

$$Y_t = \mu_t + v_t = d + X_{t-1} + v_t$$

where μ_t is the mean of the process, X_t the level of a controllable factor that is able to adjust the process, and $v_t \sim (0, \sigma_v^2)$. The problem that Grubbs defined is how to find the best adjustment weights, K_t, for an adjustment rule of the form

$$\nabla X_t = K_t Y_t$$

The setup adjustment problem can then be set in the simple state-space form

$$d = d_{t-1} = \cdots = d_0 = d$$
$$z_t = Y_t - X_{t-1} = d_{t-1} + v_t$$

Using the Kalman filter formulation at the beginning of this appendix, we have that

$$F_t = 1 \quad G_t = 1 \quad V_t = \sigma_v^2 \quad W_t = 0 \quad \hat{\theta}_t = \hat{d}_t$$

(all these quantities are scalars) and therefore $R_t = P_{t-1}$, where P_t is the posterior variance of the state estimate (given by $\hat{d}_t$). From equation (8.12), the Kalman filter weights are given by

$$K_t = \frac{P_{t-1}}{\sigma_v^2 + P_{t-1}} \tag{8.16}$$

and the residuals are given by $e_t = Y_t - X_{t-1} - \hat{d}_{t-1}$.
From equation (8.12), the estimate of d at time t is given by

$$\hat{d}_t = \hat{d}_{t-1} + \frac{P_{t-1}}{\sigma_v^2 + P_{t-1}} \left(Y_t - X_{t-1} - \hat{d}_{t-1} \right)$$

If adjustments are made following the simple rule

$$\nabla X_t = -\nabla \hat{d}_t \tag{8.17}$$

or $X_t = -\hat{d}_t$, the estimate of the offset is given by

$$\hat{d}_t = \hat{d}_{t-1} + \frac{P_{t-1}}{\sigma_v^2 + P_{t-1}} Y_t \tag{8.18}$$

From (8.13), the variance of this estimate is updated following

$$P_t = P_{t-1} - \frac{P_{t-1}^2}{\sigma_v^2 + P_{t-1}} = \frac{\sigma_v^2 P_{t-1}}{\sigma_v^2 + P_{t-1}}$$

Here the Bayesian interpretation is that, a priori, $d \sim (\hat{d}_0, P_0)$. Note that no assumption is made on the type of distribution of the offset.

One more algebraic step provides further insight into this procedure. Substituting equation (5.8}) into equation (8.16), the following is obtained after simplification:

$$K_t = \frac{1}{\left(\sigma_v^2/P_0\right) + t}$$

Using this value of K_t in equation (8.18) yields

$$\hat{d}_t = \hat{d}_{t-1} + \frac{1}{\left(\sigma_v^2/P_0\right) + t} Y_t \qquad (8.19)$$

It can be shown that the adjustment rule $X_t = -\hat{d}_t$ solves Grubbs's problem. See del Castillo and Pan (2001) for more details.   □

APPENDIX 8B: CERTAINTY EQUIVALENCE, CAUTIOUS, AND DUAL CONTROL

Suppose that a process is given in the state-space form:

$$Z_{t+1} = aZ_t + bX_t + v_t$$
$$Y_t = Z_t + \varepsilon_t \qquad t = 0, 1, 2, \ldots, N - 1 \qquad (8.20)$$

with $v_t \sim (0, \sigma_v^2)$ and $\varepsilon_t \sim (0, \sigma_\varepsilon^2)$. The parameters a and b may be known or unknown, constant or random. The criterion to optimize is the variance of the final state:

$$\min J = E\left[Z_N^2\right]$$

We now look at some cases of interest.

1. *Deterministic case.* If a and b are known constants and $\sigma_v = \sigma_\varepsilon = 0$, we just need to minimize

$$J = Z_N^2 = \left(aZ_{N-1} + bX_{N-1}\right)^2$$

with obvious solution

$$X_{N-1} = -\frac{a}{b}Z_{N-1}$$

and $X_0, X_1, \ldots, X_{N-2}$ arbitrary. We can always use the same type of control rule for all t, so

$$X_t = -\frac{a}{b}Z_t \qquad \text{for } t = 0, 1, 2, \ldots, N-1$$

which complies with the optimal solution.

2. *i.i.d. parameters varying between sampling instances.* Assume that $\sigma_v = \sigma_\varepsilon = 0$, b is a known constant, Z_0 is given, but assume now that the sequence $\{a_t\}$, where a_t is the random value of parameter a at time t, is made of i.i.d. random variables with mean μ_a and variance σ_a^2. We consider criterion (8.21), which can be rewritten as

$$J = E[Z_N^2] = E\left[E\left[Z_N^2 \mid \mathcal{Z}^{N-1}, \mathcal{X}^{N-1}\right]\right]$$

where the outer expectation is with respect to

$$\mathcal{Z}^{N-1} = \{Z_0, Z_1, \ldots, Z_{N-1}\}$$

and where $\mathcal{X}^{N-1} = \{X_0, X_1, \ldots, X_{N-1}\}$. Now the inner expectation is

$$E\left[Z_N^2 \mid \mathcal{Z}^{N-1}, \mathcal{X}^{N-1}\right] = E\left[(a_{N-1}Z_{N-1} + bX_{N-1})^2 \mid \mathcal{Z}^{N-1}, \mathcal{X}^{N-1}\right]$$

$$= (\mu_a Z_{N-1} + bX_{N-1})^2 + \sigma_a^2 Z_{N-1}^2 \qquad (8.22)$$

Clearly, the optimal controller at time $N-1$ is

$$X_{N-1} = -\frac{\mu_a}{b}Z_{N-1}$$

which from (8.22) yields the performance index value

$$E[Z_N^2] = E\left[E\left[Z_N^2 \mid \mathcal{Z}^{N-1}, \mathcal{X}^{N-1}\right]\right] = E[\sigma_a^2 Z_{N-1}^2] = \sigma_a^2 E[Z_{N-1}^2]$$

To get $E[Z_{N-1}^2]$, we can repeat the same procedure and get $X_{N-2} = -\mu_a/bZ_{N-2}$ and $E[Z_{N-1}^2] = \sigma_a^2 E[Z_{N-2}^2]$. Therefore, we have shown that at each stage the optimal control is

$$X_t = -\frac{\mu_a}{b}Z_t \qquad t = 0, 1, \ldots, N-1$$

This control rule is the same as the one obtained from the deterministic process (case 1 above) by replacing the random parameters by their expectations. Therefore, this is a case where the certainty equivalence principle holds.

3. *Measurement error.* Here a and b are both known constants, $\sigma_\varepsilon > 0$, $\sigma_v = 0$, and the performance measure is the same as before. In contrast to the previous case, in this case the control value X_t must be a function of the measurements[9] $\mathcal{Y}^t = \{Y_0, Y_1, \ldots, Yt\}$ because the state is not observable. To minimize J, we do as follows:

$$J = E\left[Z_N^2\right] = E\left[E\left[Z_N^2 | \mathcal{Y}^{N-1}\right]\right]$$

with outer expectation taken with respect to $\mathcal{Y}^{N-1}$. Then the inner expectation is

$$E\left[Z_N^2 | \mathcal{Y}^{N-1}\right] = E\left[(a_{N-1}Z_{N-1} + bX_{N-1})^2 | \mathcal{Y}^{N-1}\right]$$

$$= \left(aE\left[Z_{N-1} | \mathcal{Y}^{N-1}\right] bX_{N-1}\right)^2 + a^2 \operatorname{Var}\left(Z_{N-1} | \mathcal{Y}^{N-1}\right)$$

It turns out that the conditional mean and variance of the state given the observations are given by a Kalman filter. Iterating this reasoning for all stages, the optimal control law is

$$X_t = -\frac{a}{b}E\left[Z_t | \mathcal{Y}^t\right] \qquad t = 0, 1, 2, \ldots, N-1$$

Thus the certainty equivalence principle holds in this case since the optimal control law is just the one we would obtain from the deterministic case if we substitute the unknown random variable (the state) by its expectation.

4. *Unknown gain.* Suppose now that $\sigma_v > 0$, $\sigma_\varepsilon = 0$, a is a known constant, X_0 is given, but now assume that we have that $b \sim (\mu_b, \sigma_b^2)$, so b does not change with t. That is, with each realization of the process we observe only a single realization of b. This differs from case 2 above where one different realization of the random variable a is observed at each sample instant (that is why we used the notation a_t).

If we were to try a certainty equivalence controller, we would use

$$X_t = -\frac{a}{\hat{b}_t}Z_t$$

but as it turns out, this is *not* the optimal regulator that minimizes $J = E[Z_N^2]$. Thus the certainty equivalence principle does not hold in this case. To prove

---

[9]If $\sigma_\varepsilon > 0$, the term *incomplete state information* is frequently used in the area of stochastic control. If $\sigma_v > 0$, the term *complete state information* is used.

this assertion, we can find the optimal controller and realize it is not equal to the certainty equivalence controller. This requires dynamic programming techniques and is outside the scope of this book. Alternatively, we can find some other controller and show that it gives a better performance than the certainty equivalence controller. We will do this and we will introduce, as a by-product, the *cautious controller*.

Let us evaluate first the performance of the certainty equivalence controller. Substituting the control rule in the performance index, we obtain

$$E\left[Z_N^2 | \mathcal{Z}^{N-1}\right] = E\left[\left(\underbrace{aZ_{N-1} - \frac{ba}{\hat{b}_{N-1}}Z_{N-1}}_{1\text{st}} + \underbrace{v_{N-1}}_{2\text{nd}}\right)^2 | \mathcal{Z}^{N-1}\right]$$

Taking the square according to the two terms indicated, we have that

$$E\left[Z_N^2 | \mathcal{Z}^{N-1}\right] = E\left[a^2 Z_{N-1}^2 - 2Z_{N-1}^2 \frac{ba^2}{\hat{b}_{N-1}} + \frac{b^2 a^2}{\hat{b}_{N-1}^2}Z_{N-1}^2 | \mathcal{Z}^{N-1}\right] + \sigma_v^2$$

since the cross-term is zero. Simple algebra shows that

$$E\left[Z_N^2 | \mathcal{Z}^{N-1}\right] = E\left[(b - \hat{b}_{N-1})^2 | \mathcal{Z}^{N-1}\right]\left(\frac{aZ_{N-1}}{\hat{b}_{N-1}}\right)^2 + \sigma_v^2$$

$$\equiv p_{N-1}\left(\frac{aZ_{N-1}}{\hat{b}_{N-1}}\right)^2 + \sigma_v^2 \tag{8.23}$$

where p_t is the conditional variance of our estimate $\hat{b}_t \equiv E[b|\mathcal{Z}^t]$ and can be thought of (using a Bayesian interpretation) as a degree of confidence we have in our estimate. Let us now try a different controller:

$$X_t = -a\frac{\hat{b}_t}{\hat{b}_t^2 + p_t}Z_t$$

which is called a *cautious* controller since it considers the uncertainty in the estimate of b into account. When $\hat{b}_t$ is very uncertain (or when we have little confidence in it) the controller gain

$$\frac{\hat{b}_t}{\hat{b}_t^2 + p_t}$$

is reduced so that the values of the controllable factor are more moderate or cautious. The value of the performance index under this control rule is

$$E\left[Z_N^2 | \mathcal{Z}^{N-1}\right] = E\left[\left(Z_{N-1} - \frac{ba\hat{b}_{N-1}^2 Z_{N-1}}{\hat{b}_{N-1}^2 + p_{N-1}} + v_{N-1}\right)^2 | \mathcal{Z}^{N-1}\right]$$

where we treat b as a random variable when taking expectation because Z depends on b. Since $E[b | \mathcal{Z}^{N-1}] = \hat{b}_{N-1}$ by definition, and the cross-term is zero, we have that

$$E\left[Z_N^2 | \mathcal{Z}^{N-1}\right]$$

$$= a^2 Z_{N-1}^2 + \frac{-2a^2 Z_{N-1}^2 \hat{b}_{N-1}^2 \left(\hat{b}_{N-1}^2 + p_{N-1}\right) + a^2 \hat{b}_{N-1}^4 Z_{N-1}^2}{\left(\hat{b}_{N-1}^2 + p_{N-1}\right)^2} + \sigma_v^2$$

$$= a^2 Z_{N-1} - \frac{a^2 Z_{N-1}^2 \hat{b}_{N-1}^2 \left(\hat{b}_{N-1}^2 + p_{N-1}\right)}{\left(\hat{b}_{N-1}^2 + p_{N-1}\right)^2} + \sigma_v^2$$

$$= p_{N-1} \frac{(aZ_{N-1})^2}{\hat{b}_{N-1}^2 + p_{N-1}} + \sigma_v^2$$

Since $p_{N-1} \geq 0$, comparing this result with (8.23) it is evident that the cautious controller gives lower cost, so the certainty equivalence controller is not optimal (i.e., the certainty equivalence principle does not hold).

The cases above illustrate when the certainty equivalence principle holds. In summary, this principle holds when all the following conditions are true:

1. When the performance index is quadratic
2. When the parameters are known constants *or* when they are random variables that are independent between sample times

Notice that if the model is in state-space form, there can be measurement error, so both the complete and incomplete state information cases can be CE-optimal as long as they satisfy the conditions above. Also note that nothing is said about the distribution or correlation of the error terms, so they do not need be white or normal. In a self-tuning controller with a minimum variance objective, we have that condition 1 is satisfied but condition 2 is not. The parameters are unknown constants, so for the typical situation addressed by a self-tuning, minimum variance controller, the certainty equivalence principle does *not* hold.

Dual Controller

Adaptive certainty equivalence controllers have a "passive" attitude since the model provided by the estimator scheme is used no matter whether the model is good or bad. A more "active" approach, in which the controller tries to (1) probe the process in order to "learn" (i..e, get better parameter estimates), and (2) bring the process back to target (minimizing some relevant performance index), is called a *dual controller*. The dual control problem was first stated by Feld'baum (1965). It is usually stated for ARMAX processes with $C(\mathcal{B}) = 1$ and parameters $\theta' = (a_1, a_2, \ldots; b_1, b_2, \ldots)$ that vary at each sample according to a Markov (AR(1)) model:

$$\theta_t = \Phi\theta_{t-1} + \mathbf{W}_t.$$

The goal is to minimize the finite-horizon objective

$$E\left[\frac{1}{N}\sum_{t=1}^{N} Y_t^2\right]$$

where Y_t denotes deviations from target. Except for the simplest cases, the dynamic programming solution rapidly encounters Bellman's *course of dimensionality*, so it cannot be solved in general. Heuristic approaches have been proposed that emulate the dual goals of probing and control by using simpler objective functions.

The cautious controller is actually the optimal dual controller for $N = 1$. The optimal dual controller will not be "myopic" (as the certainty equivalence controller is) in the sense that it will incur in short-term costs due to probing for longer-term benefits in the form of better parameter estimates. As was shown in the examples of Section 8.3, a certainty equivalence self-tuning controller has built-in dual action if the initial precision matrix P_0 has large entries. This will result in large transient fluctuations in the controllable factors, which will decrease the value of P_t (the parameter estimate uncertainties) rapidly with t. This can be thought of a "probing" action. Eventually, the better parameter estimates so obtained assure better control.

CHAPTER 9

Analysis and Adjustment of Multivariate Processes

The discrete-time stochastic processes analyzed in preceding chapters[1] correspond to univariate processes, that is, processes in which there is only one response or quality characteristic and there is at most one controllable factor in case we are dealing with a transfer function system. In this chapter we analyze *multivariate* processes, that is, processes in which there are either multiple responses and/or multiple controllable factors. The most general and interesting case for most real-life situations is when we deal with a process with multiple controllable factors (inputs) and multiple responses (or outputs) that define the quality of a product. These are called MIMO processes in the control engineering literature. Here we discuss some of their properties, some practical techniques for how to identify and fit them from data using the SAS and Matlab software systems, and how to adjust them with a view to achieving some optimality goal. By necessity, this chapter is just a brief introduction to a deep and difficult topic for which very complete and more advanced texts have been written (see the "Bibliography and Comments" section). The emphasis is on providing the minimum number of results necessary to undertake the analysis and adjustment of simple multivariate dynamic processes, highlighting how to accomplish these tasks using modern software tools.

9.1 PROPERTIES OF MULTIVARIATE ARMA AND ARMAX PROCESSES

Consider again the semiconductor manufacturing process described in Example 7.4. In this process it is of interest to planarize, or make as flat as possible, silicon wafers, and it is important to do this as rapidly as possible,

---

[1]This chapter contains relatively more advanced material than the rest of the text.

due to the very expensive machines in which this process take place. For these reasons, process engineers define the nonuniformity of a wafer (Y_1) as a measure related to the standard deviation of thickness measurements taken over the wafer, and removal rate (Y_2) as the rate at which silicon oxide is removed from the surface of the wafer. Evidently, manufacturers want to maximize removal rate and minimize nonuniformity, and these goals are usually conflicting. This is an instance of a run-to-run process, as discussed at the beginning of Chapter 7, in which control actions are required from batch to batch of wafers (usually, a batch equals one to three wafers in most planarization machines). In this process, a variety of variables can be controlled, for example, the speed at which the polishing plate rotates (X_1), the forces with which the wafers are pressed down to the plate (X_2) or with which the plate presses up (X_3), the amount of a slurry abrasive that is poured on the plate (X_4), and so on. If we settle with these responses and controllable factors, we have two responses and four factors, or a 4×2 process. Our first goal in this chapter is to show how to model the simultaneous variation in time of the responses Y_i as a function of the dynamic effects of the controllable factors X_j. We then address how to set the controllable factors to achieve a particular goal.

To study a more general process, let

$$\mathbf{Y}_t = \begin{pmatrix} Y_{1,t} \\ Y_{2,t} \\ \vdots \\ Y_{N,t} \end{pmatrix}$$

denote a vector of N random variables (responses or outputs) which define the quality of a product manufactured at time t (as in previous chapters, t can also be thought of as the product or part number, if that is the case). Many properties of multivariate, discrete-time stochastic processes are analogous to their univariate counterparts, as what follows will show. Readers might wish to review the material in Section 1.2 and in Chapter 3 to contrast the univariate and multivariate cases.

We say that the process $\{\mathbf{Y}_t\}$ is *strictly stationary* (or just *stationary*) if the probability distribution of $(\mathbf{Y}_{t_1}, \mathbf{Y}_{t_2}, \ldots, \mathbf{Y}_{t_k})$ is identical to the distribution of $(\mathbf{Y}_{t_1+\tau}, \mathbf{Y}_{t_2+\tau}, \ldots, \mathbf{Y}_{t_k+\tau})$ for any $t_1, t_2, \ldots, t_k$ and any lag τ. This implies, in particular, that a stationary multivariate time-series process has $E[\mathbf{Y}_t] = \boldsymbol{\mu}$ (a vector of constants) and that

$$\Sigma_Y = \text{Var}(\mathbf{Y}_t) \equiv \Gamma(0) = E\big[(\mathbf{Y}_t - \boldsymbol{\mu})(\mathbf{Y}_t - \boldsymbol{\mu})'\big]$$

is a $N \times N$ symmetric matrix of constants.

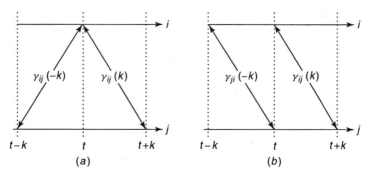

Figure 9.1 Cross-covariance function: (a) $\gamma_{ij}(k) \neq \gamma_{ij}(-k)$; (b) $\gamma_{ij}(k) = \gamma_{ji}(-k)$.

9.1.1 Autocovariance and Autocorrelation Functions

For a stationary process, the covariance between $\mathbf{Y}_t$ and $\mathbf{Y}_{t+\tau}$ is defined as

$$\Gamma(k) \equiv E\big[(\mathbf{Y}_t - \boldsymbol{\mu})(\mathbf{Y}_{t+k} - \boldsymbol{\mu})'\big]$$

(an $N \times N$ matrix), where the $(i - j)$th element equals

$$\gamma_{ij}(k) = E\big[(Y_{i,t} - \mu_i)(Y_{j,t+k} - \mu_j)\big] = E\big[(Y_{j,t} - \mu_j)(Y_{i,t-k} - \mu_i)\big] = \gamma_{ji}(-k)$$

which we can recall from Chapter 4 is the cross-covariance between response i and response j at lag k. This last relation is illustrated in Figure 9.1. Therefore, the (i, j) element of $\Gamma(k)$ is equal to the (j, i)th element of $\Gamma(-k)$ and we have the relation

$$\Gamma(k) = \Gamma(-k)' \tag{9.1}$$

An estimator of the autocovariance function is made of the sequence of $N \times N$ matrices:

$$\mathbf{C}(k) = \hat{\Gamma}(k) = \frac{1}{n} \sum_{t=1}^{n-k} (\mathbf{Y}_t - \overline{\mathbf{Y}})(\mathbf{Y}_{t+k} - \overline{\mathbf{Y}})' \qquad k = 0, 1, 2, \ldots$$

Given property (9.1), the estimated autocovariance matrices are usually computed only for positive lags. The individual elements of the $\mathbf{C}(k)$ matrix are

$$c_{ij}(k) = \hat{\gamma}_{ij}(k) = \frac{1}{n} \sum_{t=1}^{n-k} (Y_{i,t} - \overline{Y}_i)(Y_{j,t+k} - \overline{Y}_j)$$

The autocorrelation matrix at lag k is given by

$$\boldsymbol{\rho}(k) = S^{-1}\Gamma(k)S^{-1}$$

an $N \times N$ matrix with elements i, j equal to

$$\rho_{ij}(k) = \frac{\gamma_{ij}(k)}{\sqrt{\gamma_{ii}(0)}\,\sqrt{\gamma_{jj}(0)}}$$

Thus the S matrix is such that

$$S = \begin{bmatrix} \sqrt{\gamma_{11}(0)} & 0 & \cdots & 0 \\ 0 & \sqrt{\gamma_{22}(0)} & \cdots & 0 \\ \vdots & \vdots & \ddots & \vdots \\ 0 & 0 & \cdots & \sqrt{\gamma_{NN}(0)} \end{bmatrix}$$

or

$$S^{-1} = \begin{bmatrix} \dfrac{1}{\sqrt{\gamma_{11}(0)}} & 0 & \cdots & 0 \\ 0 & \dfrac{1}{\sqrt{\gamma_{22}(0)}} & \cdots & 0 \\ \vdots & \vdots & \ddots & \vdots \\ 0 & 0 & \cdots & \dfrac{1}{\sqrt{\gamma_{NN}(0)}} \end{bmatrix}$$

The diagonal elements of $\boldsymbol{\rho}(0)$ are equal to 1, but the off-diagonal elements are not zero, so $\boldsymbol{\rho}(0) \neq I$, contrary to the univariate case. The sample autocorrelations are defined as

$$r_{ij}(k) = \hat{\rho}_{ij}(k) = \frac{c_{ij}(k)}{\sqrt{c_{ii}(0)}\,\sqrt{c_{jj}(0)}} \qquad i, j = 1, 2, \ldots .$$

If one of the series (i or j) is white noise, then $\mathrm{Var}(r_{ij}(k)) \approx 1/n$. It can be shown that for a vector AR(p) process the autocovariance and autocorrelation functions tail off, whereas for a vector MA(q) process they will cut off after lag q, just as in the univariate case. Examples of these types of patterns will be given shortly in this section.

9.1.2 Partial Autocorrelations

Recall (Section 3.6) that in the univariate case the partial autocorrelation function ϕ_{kk} measures the correlation between observations separated k lags once the effect of the intermediate variables is discounted. An analogous concept exists in the multivariate case. The partial autocorrelation matrix P_k represents the cross-correlation between elements of $\mathbf{Y}_{t-k}$ and $\mathbf{Y}_t$ once we have eliminated the effect of $\mathbf{Y}_{t-k+1}, \ldots, \mathbf{Y}_{t-1}$. P_k has the property that for vector AR(p) processes it "cuts off" after lag p and for vector MA processes it "tails off."

Tiao and Box (1981) proposed the following graphical device to determine the tentative order of a vector ARMA process. They suggest writing the sample autocorrelation matrices $\mathbf{r}(k) = \hat{\rho}(k)$ ($k = 1, 2, \ldots$) and sample partial autocorrelation matrices $\hat{P}_k$, $k = 1, 2, \ldots$ with the following symbols substituting for an entry of either $\hat{\rho}_{ij}(k)$ or $P_{ij}(k)$:

1. A plus sign if the (i,j)th entry is greater than $2/\sqrt{n}$
2. A minus sign if the (i,j)th entry is smaller than $2/\sqrt{n}$
3. A dot otherwise

Example 9.1. To illustrate the graphical device proposed by Box and Tiao (1981), let us simulate the vector AR(1) process:

$$\left(I - A_1 \mathcal{B}\right)\mathbf{Y}_t = \boldsymbol{\varepsilon}_t$$

with

$$A_1 = \begin{pmatrix} 0.8 & 0.2 \\ 0.3 & -0.4 \end{pmatrix}$$

and $\boldsymbol{\varepsilon}_t$ is a multivariate normal variable with zero mean and variance

$$\Sigma = \begin{pmatrix} 25 & 7.5 \\ 7.5 & 25 \end{pmatrix}$$

Two hundreds observations of this process were simulated (the data can be consulted in the file *MultiAR.txt*). The schematic representation of the sample autocorrelation and partial autocorrelation functions are as follows:

Schematic Representation of Correlations

Name/Lag	0	1	2	3	4	5	6	7	8	9	10
Y1	++	++	++	++	++	++	++	++	++	++	+.
Y2	++	+-	++	+.	+.	+.	++	+.	++	..	++

Schematic Representation of Partial Autocorrelations

Name/Lag	1	2	3	4	5	6	7	8	9	10
Y1	++	..	..	..	..	..	..	..	..	..
Y2	+-	..	..	..	..	..	..	..	..	..

These were computed using SAS, as will be described shortly. As can be seen, the sample autocorrelations "tail off," while the sample partial autocorrelation matrices "cut off" after lag 1. This is in agreement with a vector AR(1) process, as simulated.   □

Example 9.2: Bivariate Modeling of Single-Input, Single-Output Transfer Functions (Tiao and Box, 1981). The Box–Jenkins methodology for fitting single-input, single-output (SISO) transfer functions (Chapter 4) is quite involved. An alternative way of identifying and estimating a SISO (single-output, single-input) transfer function is to consider the input–output data as a bivariate time-series process. To illustrate, consider the gas furnace data (series J) in Box and Jenkins (1976). The data can be found in the file *BJ-J.txt*. The input, X_t, is the input gas rate, and the output, Y_t, is the concentration of CO_2 in the outlet of a furnace. The schematic representations of the sample autocorrelation and partial autocorrelation of the bivariate time series

$$\mathbf{Z}_t = \begin{pmatrix} X_t \\ Y_t \end{pmatrix}$$

are:

Schematic Representation of Correlations

Name / Lag	0	1	2	3	4	5	6	7	8	9	10
X	+-	+-	+-	+-	+-	+-	+-	+-	+-	+-	+-
Y	-+	-+	-+	-+	-+	-+	-+	-+	-+	-+	-+

Schematic Representation of Partial Autocorrelations

Name / Lag	1	2	3	4	5	6	7	8	9	10
X	+.	-.	+.	..	..	-.	..	..	..	..
Y	-+	-.	.+	..	..	..	..	..	.+	

These functions indicate that the process is some type of AR process, perhaps of order 3 or 4 (notice how the sample PACF cuts off). The

Yule–Walker estimates of a vector AR(4) process are

| | ---- Lag = 1 ---- | | ---- Lag = 2 ---- | | ---- Lag = 3 ---- | | ---- Lag = 4 ---- | |
	X	Y	X	Y	X	Y	X	Y
X	1.9258	-0.001	-1.201	0.0042	0.1169	-0.008	0.1042	0.0032
Y	0.0504	1.2997	-0.020	-0.327	-0.711	-0.257	0.1954	0.1334

Each lag k corresponds to a 2×2 matrix $\hat{A}_k$. Considering the statistically significant estimates only, we can entertain the model

$$\left[I - \begin{pmatrix} 1.95 & 0 \\ 0 & 1.29 \end{pmatrix} \mathcal{B} - \begin{pmatrix} -1.20 & 0 \\ 0 & -0.32 \end{pmatrix} \mathcal{B}^2 - \begin{pmatrix} 0.11 & 0 \\ -0.71 & 0.25 \end{pmatrix} \mathcal{B}^3 \right.$$

$$\left. - \begin{pmatrix} 0.10 & 0 \\ 0.19 & 0.13 \end{pmatrix} \mathcal{B}^4 \right] \begin{pmatrix} X_t \\ Y_t \end{pmatrix} = \begin{pmatrix} \varepsilon_{1,t} \\ \varepsilon_{2,t} \end{pmatrix}$$

or, in more simple terms,

$$\left[\begin{pmatrix} \hat{\phi}_{11}(\mathcal{B}) & 0 \\ \hat{\phi}_{21}(\mathcal{B}) & \hat{\phi}_{22}(\mathcal{B}) \end{pmatrix} \right] \begin{pmatrix} X_t \\ Y_t \end{pmatrix} = \begin{pmatrix} \varepsilon_{1,t} \\ \varepsilon_{2,t} \end{pmatrix}$$

The second row gives the estimated SISO transfer function model:

$$Y_t = - \frac{-\hat{\phi}_{21}(\mathcal{B})}{\hat{\phi}_{22}(\mathcal{B})} X_t + \frac{\varepsilon_{2,t}}{\hat{\phi}_{22}(\mathcal{B})}$$

or

$$Y_t = \frac{-0.71 \mathcal{B}^3 + 0.19 \mathcal{B}^4}{1 - 1.29 \mathcal{B} + 0.32 \mathcal{B}^2 + 0.25 \mathcal{B}^3 - 0.13 \mathcal{B}^4} X_t$$

$$+ \frac{\varepsilon_{2,t}}{1 - 1.29 \mathcal{B} + 0.32 \mathcal{B}^2 + 0.25 \mathcal{B}^3 - 0.13 \mathcal{B}^4} \qquad (9.2)$$

As it can be seen, the input–output delay is 3 time units. For these data, using the procedure described in Chapter 4, Box and Jenkins (1976) identified and fit the following $(r, s, k) = (2, 2, 3)$ transfer function with AR(2) noise:

$$Y_t = \frac{-0.53 + 0.37 \mathcal{B} + 0.51 \mathcal{B}^2}{1 - 0.57 \mathcal{B} - 0.01 \mathcal{B}^2} X_{t-3} + \frac{\varepsilon_t}{1 - 1.53 \mathcal{B} + 0.63 \mathcal{B}^2} \qquad (9.3)$$

The two models identified, equations (9.2) and (9.3), look different, but as pointed out by Tiao and Box (1981) we should compare their impulse

Table 9.1 Impulse Response Weights for the Two Models Fitted to the Box–Jenkins Data

	Lag												
Model	0	1	2	3	4	5	6	7	8	9	10	11	12
Eq. (9.2)	0	0	0	−0.71	−0.72	−0.71	−0.51	−0.34	−0.19	−0.10	−0.05	−0.03	−0.02
Eq. (9.3)	0	0	0	−0.53	−0.67	−0.89	−0.51	−0.29	−0.17	−0.09	−0.05	−0.03	−0.02

response weights v_j (see Chapter 4). Table 9,1 compares the impulse response weights for these two models. As can be seen, both functions have captured essentially the same dynamic behavior.

Notice that no $(1, 2)$ entry in the $\hat{A}_j$ matrices was significant. This implies that there is no relation from Y to X; that is, there were no (linear) feedback adjustments taking place while the data were being collected. □

9.1.3 General Form of a Multivariate ARMA Process

A vector ARMA(p, q) time-series model has the form

$$A(\mathcal{B})\tilde{\mathbf{Y}}_t = C(\mathcal{B})\boldsymbol{\varepsilon}_t$$

where $\tilde{\mathbf{Y}}_t = \mathbf{Y}_t - \boldsymbol{\mu}$ and

$$A(\mathcal{B}) = I_N - A_1\mathcal{B} - A_2\mathcal{B}^2 - \cdots - A_p\mathcal{B}^p$$

$$C(\mathcal{B}) = I_N - C_1\mathcal{B} - C_2\mathcal{B}^2 - \cdots - C_q\mathcal{B}^q$$

Here the A_i and C_i are $N \times N$ matrices, where N is the dimension of the vector $\mathbf{Y}_t$. This is sometimes called a VARMA(p, q) process in the literature. This process is *stationary* if the roots of

$$\det(A(\mathcal{B})) = 0$$

lie all outside the unit circle on the complex plane. Similarly, the process is *invertible* if all the roots of

$$\det(C(\mathcal{B})) = 0$$

lie outside the unit circle. It should be pointed out that two multivariate ARMA(p, q) models with different matrices $\{A_j, C_j\}$ can have the same autocovariance function $\Gamma(k)$. For this reason, some authors suggest fitting only vector AR models, but this may lead to nonparsimonious models. See Hannan and Diestler (1988) and Reinsel (1997, Chap. 2) for details.

Example 9.3: Invertibility of a Vector MA(1) Process. Consider a bivariate (i.e., $N = 2$) MA(1) process with

$$C_1 = \begin{pmatrix} 0.8 & 0.7 \\ -0.4 & 0.6 \end{pmatrix}$$

The $C(\mathcal{B})$ matrix polynomial is

$$I - C_1\mathcal{B} = \begin{pmatrix} 1 - 0.8\mathcal{B} & -0.7\mathcal{B} \\ 0.4\mathcal{B} & 1 - 0.6\mathcal{B} \end{pmatrix}$$

from which we obtain that $\det(I - C_1\mathcal{B}) = 1 - 1.4\mathcal{B}^2 + 0.76\,\mathcal{B}$. This has roots at $0.9210 \pm 0.6837\,i$, which have magnitude equal to 1.1470. Therefore, this is an invertible process, in the same sense as discussed in Chapter 3 for univariate processes. An alternative characterization of invertibility for a vector MA(1) process is that for invertibility the eigenvalues of the C_1 matrix must be less than 1 in magnitude. In the example, the eigenvalues are $0.7 \pm 0.5196\,i$ with magnitude $0.87178 < 1$, so this implies that the process is invertible. Similarly, for a vector AR(1) process $[I - A_1(\mathcal{B})]Y_t = \varepsilon_t$, if the eigenvalues of A_1 are all less than 1 in magnitude, the process is stationary.[2]

□

9.1.4 Correlation Patterns of Vector AR(1) and Vector MA(1) Processes

Multivariate ARMA(p, q) processes have covariance functions that resemble their univariate counterparts. For example, a multivariate AR(1) process has an autocovariance function as follows:

$$\Gamma(0) = A_1\Gamma(0)A_1' + \Sigma$$
$$\Gamma(1) = A_1\Gamma(0)$$
$$\Gamma(k) = A_1^k\Gamma(0) \qquad k \geq 1$$

The autocorrelation matrices have the form

$$\begin{aligned} \rho(k) &= S^{-1}\Gamma(k)S^{-1} \\ &= S^{-1}A_1^k\Gamma(0)S^{-1} \\ &= \rho(0)SA_1^kS^{-1} \end{aligned}$$

so this function decays or tails off if the eigenvalues of A_1 are less than 1 in magnitude. Similarly, for a multivariate MA(1) process, the autocovariance

---

[2] From now on, we drop the tilde with the understanding that the series are modeled as deviations from their mean.

function is

$$\Gamma(0) = \Sigma + C_1 \Sigma C_1'$$
$$\Gamma(1) = -C_1 \Sigma$$
$$\Gamma(k) = \mathbf{0} \quad k \geq 1$$

In this case, the function cuts off after lag 1. These behaviors are identical to those of univariate AR(1) and MA(1) processes, as explained in Sections 3.1 and 3.2, respectively. Given sample estimates of $\Gamma(0), \Gamma(1), \ldots$, or of $\boldsymbol{\rho}(0), \boldsymbol{\rho}(1) \ldots$ one can solve for the matrices of parameters A_1 and C_1 and get moment estimates. (These are the *Yule–Walker estimates*, as discussed for the univariate case in Chapter 3 and shown for the vector case in Example 9.2.)

9.1.5   Vector ARIMA and ARMAX Processes

In a similar way as in univariate time-series models, nonstationary but nonexplosive behavior is a very useful model feature in applications. If in a vector ARMA(p, q) process some of the roots of $\det(A(\mathcal{B})) = 0$ equal to 1 in magnitude, homogeneous nonstationary behavior will result. A vector ARIMA process is defined by

$$A_I(\mathcal{B}) \, D(\mathcal{B}) \mathbf{Y}_t = C(\mathcal{B}) \boldsymbol{\varepsilon}_t$$

where $\det(A_I(\mathcal{B})) = 0$ has all roots outside the unit circle and $D(\mathcal{B}) = \mathrm{Diag}\{\nabla^{d_1}, \nabla^{d_2}, \ldots, \nabla^{d_N}\}$. This allows each series to have a different nonstationary order. For example, a vector IMA$(1, 1)$ process is given by

$$(1 - \mathcal{B})\mathbf{Y}_t = \left(I - C_1 \mathcal{B}\right) \boldsymbol{\varepsilon}_t$$

where the eigenvalues of C_1 are less than 1 in magnitude.

Finally, let us denote by r_1 the number of controllable factors and r_2 the number of responses. A multivariate ARMAX model is of the form

$$A(\mathcal{B})\mathbf{Y}_t = B_{N_D}(\mathcal{B})\mathbf{X}_{t-k} + C(\mathcal{B})\boldsymbol{\varepsilon}_t \tag{9.4}$$

which can also be written as

$$A(\mathcal{B})\mathbf{Y}_t = B(\mathcal{B})\mathbf{X}_t + C(\mathcal{B})\boldsymbol{\varepsilon}_t \tag{9.5}$$

where

$$A(\mathcal{B}) = I_{r_2} - A_1 \mathcal{B} - A_2 \mathcal{B}^2 - \cdots - A_p \mathcal{B}^p$$

is a matrix polynomial of order p in which the A_i are $r_2 \times r_2$ matrices,

$$B(\mathcal{B}) = B_0 \mathcal{B} + B_1 \mathcal{B}^2 + B_2 \mathcal{B}^3 + \cdots + B_l \mathcal{B}^{l+1}$$

is an $(l + 1)$-order matrix polynomial of $r_2 \times r_1$ matrices, and

$$C(\mathcal{B}) = I_{r_2} - C_1\mathcal{B} - C_2\mathcal{B}^2 - \cdots - C_q\mathcal{B}^q$$

is a matrix polynomial of order q in which the C_i are $r_2 \times r_2$ matrices. In the discussion in Sections 9.2 and 9.3, model (9.5), which assumes that $k \geq 1$, will be used for *model estimation* purposes. However, in Section 9.4 the form (9.4) will be used for forecasting and *to derive process adjustment rules*, as it is the more natural form for this purpose. If we fit (9.5) and the delay k is greater than one period, this will result in nonsignificant estimates for the matrices $B_0, B_1, \ldots, B_{k-2}$ from which we can infer k and write the model back in the (9.4) form.

In a way analogous to the univariate case, the k-step-ahead minimum mean square error forecast of a vector ARMAX process is

$$\hat{\mathbf{Y}}_{t+k|t} = F(\mathcal{B})C^{-1}(\mathcal{B})B_{N_D}(\mathcal{B})\mathbf{X}_t + G(\mathcal{B})C^{-1}(\mathcal{B})\mathbf{Y}_t$$

where $F(\mathcal{B})$ and $G(\mathcal{B})$ are matrix polynomials (of $r_2 \times r_2$ matrices) of orders $k - 1$ and $\max(p, n_{B_{N_D}}, q) - 1$, respectively. Notice that we are using the form (9.4), with $n_{B_{N_D}}$ equal to the highest power of $\mathcal{B}$ in the $B_{N_D}(\mathcal{B})$ polynomial. In the same way as in the univariate case (see Chapter 5), the G and F polynomials are obtained by equating matrix coefficients of like powers of $\mathcal{B}$ in the matrix Diophantine identity:

$$C(\mathcal{B}) = F(\mathcal{B})A(\mathcal{B}) + \mathcal{B}^k G(\mathcal{B})$$

Examples of the use of this equation are given in Section 9.4 when we apply them to process adjustment problems.

9.2 IDENTIFICATION AND FITTING OF ARMA AND ARMAX PROCESSES USING SAS PROC STATESPACE

SAS PROC STATESPACE helps identify and fit the parameters F, G, and Σ in the state-space model[3]

$$\mathbf{Z}_{t+1} = F\mathbf{Z}_t + G\boldsymbol{\varepsilon}_{t+1} \tag{9.6}$$

$$\mathbf{W}_t = [I_r \quad 0]\mathbf{Z}_t \tag{9.7}$$

where $\mathbf{Z}$ is a $n \times 1$ vector of state variables, $\{\varepsilon_t\}$ is a sequence of $r \times 1$ i.i.d. multivariate normal $(\mathbf{0}, \Sigma)$ random variables, F is a $n \times n$ matrix called the *transition matrix*, and G is an $n \times r$ matrix called the *input matrix*. Clearly, G

---

[3]Readers should consult Appendix 4A for an introduction to state-space models.

determines the variance of the state variable. In this formulation, the first r ($< n$) state variables are assumed directly observable, so equation (9.7) simply retrieves them. The last $n - r$ state variables are predicted values of the observable variables. This model differs from conventional state-space models in that some of the state variables can be observed directly. Also, the controllable factors, if any, are included in the state vector, and this can complicate the state-space representation. The model is explained in Akaike (1976) and in Akaike and Nakagawa (1988), which SAS PROC STATES-PACE follows closely. This formulation can be used to fit vector ARMA and ARMAX models.

For example, consider the multivariate ARMA(p,q) model

$$\left(I - A_1\mathcal{B} - A_2\mathcal{B}^2 - \cdots -A_p\mathcal{B}^p\right)\mathbf{Y}_t = \left(I - C_1\mathcal{B} - C_2\mathcal{B}^2 \cdots C_q\mathcal{B}^q\right)\boldsymbol{\varepsilon}_t$$

This can be fitted by considering the model

$$\mathbf{Z}_{t+1} = \begin{pmatrix} \mathbf{Y}_{t+1} \\ \mathbf{Y}_{t+2\,|t+1} \\ \vdots \\ \mathbf{Y}_{t+p|t+1} \end{pmatrix}$$

$$= \begin{bmatrix} \mathbf{0} & I_r & \cdots & & \mathbf{0} \\ \mathbf{0} & \mathbf{0} & I_r & \cdots & \mathbf{0} \\ \vdots & \vdots & \vdots & \ddots & \vdots \\ \mathbf{0} & \mathbf{0} & \cdots & \cdots & I_r \\ A_p & A_{p-1} & \cdots & \cdots & A_1 \end{bmatrix} \begin{pmatrix} \mathbf{Y}_t \\ \mathbf{Y}_{t+1|t} \\ \vdots \\ \mathbf{Y}_{t+p-1|t} \end{pmatrix} + \begin{pmatrix} I_r \\ \boldsymbol{\Psi}_1 \\ \vdots \\ \boldsymbol{\Psi}_{p-1} \end{pmatrix} \boldsymbol{\varepsilon}_{t+1} \quad (9.8)$$

where $\boldsymbol{\varepsilon}_{t+1}$ has the same dimensions as the $\mathbf{Y}$'s (i.e., $r \times 1$). Here $\mathbf{Y}_{t+j|t}$ denotes a j-steps-ahead forecast or prediction computed at time t (we omit the "hat" notation in this section for simplicity of notation). The matrices $\boldsymbol{\Psi}_j$ ($j = 1, 2, \ldots, p - 1$) are the impulse response weight matrices of the ARMA process, the multivariate counterpart of the v_j weights described in Chapter 4. Note that to represent a model with an AR(p) component, the state vector must contain p future values of the multivariate series $\mathbf{Y}_t$. More details about the equivalence between vector ARMA and ARMAX processes and the state-space formulation used by PROC STATESPACE will be given later in this chapter.

SAS uses an approximate maximum likelihood method to fit the parameters in (9.8). The C_j matrices of the MA part of the model can be obtained simply from the $\boldsymbol{\Psi}_j$ and A_j matrices by equating matrix coefficients of like powers of $\mathcal{B}$ in

$$C(\mathcal{B}) = A(\mathcal{B}) \sum_{j=0}^{\infty} \Psi_j \mathcal{B}^j$$

Each individual series Y_i is assumed stationary, so differencing may be necessary prior to fitting an ARMA or ARMAX model using SAS.

A few examples illustrate this procedure. We first illustrate the use of PROC STATESPACE for fitting vector AR(p) processes and then show how to deal with vector ARMA processes. The case of vector ARMAX processes is explained next. In the following examples, simulated time-series models with known structure were used to observe clearly how the state-space approach described in this section is successful in identifying the underlying ARMA or ARMAX process.

Example 9.4: Identification and Estimation of a Vector AR(1) Process. Consider the same process and data as in Example 9.1. After reading the data into a SAS file called *vecAR1*, the statements

```
proc statespace data = vecAR1 out = out;
   var Y1 Y2;
 run;
```

will generate output that is partially shown in Figure 9.2.

As discussed in Example 9.1, the sample ACF and sample PACF indicate an AR(1) process. From Figure 9.2, we observe that the fitted model is

$$\begin{pmatrix} Y_{1,t+1} - 2.44 \\ Y_{2,t+1} - 0.42 \end{pmatrix} = \begin{bmatrix} 0.813 & 0.164 \\ 0.330 & -0.486 \end{bmatrix} \begin{pmatrix} Y_{1,t} - 2.44 \\ Y_{2,t} - 0.42 \end{pmatrix} + \varepsilon_{t+1} \quad (9.9)$$

where

$$\hat{\Sigma} = \begin{pmatrix} 29.81 & 10.41 \\ 10.41 & 28.17 \end{pmatrix}$$

from where we can see that $\hat{F} = \hat{A}_1$ and the state-space model is already in AR(1) from. The constants 2.44 and 0.42 correspond to the mean of the two time series. PROC STATESPACE always centers the data by default.

To look at the residuals of this model, one can use again PROC STATESPACE applied to RES1 and RES2, the internal names SAS uses for the two residual series. The following statements accomplish such a task:

```
proc statespace data = out out = out2;
var RES1 RES2;
run;
```

Selected statespace form and fitted model

State vector

Y1 (T;T)   Y2 (T;T)

Estimate of transition matrix

0.813	0.164595
0.330702	−0.48689

Input matrix for innovation

1	0
0	1

Variance matrix for innovation

29.81443	10.41516
10.41516	28.17115

Parameter estimates

Parameter	Estimate	Standard Error	t Value
F (1,1)	0.813000	0.039697	20.48
F (1,2)	0.164595	0.065711	2.50
F (2,1)	0.330702	0.038588	8.57
F (2,2)	−0.48689	0.063876	−7.62

Schematic representation of correlations

Name/Lag	0	1	2	3	4	5	6	7	8	9	10
RES1	++	..	..	..	..	..	..	..	..	..	..
RES2	++	..	..	..	..	..	..	..	..	..	..

+ is > 2*std error, − is < −2*std error, . is between

Schematic representation of partial autocorrelations

Name/Lag	1	2	3	4	5	6	7	8	9	10
RES1	..	..	..	..	..	..	..	..	..	..
RES2	..	..	..	..	..	..	..	..	..	..

+ is > 2*std error, − is < −2*std error, . is between

Figure 9.2 Partial output generated by SAS PROC STATESPACE in Example 9.4. Top, final estimates; bottom, sample ACF and sample PACF of residuals.

Part of the output generated by these statements is shown at the bottom of Figure 9.2. The schematic representation of the sample ACF and sample PACF of the residuals indicates that they are multivariate white noise, and therefore model (9.9) is deemed appropriate.   □

The next example illustrates how to fit a vector ARMA model.

Example 9.5: Identification and Estimation of a Vector ARMA(1, 1) Process.
Two hundred observations were simulated from a vector ARMA(1, 1) process
with parameters

$$A_1 = \begin{pmatrix} 0.8 & 0.2 \\ 0.3 & -0.4 \end{pmatrix} \quad C_1 = \begin{pmatrix} 0.7 & 0.2 \\ -0.3 & 0.6 \end{pmatrix} \quad \Sigma = \begin{pmatrix} 25 & 7.5 \\ 7.5 & 25 \end{pmatrix}$$

The data can be found in the file *MultiARMA.txt*. The schematic sample ACF
and PACF are as follows:

```
                 Schematic Representation of Correlations
Name / Lag    0    1    2    3    4    5    6    7    8    9    10
Y1           ++   .+   ..   ..   ..   ..   ..   ..   ..   ..   ..
Y2           ++   +-   -.   ..   ..   ..   ..   ..   ..   ..   ..

              Schematic Representation of Partial Autocorrelations
Name / Lag    1    2    3    4    5    6    7    8    9    10
Y1           ..   ..   ..   .-   ..   ..   ..   ..   ..   ..
Y2           +-   .-   ..   ..   ..   ..   ..   ..   ..   ..
```

These do not reveal an AR − like (tailing off) pattern, and a mixed model is
a natural candidate.[4] To fit a state-space model from where we can extract
the tentative ARMA(1, 1) parameter estimates, the following statements were
entered after loading the data set into the SAS file *vecARMA*:

```
proc statespace data = vecARMA out = out;
      var Y1 Y2;
      form Y1 2 Y2 2;
run;
```

The optional statement `form Y1 2 Y2 2;` forces SAS to include $Y_{1, t+1|t}$ and
$Y_{2, t+1|t}$ as part of the state vector in addition to the other state variables $Y_{i, t}$,
which are always included. This is important if we suspect that an AR(1)
component should be in the model. Figure 9.3 shows a partial listing of the
output generated by SAS.

[4]SAS PROC STATESPACE uses an automatic identification procedure based on canonical
correlations and an information statistic proposed by Akaike (1976, this is *not* his celebrated
AIC). In essence, if the current state vector seems to contain all the information needed to
predict a future value of some variable currently in the state vector, (say $\mathbf{Y}_t$), the predictor $\mathbf{Y}_{t+k|t}$
is not added to the state-space vector. We advise against blind use of these automatic routines,
which although they can give insight into a given problem, should not be used as a substitute for
an iterative identification approach based on inspection of the sample ACF and PACF and a
residual analysis. This is the same approach as explained in Chapters 3 and 4 and recommended
in Box and Jenkins (1976) and Box and Tiao (1981). Automatic identification routines become
more useful as r_1 and r_2 get large. Another commercial software for automatic identification
which follows Akaike's method is ADAPT$_X$ (Larimore, 1999).

State vector

Y1 (T;T)	Y2 (T;T)	Y1 (T+1;T)	Y2 (T+1;T)

Estimate of transition matrix

0	0	1	0
0	0	0	1
−0.00409	−0.02479	−0.75725	0.300663
−0.039	−0.10137	0.193588	−0.43804

Input matrix for innovation

1	0
0	1
0.026779	0.054677
0.676872	− 0.83162

Variance matrix for innovation

26.08026	9.339762
9.339762	25.22798

Parameter estimates

Parameter	Estimate	Standard Error	t Value
F (3, 1)	−0.00409	0.108387	−0.04
F (3, 2)	−0.02479	0.133237	−0.19
F (3, 3)	0.757250	0.152696	4.96
F (3, 4)	0.300663	0.174780	1.72
F (4, 1)	−0.03900	0.104052	−0.37
F (4, 2)	−0.10137	0.117774	−0.86
F (4, 3)	0.193588	0.267685	0.72
F (4, 4)	−0.43804	0.167533	−2.61
G (3, 1)	0.026779	0.075363	0.36
G (3, 2)	0.054677	0.075520	0.72
G (4, 1)	0.676872	0.074555	9.08
G (4, 2)	−0.83162	0.075202	−11.06

Schematic representation of correlations

Name/Lag	0	1	2	3	4	5	6	7	8	9	10
RES1	++	..	..	..	..	..	..	..	..	..	..
RES2	++	..	..	..	..	..	..	..	..	..	..

+ is > 2*std error, − is < −2*std error, . is between

Schematic representation of partial autocorrelations

Name/Lag	1	2	3	4	5	6	7	8	9	10
RES1	..	..	..	..	..	..	..	..	..	..
RES2	..	..	..	..	..	..	..	..	..	..

Figure 9.3 Partial output generated by SAS PROC STATESPACE in Example 9.5. Top, final estimates; bottom, sample ACF and sample PACF of residuals.

Following the state-space formulation given by equation (9.8), the $\hat{A}_1$ matrix is extracted from the transition matrix $\hat{F}$ taking the lower right 2×2 submatrix:

$$\hat{A}_1 = \begin{pmatrix} 0.7572 & 0.3006 \\ 0.1935 & -0.4380 \end{pmatrix}$$

Note how the entries of this submatrix have relative large t-statistics (Figure 9.3). These t-values test the null hypothesis that the corresponding coefficient is zero, a hypothesis that should be rejected when the t-values are large. In comparison, the lower-left submatrix in $\hat{F}$ gives an estimate of A_2, the second-order AR matrix of coefficients. The t-values associated with this matrix are quite small, so the estimates are not significant and the AR part of the model appears to be of first order only.[5]

To get the estimates of the MA matrix, C_j, we first extract the $\hat{\mathbf{\Psi}}_j$ weights from the estimated input matrix $\hat{G}$, following equation (9.8). From the output in Figure 9.3, we have that

$$\mathbf{\Psi}_0 = I \qquad \hat{\mathbf{\Psi}}_1 = \begin{pmatrix} 0.0267 & 0.0546 \\ 0.6768 & -0.8316 \end{pmatrix}$$

In general, the relation

$$I - C_1 \mathcal{B} - C_2 \mathcal{B}^2 - \cdots - C_q \mathcal{B}^q = \left(I - A_1 \mathcal{B} - A_2 \mathcal{B}^2 - \cdots - A_p \mathcal{B}^p \right) \sum_{j=0}^{\infty} \mathbf{\Psi}_j \mathcal{B}^j$$

holds, so equating matrices of coefficients associated to like powers of $\mathcal{B}$, we find that

$$C_j = \sum_{i=1}^{j} A_i \mathbf{\Psi}_{j-i} - \mathbf{\Psi}_j$$

From this last result, we can get an estimate of C_1 in the example

$$\hat{C}_1 = \hat{A}_1 \mathbf{\Psi}_0 - \hat{\mathbf{\Psi}}_1 = \hat{A}_1 - \hat{\mathbf{\Psi}}_1 = \begin{pmatrix} 0.7312 & 0.246 \\ -0.483 & 0.7872 \end{pmatrix}$$

These estimates are very close to the true ones which were simulated. Performing a residual analysis in exactly the same way as in example 9.4, we find that the residuals behave like white noise (Figure 9.3, bottom), and we can conclude that the fitted ARMA(1, 1) model is appropriate. The final

---

[5]The optional statement `restrict` could be used to fix the nonsignificant parameters to zero and then the model could be refit.

model is

$$\left(I - \hat{A}_1 \mathcal{B}\right)\begin{pmatrix} Y_{1,t} - 0.6116 \\ Y_{2,t} - 0.2866 \end{pmatrix} = \left(I - \hat{C}_1 \mathcal{B}\right)\varepsilon_t$$

with

$$\hat{\Sigma} = \begin{pmatrix} 26.08 & 9.33 \\ 9.33 & 25.22 \end{pmatrix} \qquad \square$$

Fitting ARMAX Models with SAS PROC STATESPACE

PROC STATESPACE can also be used to fit multivariate ARMAX models, which are of obvious concern for process adjustment purposes. In this case, the state vector will contain all observed controllable factors $X_{i,t}$, $i = 1, 2, \ldots, r_1$, followed by all the observed outputs $Y_{i,t}$, $i = 1, 2, \ldots, r_2$. Similar to vector ARMA models, the state vector may also contain $n - r$ forecasted variables in addition to the aforementioned $r \, (= r_1 + r_2)$ directly observable variables. With this state-space formulation, we wish to find estimates of the parameters in the multivariate ARMAX process:

$$\left(I - A_1 \mathcal{B} - \cdots - A_p \mathcal{B}^p\right)\mathbf{Y}_t$$

$$= \left(B_0 \mathcal{B} + B_1 \mathcal{B}^2 + \cdots + B_l \mathcal{B}^{l+1}\right)\mathbf{X}_t + \left(I - C_1 \mathcal{B} - \cdots - C_q \mathcal{B}^q\right)\varepsilon_t$$

where $\varepsilon_t \sim N(\mathbf{0}, \Sigma)$ is a $r_2 \times 1$ vector of errors.

We would like to represent the multivariate ARMAX model in a form similar to the state-space model given by equation (9.8). To achieve this end, notice that the ARMAX model can be written as

$$\mathbf{Y}_t = \sum_{j=0}^{\infty} V_j \mathbf{X}_{t-j} + \sum_{j=0}^{\infty} \mathbf{\Psi}_j \varepsilon_{t-j}$$

where the V_j matrices are the impulse response weights of the input–output transfer function (see Chapter 4) and the $\mathbf{\Psi}_i$ matrices are the impulse response weights of the noise model. The above implies that

$$\mathbf{Y}_{t+i|t} = \sum_{j=1}^{\infty} V_j \mathbf{X}_{t+i-j} + \sum_{j=1}^{\infty} \mathbf{\Psi}_j \varepsilon_{t+i-j}$$

$$= V_i \mathbf{X}_t + V_{i+1} \mathbf{X}_{t+1} + \cdots + \mathbf{\Psi}_i \varepsilon_t + \mathbf{\Psi}_{i+1} \varepsilon_{t+1} + \cdots \qquad (9.10)$$

To find $\mathbf{Y}_{t+i|t+1}$ ($i \geq 2$), which gives potential elements of the state vector, we do as follows:

$$\mathbf{Y}_{t+i|t+1} = \mathbf{Y}_{(t+1)+(i-1)|t+1} = \sum_{j=i-1}^{\infty} V_j \mathbf{X}_{(t+1)+(i-1)-j}$$

$$+ \sum_{j=i-1}^{\infty} \boldsymbol{\Psi}_j \boldsymbol{\varepsilon}_{(t+1)+(i-1)-j}$$

$$= V_{i-1}\mathbf{X}_{t+1} + V_i\mathbf{X}_t + V_{i+1}\mathbf{X}_{t+1} + \cdots$$

$$+ \boldsymbol{\Psi}_{i-1}\boldsymbol{\varepsilon}_{t+1} + \boldsymbol{\Psi}_i\boldsymbol{\varepsilon}_t + \boldsymbol{\Psi}_{i+1}\boldsymbol{\varepsilon}_{t+1} + \cdots$$

$$= \mathbf{Y}_{t+i|t} + V_{i-1}\mathbf{X}_{t+1} + \boldsymbol{\Psi}_{i-1}\boldsymbol{\varepsilon}_{t+1} \tag{9.11}$$

where the last expression follows from (9.10). Expression (9.11) gives us equations that can be written using a recursion involving entries in the F and G matrices. If we let $i = m = \max(p, l + 1, q + 1)$, we have that

$$\mathbf{Y}_{t+m|t+1} = \sum_{j=1}^{m} A_j \mathbf{Y}_{t+m-j|t} + \sum_{j=1}^{m} B_{j-1} \mathbf{X}_{t+m-j|t} + V_{m-1}\mathbf{X}_{t+1} + \boldsymbol{\Psi}_{m-1}\boldsymbol{\varepsilon}_{t+1} \tag{9.12}$$

where the two terms with summations equal the prediction $\mathbf{Y}_{t+m|t}$, obtained directly from the ARMAX model definition. If there are r_1 controllable factors and r_2 responses (so $r_1 + r_2 = r$), we can define the $n \times 1$ state vector

$$\mathbf{Z}_{t+1} = \begin{pmatrix} \mathbf{X}_{t+1} \\ \mathbf{Y}_{t+1} \\ \mathbf{X}_{t+2|t+1} \\ \mathbf{Y}_{t+2|t+1} \\ \vdots \\ \mathbf{X}_{t+m|t+1} \\ \mathbf{Y}_{t+m|t+1} \end{pmatrix}$$

The state vector thus contains m pairs of $\mathbf{X}, \mathbf{Y}$ vectors, with $n = m(r_1 + r_2)$. Model (9.6) can then be written as

$$
\mathbf{Z}_{t+1} =
\begin{bmatrix}
\mathbf{0} & I & \mathbf{0} & \cdots & \mathbf{0} \\
\mathbf{0} & \mathbf{0} & I & \cdots & \mathbf{0} \\
\vdots & \vdots & \vdots & \ddots & \vdots \\
\mathbf{0} & \mathbf{0} & \mathbf{0} & \cdots & I \\
* & * & * & * & * \\
B_{m-1},\,A_m & B_{m-2},\,A_{m-1} & \cdots & \cdots & B_0,\,A_1
\end{bmatrix}
\begin{pmatrix}
\mathbf{X}_t \\
\mathbf{Y}_t \\
\mathbf{X}_{t+1|t} \\
\mathbf{Y}_{t+1|t} \\
\vdots \\
\mathbf{X}_{t+m-1|t} \\
\mathbf{Y}_{t+m-1|t}
\end{pmatrix}
$$

$$
+
\begin{bmatrix}
I & \mathbf{0} \\
\mathbf{0} & I \\
* & * \\
V_1 = B_0 & \Psi_1 \\
* & * \\
V_2 = B_1 + A_1 B_0 & \Psi_2 \\
\vdots & \vdots \\
* & * \\
V_{m-1} & \Psi_{m-1}
\end{bmatrix}
\begin{pmatrix}
\boldsymbol{\varepsilon}_{X,t+1} \\
\boldsymbol{\varepsilon}_{t+1}
\end{pmatrix}
\tag{9.13}
$$

where the entries labeled with an asterisk ($*$) are not relevant for our purposes. The first r_1 equations imply that

$$
X_{i,t+1} = \varepsilon_{X_i,t+1} \qquad i = 1, 2, \ldots, r_1
$$

where $\varepsilon_{X_i,t+1}$ is the error assumed associated with the ith controllable variable. If this assumption holds, equations (9.11) and (9.12) hold in the formulation (9.13). This formulation thus assumes that the process is operated in open-loop mode, with the controllable factors varying as i.i.d. variables. The second set of r_2 variables imply the vector equation

$$
\mathbf{Y}_{t+1} = \mathbf{Y}_{t+1|t} + \boldsymbol{\varepsilon}_{t+1} \quad \text{or} \quad \boldsymbol{\varepsilon}_{t+1} = \mathbf{Y}_{t+1} - \mathbf{Y}_{t+1|t}
$$

which are the one-step-ahead forecast errors or innovations.

Example 9.6: Identification and Fitting of a Vector ARMAX Process. As an illustration, consider the 500 observations in the file *MultiARMAX.txt*. These were simulated from an uncontrolled 3×2 ARMAX process of the form

$$
(I - A_1 \mathcal{B})\mathbf{Y}_t = (B_0 \mathcal{B} + B_1 \mathcal{B}^2 + B_2 \mathcal{B}^3)\mathbf{X}_t + (I - C_1 \mathcal{B})\boldsymbol{\varepsilon}_t
$$

where

$$A_1 = \begin{pmatrix} 0.8 & -0.2 \\ 0.5 & -0.7 \end{pmatrix} \qquad B_0 = \begin{pmatrix} -0.5 & 0.4 & 0.8 \\ 0.9 & -0.7 & -0.4 \end{pmatrix}$$

$$B_1 = \begin{pmatrix} 0.6 & 0.3 & -0.3 \\ 0.6 & 0.5 & 0.9 \end{pmatrix} \qquad B_2 = \begin{pmatrix} 0.8 & 0.4 & 0.9 \\ -0.6 & -0.7 & -0.3 \end{pmatrix}$$

$$C_1 = \begin{pmatrix} 0.7 & 0.2 \\ -0.3 & 0.6 \end{pmatrix} \qquad \varepsilon_t \sim N(\mathbf{0}, I_2)$$

and $\mathbf{X} \sim N(\mathbf{0}, I_2)$. Therefore, in this case we have $r_1 = 3$ and $r_2 = 2$. From the estimated autocorrelation and partial autocorrelations, the process seems to follow some type of mixed ARMA process as far as Y_1 and Y_2 go (last two rows of the matrices):

```
              Schematic Representation of Correlations
Name     0      1      2      3      4      5      6      7      8      9
X1    +.-..  .....  .....  .....  .....  .....  .-...  +....  .....  .....
X2    .+...  .....  .....  .....  .....  .....  .....  ..-..  .....  .....
X3    -.+..  .+...  .....  .....  .+...  .....  .....  .....  .....  .....
Y1    ...++  -+++.  .++++  ++++.  +++++  ++++.  .+++.  .+.+.  .+.+.  .....
Y2    ...++  +--+-  -++.+  ---+-  +++.+  -..+.  +++..  .....  .....  ....-

          Schematic Representation of Partial Autocorrelations
Name     1      2      3      4      5      6      7      8      9
X1    .....  .....  .....  -....  .....  .....  .....  .....  .....
X2    .....  .....  .....  .....  .....  .....  .....  .....  .....
X3    .+...  ....-  .....  .+...  .....  .....  .....  .....  .....
Y1    -+++-  ++..+  ++++.  +++..  +.++.  ...+..  ++...  .....  .....
Y2    +--+-  ++++-  ---.-  ---.-  ---..  -.--+  .....  ..-..  .....
```

From these functions it appears that the three controllable variables (the first three elements in state vector) behave as a white noise series, in accordance to the open-loop simulation used to collect the data set.

Building multivariate ARMAX models is an iterative process, in just the same way that we build univariate transfer function models. To handle the additional complexities due to the exogenous variables, it is wise to start with a large enough initial model and proceed by eliminating terms that seem insignificant. At the same time, we should check the residuals of any tentative model, just as we did for vector ARMA models. Given the large number of parameters involved, large data sets will be required for an adequate identification.

In this example, suppose that we start by considering an order of the $B(\mathcal{B})$ polynomial equal to $l + 1 = 3$, so we should have $m = \max(p, l + 1, q + 1) = \max(1, 3, 2) = 3$ for an adequate state-space representation. This implies that the state vector should contain $m(r_1 + r_2) = 3(3 + 2) = 15$ elements. The following SAS statements fit model (9.13):

```
proc statespace data = vecARMAX dimmax = 15 out = out;
      var X1 X2 X3 Y1 Y2;
      form X1 3 X2 3 X3 3 Y1 3 Y2 3;
run;
```

The `dimmax` option is important since the default maximum value for the dimension of the state vector (n) is 10. Table 9.2 shows the F matrix estimated by SAS. The standard errors and t-statistics given by SAS (not shown here) indicate that only $\hat{B}_2$ and $\hat{A}_1$ are significant.[6] Note how these two matrices of estimates are quite close to the true (simulated) ones. This is not the case for $\hat{B}_1$ and $\hat{B}_0$. Table 9.3 shows the corresponding G matrix estimated by SAS. The B_j matrices can be obtained from the relation

$$B_0\mathcal{B} + B_1\mathcal{B}^2 + \cdots + B_l\mathcal{B}^{l+1}$$
$$= \left(I - A_1\mathcal{B} - \cdots - A_p\mathcal{B}^p\right)\left(V_0 + V_1\mathcal{B} + V_2\mathcal{B}^2 + V_3\mathcal{B}^3 + \cdots\right)$$

Equating matrix coefficients of like powers of $\mathcal{B}$, we get

$$B_0 = V_1$$

$$B_j = V_{j+1} - \sum_{i=0}^{j} A_{j+1-i} V_i$$

with $V_0 = 0$. Applying these results to the data in the problem (Table 9.3), we obtain

$$\hat{B}_0 = \hat{V}_1 = \begin{pmatrix} -0.549 & 0.399 & 0.801 \\ 0.874 & -0.722 & -0.414 \end{pmatrix}$$

Similarly, we obtain that $B_1 = V_2 - A_2V_0 - A_1V_1 = V_2 - A_1V_1$. From the table we have that

$$\hat{V}_2 = \begin{pmatrix} 0.021 & 0.794 & 0.504 \\ -0.304 & 1.150 & 1.594 \end{pmatrix} = \hat{B}_1 + \hat{A}_1\hat{B}_0.$$

so we get

$$\hat{B}_1 = \hat{V}_2 - \hat{A}_1\hat{B}_0 = \begin{pmatrix} 0.661 & 0.314 & -0.288 \\ 0.559 & 0.464 & 0.904 \end{pmatrix}$$

[6]As a rule of thumb, an estimated matrix with *any* entry resulting in a t-statistic of at least 3.5 was considered significant.

Table 9.2 $\hat{F}$ Matrix Estimate for Example 9.6[a]

0	0	0	0	0	0	0	0	0	0	1	0	0	0	0	0	0	0
0	0	0	0	0	0	0	0	0	0	0	1	0	0	0	0	0	0
0	0	0	0	0	0	0	0	0	0	0	0	1	0	0	0	0	0
0	0	0	0	0	0	0	0	0	0	0	0	0	1	0	0	0	0
0	0	0	0	0	0	0	0	0	0	0	0	0	0	1	0	0	0
0	0	0	0	0	0	0	0	0	0	0	0	0	0	0	1	0	0
0	0	0	0	0	0	0	0	0	0	0	0	0	0	0	0	1	0
0	0	0	0	0	0	0	0	0	0	0	0	0	0	0	0	0	1
0.030	0.265	0.162	0	0	0.041	0.033	−0.316	0.571	−0.348	0	0.129	−0.078	−0.359	−0.062	0.883	−0.129	−0.155
−0.251	−0.013	−0.196	0	0	−0.145	0.157	−1.014	−0.536	0.456	0	0.037	0.248	−0.836	−0.392	1.944	0.046	0.114
−0.066	0.158	0.130	0	0	0.055	−0.008	0.086	0.217	−0.524	0	−0.035	−0.073	0.037	−0.068	0.383	0.004	−0.100
0.822	**0.204**	**0.827**	0	0	−0.004	−0.045	0.767	−0.956	−0.262	0	−0.049	−0.008	−0.101	0.412	0.842	**0.904**	**−0.165**
−0.613	**−0.889**	**−0.507**	0	0	−0.059	0.013	1.070	0.398	1.223	0	0.023	0.006	1.241	0.035	−0.921	**0.518**	**−0.663**

[a] Bold items correspond to the statistically significant matrices: $\hat{B}_2$ (lower left) and $\hat{A}_1$ (lower right).

Table 9.3 Estimated $\hat{G}$ Matrix for Example 9.6[a]

1	0	0	0	0
0	1	0	0	0
0	0	1	0	0
0	0	0	1	0
0	0	0	0	1
−0.039	−0.012	−0.073	0.001	−0.056
−0.025	−0.062	0.035	−0.020	−0.041
−0.012	0.141	0.061	0.043	0.007
−0.549	**0.399**	**0.802**	**0.110**	**−0.366**
0.875	**−0.722**	**−0.414**	**0.792**	**−1.301**
−0.037	0.059	−0.004	−0.046	0.039
0.088	−0.004	0.014	0.009	0.036
0.024	−0.023	0.000	−0.046	0.052
0.021	**0.794**	**0.505**	−0.006	0.005
−0.304	**1.151**	**1.594**	−0.430	0.701

[a]Bold items correspond to the statistically significant matrices: $\hat{V}_1$ (upper left), $\hat{\Psi}_1$ (upper right), and $\hat{V}_2$ (lower left).

These estimates are much closer to the true (simulated) ones than the estimates obtained from the transition matrix F (which were not even significant).[7] Both $\hat{V}_1$ and $\hat{V}_2$ are significant, so we have that the delay should be one period, according to equation (9.5). Also from the estimated G matrix, we find that

$$\hat{\Psi}_1 = \begin{pmatrix} 0.109 & -0.365 \\ 0.559 & -1.301 \end{pmatrix}$$

This, together with the relation $C(\mathcal{B}) = A(\mathcal{B})\sum_{j=0}^{\infty}\Psi_j\mathcal{B}^j$, results in

$$\hat{C}_1 = \hat{A}_1 - \Psi_1 = \begin{pmatrix} 0.793 & 0.200 \\ -0.274 & 0.638 \end{pmatrix}$$

This is in close agreement with the true (simulated) value. The estimated variance matrix of the errors is

$$\hat{\Sigma} = \begin{pmatrix} 1.082 & 0.021 \\ 0.021 & 1.069 \end{pmatrix}$$

From the standard errors given by SAS, $\hat{V}_1$, $\hat{V}_2$, and $\hat{\Psi}_1$ are significant.

[7]This indicates that there are redundancies which can be eliminated, as shown in Example 9.7.

The residuals of the fitted model can be studied using the statements

```
proc statespace data = out out = out2;
var RES1 RES2 RES3 RES4 RES5;
run;
```

The SAS output, not shown here, indicates that the residuals are a multivariate white noise sequence, so the fitted model can be considered adequate. This model is, in fact, the exact state-space representation of the simulated ARMAX process put in form (9.13). In real practice, we will probably not be so lucky, and some further attempts and residual analysis with different values of l will be necessary.   □

The preceding example indicates that the B_j matrices are estimated twice in the state-space formulation (9.13); this is obviously redundant and perhaps unnecessary. A more parsimonious model representation results from removing all $\mathbf{X}_{t+i|t}$ $(i = 1, \ldots, t + m - 1)$ variables from the state vector. The state vector will have $n = r_1 + r_2 \, m$ elements, that is,

$$
\mathbf{Z}_{t+1} = \begin{pmatrix} \mathbf{X}_{t+1} \\ \mathbf{Y}_{t+1} \\ \mathbf{Y}_{t+2|t+1} \\ \vdots \\ \mathbf{Y}_{t+m|t+1} \end{pmatrix}
$$

As the reader can confirm, the columns associated with the B_j $(j < m - 1)$ matrices in matrix F of model (9.13) are eliminated. Estimates of the B_j $(j \leq m - 1)$ matrices can be obtained from the V_j impulse weight matrices, contained in the G matrix. An example illustrates this procedure.

Example 9.7: Reduced State-Space Formulation of Vector ARMAX Processes. Let us fit a less redundant state-space model to the same data as in Example 9.6 along the suggestion given in the preceding paragraph. The SAS statements that will do this are:

```
proc statespace data = vecARMAX dimmax = 15 out = out;
      var X1 X2 X3 Y1 Y2;
      form Y1 3 Y2 3;
run;
```

Excluding the X_j variables from the form statement will not force SAS to include prediction of these variables in the state vector. Tables 9.4 and 9.5 give the estimated $\hat{F}$ and $\hat{G}$ matrices, respectively, obtained with SAS.

Table 9.4   Estimated $\hat{F}$ Matrix for Example 9.7[a]

-0.050	-0.016	-0.082	0.009	0.002	0	0	0	0
-0.037	-0.078	0.022	-0.005	0.010	0	0	0	0
0.027	0.110	0.079	0.006	-0.011	0	0	0	0
0	0	0	0	0	1	0	0	0
0	0	0	0	0	0	1	0	0
0	0	0	0	0	0	0	1	0
0	0	0	0	0	0	0	0	1
0.830	**0.332**	**0.904**	0.141	-0.121	-0.089	-0.182	**0.875**	**-0.303**
-0.752	**-0.824**	**-0.525**	-0.021	-0.005	-0.092	0.067	**0.613**	**-0.606**

[a]Bold items correspond to the statistically significant matrices: $\hat{B}_2$ (lower left) and $\hat{A}_1$ (lower right).

Table 9.5   Estimated $\hat{G}$ Matrix for Example 9.7

1	0	0	0	0
0	1	0	0	0
0	0	1	0	0
0	0	0	1	0
0	0	0	0	1
-0.538	**0.420**	**0.816**	**0.110**	**-0.386**
0.882	**-0.698**	**-0.401**	**0.807**	**-1.326**
0.070	**0.743**	**0.507**	-0.093	-0.018
-0.287	**1.153**	**1.593**	-0.471	0.761

[a]Bold items correspond to the statistically significant matrices: $\hat{V}_1$ (upper left 2 × 3 submatrix), $\hat{V}_2$, (lower left 2 × 3 submatrix), and $\hat{\Psi}_1$ (upper right 2 × 2 submatrix).

From the standard errors (computed by SAS and not shown here) only $\hat{B}_2$, $\hat{A}_1$, $\hat{V}_1(=\hat{B}_0)$, $\hat{V}_2 = \hat{B}_1 + \hat{A}_1\hat{B}_0$ and $\hat{\Psi}_1$ are significant and equal to:

$$\hat{B}_2 = \begin{pmatrix} 0.829 & 0.332 & 0.903 \\ -0.752 & -0.823 & -0.525 \end{pmatrix} \qquad \hat{A}_1 = \begin{pmatrix} 0.875 & -0.303 \\ 0.613 & -0.606 \end{pmatrix},$$

$$\hat{V}_1 = \hat{B}_0 = \begin{pmatrix} -0.538 & 0.419 & 0.816 \\ 0.882 & -0.697 & -0.400 \end{pmatrix}$$

$$\hat{B}_1 = \hat{V}_2 - \hat{A}_1\hat{B}_0 = \begin{pmatrix} 0.808 & 0.165 & -0.328 \\ 0.578 & 0.473 & 0.849 \end{pmatrix},$$

and

$$\hat{C}_1 = \hat{A}_1 - \hat{\Psi}_1 = \begin{pmatrix} 0.765 & 0.083 \\ -0.194 & 0.72 \end{pmatrix} \qquad \hat{\Sigma} = \begin{pmatrix} 1.129 & 0.047 \\ 0.047 & 1.110 \end{pmatrix}$$

Comparing this estimate of the variance–covariance matrix of the errors with that obtained using the full state-space model (Example 9.6), we see that this estimate contains slightly larger entries, indicating a somewhat worse model. This should be expected since once we drop the $\mathbf{X}_{t+i|t}$ ($i = 1, \ldots, t + m - 1$) variables from the state vector, the resulting formulation will *not* satisfy exactly equations (9.12). Analysis of the residuals of the fitted model, however, shows that the reduced model achieves an adequate representation of the underlying ARMAX process. As a rule of thumb, one should try fitting the reduced model and if the residuals do not look white, the full model based on tentative values of p, l, and q should be fitted instead (and the residuals should be checked again, etc.). Fitting and checking the model suggested automatically by PROC STATESPACE is also useful sometimes, particularly when the number of variables is large.   □

Examples 9.6 and 9.7 show how to identify and estimate a vector ARMAX process operating open loop, that is, while the controllable factors $X_{i,t}$ are simply varied according to an i.i.d. sequence as opposed to varying them according to a feedback adjustment rule. Identification and estimation in closed loop were discussed for univariate processes in Chapter 8. The problems discussed there also applied to the multivariate case. However, as the next example shows, identification and estimation of input-output processes is sometimes possible in closed-loop form provided that the feedback adjustment rule is not linear. As Akaike and Nakagawa (1988) suggest, this sometimes happens when manual control is exercised by an operator while the data are collected.

Example 9.8: Closed-Loop Identification under Manual Control. An (p, l, q) = $(1, 0, 1)$ vector ARMAX process with three inputs and two outputs was simulated. The process parameters are

$$B_0 = \begin{pmatrix} -0.5 & 0.4 & 0.8 \\ 0.9 & -0.7 & -0.4 \end{pmatrix} \qquad A_1 = \begin{pmatrix} 0.8 & 0.2 \\ 0.3 & -0.4 \end{pmatrix}$$

$$C_1 = \begin{pmatrix} 0.7 & 0.2 \\ -0.3 & 0.6 \end{pmatrix} \qquad \Sigma = \begin{pmatrix} 25 & 7.5 \\ 7.5 & 25 \end{pmatrix}$$

The process was simulated for 190 time units (data can be found in the file *MultiARMAXClosed.txt*). A deadband controller (which is nonlinear) was used to simulate adjustments by a human operator. A schematic description of the estimated ACF and PACF is

```
                  Schematic Representation of Correlations
Name      0     1     2     3     4     5     6     7     8     9
X1     +-+-.  +..--  -+-..  -+-+.  .....  .....  .....  +....  .....  -+-..
X2     -+-+.  .+.++  +-+..  +-+-.  .....  .....  .....  .+-..  .....  +-+..
X3     +-+--  ...--  -+-..  -+-+.  ...+.  .....  ...+.  +-+-.  ...-.  -+-..
Y1     -+-++  +-++.  +-+--  ...--  -+-..  -+-..  .+...  ...+.  +.+..  +-+-.
Y2     ..-++  +-++-  +-+.+  ....-  ....+  ....-  .....  .....  .....  .....


             Schematic Representation of Partial Autocorrelations
Name      1     2     3     4     5     6     7     8     9
X1     +.---  .....  .++..  .....  .....  .....  .....  .....  .....
X2     .++++  .....  .-...  .....  .....  .....  .....  .....  .....
X3     ..--.  .....  .+...  .....  .....  .....  .....  .....  .....
Y1     ..++.  ...+.  ...+.  ...+.  .....  .....  .....  .....  .....
Y2     +--+-  +.-.-  ..-.-  .....  .....  .....  .....  .....  .....
```

which indicates some type of ARMA process for Y_1 and Y_2, whereas for X_1, X_2, and X_3 the functions clearly indicate that these series are *not* white noise (contrast with similar plots in Examples 9.6 and 9.7). If we use the automatic identification routine of SAS PROC STATESPACE, a model with state vector

$$\mathbf{Z}_t = \begin{pmatrix} \mathbf{X}_t \\ \mathbf{Y}_t \\ \mathbf{Y}_{t+1|t} \end{pmatrix}$$

results. This turns out to correspond to the reduced state-space model discussed in Example 9.7, but in this case $m = \max(p, l+1, q+1) = 2$ instead. Part of the output generated by the software is shown in Figure 9.4. From the standard errors (not shown here), we see that only $\hat{A}_1$, $\hat{B}_0$, and $\hat{\mathbf{\Psi}}_1$ are significant, so we obtain, after computations similar to those performed in the previous two examples, that

$$\hat{B}_0 = \begin{pmatrix} -0.238 & 0.521 & 0.711 \\ 0.864 & -0.846 & -0.433 \end{pmatrix} \qquad \hat{A}_1 = \begin{pmatrix} 0.888 & 0.191 \\ 0.110 & -0.432 \end{pmatrix}$$

$$\hat{C}_1 = \hat{\mathbf{\Psi}}_1 - \hat{A}_1 = \begin{pmatrix} 0.655 & 0.155 \\ -0.411 & 0.644 \end{pmatrix}$$

and

$$\hat{\Sigma} = \begin{pmatrix} 30.97 & 8.03 \\ 8.03 & 23.34 \end{pmatrix}$$

The schematic plots of the sample ACF and PACF of the residuals indicate that the fitted model is adequate in general terms (see Figure 9.4, bottom).

□

State vector

X1 (T;T)	X2 (T;T)	X3 (T;T)	Y1 (T;T)	Y2 (T;T)	Y1 (T+1;T)	Y2 (T+1;T)

Estimate of transition matrix

X1 (T;T)	X2 (T;T)	X3 (T;T)	Y1 (T;T)	Y2 (T;T)	Y1 (T+1;T)	Y2 (T+1;T)
1.362148	−0.18234	−0.65412	−0.45407	−0.35053	0	0
−0.29829	0.963932	0.429268	0.301377	0.263777	0	0
1.527568	−0.86956	−1.06014	−1.34672	−0.53532	0	0
0	0	0	0	0	1	0
0	0	0	0	0	0	1
0.392797	−0.27517	−0.38635	−0.81847	−0.09091	0.888208	0.191553
0.609256	−0.32165	−0.25083	0.040421	−0.13659	0.110444	−0.43267

Input matrix for innovation

1	0	0	0	0
0	1	0	0	0
0	0	1	0	0
0	0	0	1	0
0	0	0	0	1
−0.23866	0.521982	0.711092	0.233292	0.036229
0.864461	−0.84651	−0.43378	0.521966	−1.07705

Variance matrix for innovation

84.38694	−58.9048	213.1434	−17.996	−10.0684
−58.9048	46.68138	−154.051	13.15851	7.52311
213.1434	−154.051	570.8706	−54.8649	−24.8787
−17.996	13.15851	−54.8649	30.97809	8.031312
−10.0684	7.52311	−24.8787	8.031312	23.3446

Schematic representation of correlations

Name/Lag	0	1	2	3	4	5	6	7	8	9	10
RES1	+−+−−	.....	.....	.....	.....	.....	.....	...− .	.....	.....	.....
RES2	−+−++	.....	.....	.....	.....	.....	.....	...+ .	.....	.....	.....
RES3	+−+−−	.....	.....	.....	.....	.....	.....	...− .	.....	.....	.....
RES4	−+−++	.....	.....	...− .	.....	.....	.....	−+−+ .	.....	.....	.....
RES5	−+−++	.....	.....	.....	....+	.....	.....	.+...	.....	.....	.....

Schematic representation of partial autocorrelations

Name/Lag	1	2	3	4	5	6	7	8	9	10
RES1	.....	.−...	.....	.....	.....	.....	.....	.....	.....	.....
RES2	.....	.....	.....	.....	.....	.....	.....	.....	.....	.....
RES3	.....	.....	.....	.....	.....	.....	.....	.....	.....	.....
RES4	.....	.....	.....	.....	.....	.....	.....	+....	.....	.....
RES5	.....	.....	.....	.....	.....	.....	.....	.....	.....	.....

Figure 9.4 Partial output generated by SAS PROC STATESPACE in Example 9.8. Top, final estimates; bottom, sample ACF and sample PACF of residuals.

9.3 IDENTIFICATION AND FITTING OF ARMAX PROCESSES IN STATE SPACE USING MATLAB'S SYSTEMS ID TOOLBOX

Matlab's Systems Identification Toolbox, introduced in Section 4.5, can also help in identifying and fitting multivariate state-space models that correspond to ARMAX models. The state-space representation used by Matlab is, however, different than that used by SAS PROC STATESPACE. Contrary as when using SAS, the state vector obtained when using Matlab does not contain controllable factors; it contains predicted outputs only. Using Matlab's Systems ID Toolbox, the user needs only to specify the size of the state vector, n. The state-space formulation described in this section is explained in full detail by Ljung (1999, App. 4A). The state-space form utilized by this software is

$$\mathbf{Z}_{t+1} = \mathbf{A}\mathbf{Z}_t + \mathbf{B}\mathbf{X}_t + K\boldsymbol{\varepsilon}_t$$
$$\mathbf{Y}_t = \mathbf{C}\,\mathbf{Z}_t + \boldsymbol{\varepsilon}_t \tag{9.14}$$

where the $\mathbf{Z}$'s are $n \times 1$ state vectors; $\mathbf{A}$ is an $n \times n$ vector [not to be confused with the polynomial $A(\mathcal{B})$ nor with its A_i matrix coefficients, which are not denoted in bold]; $\mathbf{B}$ is an $n \times r_1$ vector, where r_1 is the number of inputs [not to be confused with $B(\mathcal{B})$ nor with its B_i matrix coefficients]; the $\mathbf{X}$'s are $r_1 \times 1$ vectors; K is an $n \times r_2$ vector, where r_2 is the number of outputs; $\boldsymbol{\varepsilon}_t$ is an $r_2 \times 1$ vector, $\mathbf{Y}_t$ is an $r_2 \times 1$ vector; and $\mathbf{C}$ is an $r_2 \times n$ vector [not to be confused with $C(\mathcal{B})$ or with its C_i matrix coefficients]. This state-space representation is called the *innovations form* since the $\mathbf{Z}$'s can be understood as predictions of the output, and equation (9.14) can be understood as a correction factor that modifies the predictions.

The key to this representation is the structure of the state vector and of the transition matrix $\mathbf{A}$. This matrix has r_2 rows of parameters. The last row is always full of parameters, so the other $r_2 - 1$ rows of parameters can be chosen among the remaining $n - 1$ rows. The $n - r_2$ rows without parameters have 0 entries except at the entry on the superdiagonal of the matrix, where there is a 1.

The model is defined by r_2 indices $\{s_i\}$ that give the row numbers where we have parameters. The differences $d_i = s_i - s_{i-1}$ (with $s_0 = 0$) are called the *multiindex* of the state-space representation and are such that $\sum_{i=1}^{r_2} d_i = n$. With this notation, the elements of the state vector are defined according to

$$Z_{s_{i-1}+k+1,t} = \hat{Y}_{i,t+k|t-1} \qquad i = 1,2,\ldots,r_2; \quad 0 \le k < d_i \tag{9.15}$$

To illustrate this formulation, suppose that we define $n = 9$ and that we have the same transition matrix $\mathbf{A}$ as given by Ljung (1999, App. 4A):

$$
\begin{array}{ccccccccc}
0 & 1 & 0 & 0 & 0 & 0 & 0 & 0 & 0 \\
0 & 0 & 1 & 0 & 0 & 0 & 0 & 0 & 0 \\
X & X & X & X & X & X & X & X & X \\
0 & 0 & 0 & 0 & 1 & 0 & 0 & 0 & 0 \\
0 & 0 & 0 & 0 & 0 & 1 & 0 & 0 & 0 \\
X & X & X & X & X & X & X & X & X \\
0 & 0 & 0 & 0 & 0 & 0 & 0 & 1 & 0 \\
0 & 0 & 0 & 0 & 0 & 0 & 0 & 0 & 1 \\
X & X & X & X & X & X & X & X & X
\end{array}
$$

Here entries labeled with an X denote parameters; otherwise, entries have zeros, except on the superdiagonal, where we find 1's. We thus have that rows with parameters are $s_1 = 3$, $s_2 = 5$, and $s_3 = 9$. This means that we have $d_1 = s_1 - s_0 = 3$, $d_2 = s_2 - s_1 = 2$, and $d_3 = s_3 - s_2 = 4$. Note that $d_1 + d_2 + d_3 = 9 = n$. To figure out the elements of the state vector, we refer to expression (9.15), from where we have that for $i = 1$, $s_{i-1} = s_0 = 0$, and $d_1 = 3$, we get

$$k = 0: \quad Z_{0+0+1,t} = Z_{1,t} = \hat{Y}_{1,t+0|t-1} = \hat{Y}_{1,t|t-1}$$

$$k = 1: \quad Z_{2,t} = \hat{Y}_{1,t+1|t-1}$$

$$k = 2: \quad Z_{3,t} = \hat{Y}_{1,t+2|t-1}$$

For $i = 2$, $s_{i-1} = s_1 = 3$, and $d_2 = 2$, we get

$$k = 0: \quad Z_{3+0+1,t} = Z_{4,t} = \hat{Y}_{2,t+0|t-1} = \hat{Y}_{2,t|t-1}$$

$$k = 1: \quad Z_{5,t} = \hat{Y}_{2,t+1|t-1}$$

Finally, for $i = 3$, $s_{i-1} = 5$, and $d_3 = 4$, we get $Z_{6,t} = \hat{Y}_{3,t|t-1}$, $Z_{7,t} = \hat{Y}_{t+1|t-1}$, $Z_{8,t} = \hat{Y}_{3,t+2|t-1}$, and $Z_{9,t} = \hat{Y}_{3,t+3|t-1}$. In Matlab's Systems ID Toolbox, the **B** and **K** matrices are full of parameters, and the **C** matrix has zeros everywhere except that it has 1's in cells $(i, S_{i-1} + 1)$. In the example above, this means that **C** will have 1's in cells $(1, 1)$, $(2, 4)$, and $(3, 6)$. Note that the effect of these entries is to retrieve the variables $\hat{Y}_{1,t|t-1}$, $\hat{Y}_{2,t|t-1}$, and $\hat{Y}_{3,t|t-1}$ from the state vector.

Obtaining Vector ARMAX Models from the State-Space Representation in Matlab's Systems ID Toolbox

Suppose we have data that were generated by an ARMAX model such as equation (9.5). Note that if $B_0 \neq 0$, this model form assumes that the input–output delay k is one period. Given the way the state vector is assembled, it is easier to explain how to retrieve the matrices $\hat{A}_i$, $\hat{B}_i$, and $\hat{C}_i$ from a fitted state-space model such as (9.14) by means of an example.

Example 9.9. Consider the same data set as in Example 9.6. Recall that this is a 3×2 process ran in open loop while the data were collected. We import the five series into the `ident` graphical user interface provided by Matlab's Systems ID Toolbox.[8] To fit a state-space model in innovation form, select "Parametric Models," "State Space," and the dimension of the state vector, n. The same recommendation applies as when using SAS PROC STATESPACE: In multivariate models it makes sense to fit initially a model

---

[8]The `ident` GUI was introduced in Chapter 4.

large enough to accommodate a variety of models and then look at the residuals and at the significance of the parameter estimates[9] to see if a simpler model works better.

As shown below, if we want to fit a state-space model that allows an ARMAX representation of orders p, l, and q, the state dimension should be at least equal to

$$n = \max\{p \times r_2, r_2 \times (l + 1), q \times r_2\}$$

where r_2 is the number of responses.

In our case, if we want to fit a model that allows to represent an ARMAX model with $p = 1$, $l = 2$, and $q = 1$, we should use $n = \max\{1 \times 2, 2 \times 3, 1 \times 2\} = 6$. Fitting such model from the data at hand, we obtain the matrices

$$\hat{A} = \begin{bmatrix} 0 & 1 & 0 & 0 & 0 & 0 \\ 0 & 0 & 1 & 0 & 0 & 0 \\ 0.40624 & -0.63465 & 1.2354 & -0.30045 & -0.24443 & -0.19484 \\ 0 & 0 & 0 & 0 & 1 & 0 \\ 0 & 0 & 0 & 0 & 0 & 1 \\ 0.065277 & -0.31657 & 0.75709 & -0.056274 & 0.065805 & -0.55043 \end{bmatrix}$$

$$\hat{B} = \begin{bmatrix} -0.54665 & 0.411 & 0.80631 \\ 0.019296 & 0.67943 & 0.39795 \\ 0.9071 & 0.76551 & 0.75199 \\ 0.87709 & -0.71083 & -0.41681 \\ -0.28465 & 1.1806 & 1.7351 \\ -0.4104 & -1.2233 & -1.5241 \end{bmatrix}$$

$$\hat{K} = \begin{bmatrix} 0.076577 & -0.4437 \\ -0.089747 & 0.013918 \\ 0.015628 & -0.10051 \\ 0.82807 & -1.3619 \\ -0.49106 & 0.75369 \\ 0.24144 & -0.47574 \end{bmatrix}$$

We also have that

$$C = \begin{bmatrix} 1 & 0 & 0 & 0 & 0 & 0 \\ 0 & 0 & 0 & 1 & 0 & 0 \end{bmatrix}$$

[9]Matlab's Systems ID Toolbox provides the N4SID algorithm for determining the state dimension (see van Overschee and de Moor, 1996; Ljung 1999). If this method is selected, no standard error information is given for the parameters. The examples shown below assume that models were fitted by using the prediction error method (PEM), in which parameters are selected by minimizing the sum-of-squared prediction errors. By default, a robustification constant gives less weight to errors than does a pure quadratic criterion. See Ljung (1995, 1999, 2001) for details.

The structure of these matrices is as follows:

$$
\mathbf{A}_{(6\times6)} =
\begin{array}{|c|c|c|c|c|c|}
\hline
0 & 1 & 0 & 0 & 0 & 0 \\
\hline
0 & 0 & 1 & 0 & 0 & 0 \\
\hline
A_{3,(1,1)} & A_{2,(1,1)} & A_{1,(1,1)} & A_{3,(1,2)} & A_{2,(1,2)} & A_{1,(1,2)} \\
\hline
0 & 0 & 0 & 0 & 1 & 0 \\
\hline
0 & 0 & 0 & 0 & 0 & 1 \\
\hline
A_{3,(2,1)} & A_{2,(2,1)} & A_{1,(2,1)} & A_{3,(2,2)} & A_{2,(2,2)} & A_{1,(2,2)} \\
\hline
\end{array}
$$

$$
\mathbf{B}_{(6\times3)} =
\begin{array}{|c|}
\hline
V_{1,(1)} \\
\hline
V_{2,(1)} \\
\hline
V_{3,(1)} \\
\hline
V_{1,(2)} \\
\hline
V_{2,(2)} \\
\hline
V_{3,(2)} \\
\hline
\end{array}
$$

$$
K_{(6\times2)} =
\begin{array}{|c|}
\hline
\Psi_{1,(1)} \\
\hline
\Psi_{2,(1)} \\
\hline
\Psi_{3,(1)} \\
\hline
\Psi_{1,(2)} \\
\hline
\Psi_{2,(2)} \\
\hline
\Psi_{3,(2)} \\
\hline
\end{array}
$$

In the above, subscripts (i, j) denote entries of an ARMAX matrix coefficient and subscripts (i) denote a row of an impulse response weight matrix. The state vector $\mathbf{Z}_t$ contains $\hat{Y}_{1,\,t+k\,|\,t-1}$, $k = 0, \ldots, 2$, followed by $\hat{Y}_{2,\,t+k\,|\,t-1}$, $k = 0, \ldots, 2$. From the standard errors reported by Matlab's Systems ID Toolbox[10] (not shown here), all $\hat{V}_i$ matrices as well as $\hat{A}_1$ and $\hat{\Psi}_1$ have highly significant entries. This is in accordance with the simulated ARMAX model. There are some entries in $\hat{A}_2$, $\hat{A}_3$, $\hat{\Psi}_2$, and $\hat{\Psi}_3$ that seem significant, so in practice one should consider a higher-order ARMAX model with $p = 3$ and

[10] The present command will show the estimated parameters and their associated standard errors.

$q = 3$. If only $p = 1$ and $q = 1$ are considered, the estimates are

$$\hat{A}_1 = \begin{pmatrix} 1.235 & -0.194 \\ 0.757 & -0.550 \end{pmatrix} \qquad \hat{B}_0 = \hat{V}_1 = \begin{pmatrix} -0.546 & 0.411 & 0.806 \\ 0.877 & -0.710 & -0.416 \end{pmatrix}$$

$$\hat{B}_1 = \hat{V}_2 - \hat{A}_1 - \hat{B}_0 = \begin{pmatrix} 0.865 & 0.032 & -0.679 \\ 0.612 & 0.478 & 0.895 \end{pmatrix}$$

$$\hat{B}_2 = \hat{V}_3 - \hat{A}_2 \hat{B}_0 - \hat{A}_1 \hat{V}_2 = \begin{pmatrix} 0.695 & 0.243 & 1.008 \\ -0.8122 & -0.910 & -0.587 \end{pmatrix}$$

$$\hat{C}_1 = \hat{A}_1 - \hat{\Psi}_1 = \begin{pmatrix} 1.158 & 0.248 \\ -0.0709 & 0.8115 \end{pmatrix}$$

where the relations $B(\mathcal{B}) = A(\mathcal{B})(V_0 + V_1\mathcal{B} + V_2\mathcal{B}^2 + V_3\mathcal{B}^3 + \cdots)$ and $C(\mathcal{B}) = A(\mathcal{B})(\Psi_0 + \Psi_1\mathcal{B} + \Psi_2\mathcal{B}^2 + \cdots)$ were used. Plots of the autocorrelation of the residuals and crosscorrelations of the inputs with the residuals of each output (with 99% confidence limits) are shown in Figure 9.5. These

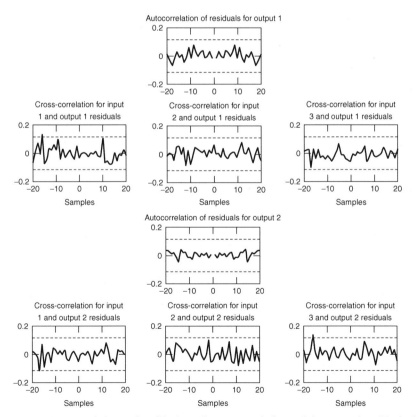

Figure 9.5 Autocorrelations of residuals and cross-correlations of inputs and residuals for Example 9.9. Top four graphs, output 1; bottom four graphs, output 2.

residuals, obtained without eliminating any nonsignificant parameter estimate, indicate that we have a good fit. Just as in the univariate case, we want the residuals to be uncorrelated with the inputs. If this is not the case (see the last part of Section 4.2), there is evidence that the multivariate input–output dynamics have not been identified correctly. □

We point out that in this state-space representation, input–output delays greater than one time period will show up as nonsignificant entries in the B_i matrices.

9.4 OPTIMAL ADJUSTMENT OF ARMAX PROCESSES

Suppose that we have fitted a vector ARMAX model as in Sections 9.2 and 9.3 and now we would like to adjust the process using our model. Let us write the fitted multivariate ARMAX model as in equation (9.4). As mentioned earlier, the minimum mean square error k-steps-ahead forecast is given by

$$\hat{\mathbf{Y}}_{t+k|t} = F(\mathcal{B})C^{-1}(\mathcal{B})B_{N_D}(\mathcal{B}) + G(\mathcal{B})C^{-1}(\mathcal{B})\mathbf{Y}_t$$

where $F(\mathcal{B})$ is a polynomial (of $r_2 \times r_2$ matrices) of order $k - 1$ and $G(\mathcal{B})$ is a polynomial (of $r_2 \times r_2$ matrices) of order $\max(p, n_{B_{N_D}}, q) - 1$ (G does not start with an identity). These polynomials are obtained from equating matrix coefficients of equal powers of $\mathcal{B}$ in

$$C(\mathcal{B}) = F(\mathcal{B})A(\mathcal{B}) + \mathcal{B}^k G(\mathcal{B}) \tag{9.16}$$

It will be assumed in this section that the input-output delay equals k in all input–output combinations. If $\mathbf{Y}_t$ denotes deviations from a vector of targets $\mathbf{T}$, then we want to set $\mathbf{Y}_t$ to a zero vector. Doing this and rearranging terms in (9.16), we get the minimum variance, or *minimum MSE multivariate controller*:

$$F(\mathcal{B})B_{N_D}(\mathcal{B})\mathbf{X}_t = -G(\mathcal{B})\mathbf{Y}_t \tag{9.17}$$

Note that this is an exact generalization of the univariate minimum MSE controller of Section 5.1.1.

If $r_1 = r_2$ (i.e., if we have a "squared" process in which the number of inputs and outputs is the same), $k = 1$, and $B_{N_D}(\mathcal{B}) = B_0$, expression (9.17) reduces to

$$\mathbf{X}_t = -B_0^{-1}G(\mathcal{B})\mathbf{Y}_t$$

Example 9.10. Suppose that we have a responsive multivariate process (i.e., no process dynamics apart from a unit delay are present) corrupted with additive multivariate IMA(1, 1) noise:

$$\mathbf{Y}_t = B_0 \mathbf{X}_{t-1} + \left(I - C_1 \mathcal{B}\right)(I - \mathcal{B})^{-1} \boldsymbol{\varepsilon}_t$$

which was fit from the equivalent ARMAX model

$$\underbrace{\nabla \mathbf{Y}_t}_{A} = \underbrace{B_0 \nabla \mathbf{X}_{t-1}}_{B} + \underbrace{\left(I - C_1 \mathcal{B}\right) \boldsymbol{\varepsilon}_t}_{C}.$$

We thus have that $k = 1$, $p = 1$, $q = 1$, and $n_{B_{ND}} = 1$, so F is of order 0 and G is of order 0. From the Diophantine identity and equating matrices of coefficients multiplying terms containing $\mathcal{B}^1$, we obtain

$$I - C_1 \mathcal{B} = I - \mathcal{B} + \mathcal{B} G_0 \Rightarrow G_0 = I - C_1$$

and the MMSE forecast is given by

$$\hat{\mathbf{Y}}_{t+1|t} = \left(I - C_1 \mathcal{B}\right)^{-1}(I - \mathcal{B}) B_0 \mathbf{X}_t + (I - C_1)\left(I - C_1 \mathcal{B}\right)^{-1} \mathbf{Y}_t$$

This can be written in difference equation form:

$$\hat{\mathbf{Y}}_{t+1|t} = C_1 \hat{\mathbf{Y}}_{t|t-1} + B_0 \nabla \mathbf{X}_t + (I - C_1) \mathbf{Y}_t$$

Note that if $\mathbf{X}_t = \mathbf{0}$ for all periods t, we end up with

$$\hat{\mathbf{Y}}_{t+1|t} = C_1 \hat{\mathbf{Y}}_{t|t-1} + (I - C_1) \mathbf{Y}_t$$

which is just a multivariate EWMA. To get the MMSE controller, from (9.17) we get

$$B_0 \nabla \mathbf{X}_t = -(I - C_1) \mathbf{Y}_t \tag{9.18}$$

If $r_2 = r_1$, we immediately obtain

$$\mathbf{X}_t = \mathbf{X}_{t-1} - B_0^{-1}(I - C_1) \mathbf{Y}_t$$

assuming that B_0 is invertible. Note that this is a pure integral controller exactly analogous to the univariate case $[X_t = X_{t-1} - (1 - c_1)Y_t/b_0]$. If we have $r_2 < r_1$ (more controllable factors than responses), it can be seen from (9.18) that we have only r_2 equations to find settings for $r_1 > r_2$ controllable factors, so evidently something else needs to be done, as described shortly below. If $r_2 = r_1$, the closed-loop equation is

$$\nabla \mathbf{Y}_t = -B_0 B_0^{-1}(I - C_1) \mathbf{Y}_{t-1} + \left(I - C_1 \mathcal{B}\right) \boldsymbol{\varepsilon}_t$$

which, after simplifying terms is $\mathbf{Y}_t = \boldsymbol{\varepsilon}_t$, a multivariate white noise process that gives the lowest possible variance (Σ_ε). $\square$

Before providing a numerical illustration of multivariate controllers, we look first at the more practical case when $r_2 < r_1$.

Case When $r_2 < r_1$

Example 9.10 raised the practical question[11] of what to do if there are more controllable factors (r_1) than quality characteristics (r_2). A usual approach in the control engineering literature is to modify the objective function from one in which attention is also paid to the variability of the controllable factors, in much the way of the Clarke–Gawthrop controller discussed in Chapter 5. Such cost functions solve the problem of having $r_2 < r_1$ but tend to result in more complicated controllers. Instead, following the general philosophy in this book for utilizing adjustment schemes as simple as possible, we introduce in what follows two multivariate adjustment procedures for ARMAX processes that are much simpler to implement.

Right Pseudoinverse Controller

From equation (9.18) it is tempting to use

$$\nabla \mathbf{X}_t = -(B_0' B_0)^{-1}(I - C_1)\mathbf{Y}_t$$

However, since B_0 is $r_2 \times r_1$ with $r_2 < r_1$, $B_0' B_0$ is $r_1 \times r_1$ but it is of rank at most r_2, implying that we cannot find its inverse. One simple alternative is to make $r_1 - r_2$ controllable variables numerically equal to zero and solve for the remaining r_1 controllable factors, but the problem is how to choose this subset of nonbasic variables (using optimization terminology). Another alternative is to use a right pseudoinverse controller:

$$\nabla \mathbf{X}_t = B_0'(B_0 \, B_0')^{-1}(I - C_1)\mathbf{Y}_t$$

where $B_0 B_0'$ is $r_2 \times r_2$ of rank r_2, so it can be inverted. We illustrate this type of control rule with the following example.

Example 9.11: Right Pseudoinverse Controller. Consider the process in Example 9.6. We had that $r_2 = 2$ responses are to be controlled by manipulating $r_1 = 3$ factors. The delay is one period and the orders of the polynomials are $p = 1$, $q = 1$, and $n_{B_{N_D}} = 2$. Thus F is of order $k - 1 = 0$ and G is of order $\max(p, n_{B_{N_D}}, q) - 1 = 1$. The Diophantine equation yields

$$I - C_1 \mathcal{B} = I - A_1 \mathcal{B} + \mathcal{B}(G_0 + G_1 \mathcal{B})$$

---

[11] The case when $r_2 > r_1$ can be handled by using a least squares solution for $\mathbf{X}_t$ since $B_0' B_0$ will always be invertible in such a case (this is the left pseudoinverse).

Table 9.6 Numerical Computations in a Right Inverse Controller, Example 9.11

Run, t	$\mathbf{Y}_t$	$\mathbf{X}_t$
0	$\begin{pmatrix} 0.0 \\ 0.0 \end{pmatrix}$	$\begin{pmatrix} 0.0 \\ 0.0 \\ 0.0 \end{pmatrix}$
1	$\begin{pmatrix} -1.356 \\ 3.529 \end{pmatrix}$	$\begin{pmatrix} 6.307 \\ -4.694 \\ 8.224 \end{pmatrix}$
2	$\begin{pmatrix} -0.973 \\ -0.538 \end{pmatrix}$	$\begin{pmatrix} -8.560 \\ 6.420 \\ -8.594 \end{pmatrix}$
3	$\begin{pmatrix} 0.368 \\ 0.646 \end{pmatrix}$	$\begin{pmatrix} 8.000 \\ -6.231 \\ -4.030 \end{pmatrix}$
$\vdots$	$\vdots$	$\vdots$

Equating matrices of coefficients of like powers of $\mathcal{B}$, we get

$$\mathcal{B}: \quad -C_1 = -A_1 + G_0 \Rightarrow G_0 = A_1 - C_1$$

$$\mathcal{B}^2: \quad 0 = G_1$$

Thus the optimal controller satisfies

$$\left(B_0 + B_1\mathcal{B} + B_2\mathcal{B}^2 \right)\mathbf{X}_t = -(A_1 - C_1)\mathbf{Y}_t$$

or

$$B_0\mathbf{X}_t = -B_1\mathbf{X}_{t-1} - B_2\mathbf{X}_{t-2} - (A_1 - C_1)\mathbf{Y}_t$$

The right pseudoinverse controller is

$$\mathbf{X}_t = -B_0'(B_0 B_0')^{-1}\left[B_1\mathbf{X}_{t-1} + B_2\mathbf{X}_{t-2} + (A_1 - C_1)\mathbf{Y}_t \right]$$

To illustrate the computations numerically, Table 9.6 shows hypothetical values of $\mathbf{Y}$ and the corresponding controllable factor vector $\mathbf{X}$ (the values of B_0, B_1, B_2, A_1, and C_1 are as in Example 9.6).

A simulation of this controller and process indicates that the behavior of the two responses (quality characteristics) is very close to minimum MSE. However, the controllable factors, and hence the adjustments, diverge with time. To illustrate, averages of five 500-time-long simulations are

$$s_{y_1} = 1.01 \qquad s_{y_2} = 0.99 \qquad s_{x_1} = 12,450 \qquad s_{x_2} = 9472 \qquad s_{x_3} = 13,654$$

It is evident that for this particular process, the right inverse controller is completely unfeasible and a different controller must be used. ☐

Ridge Controller
An alternative way of coping with a noninvertible $B_0' B_0$ matrix, common in the numerical analysis literature, is to regularize the matrix prior to attempt computing the inverse. A common trick is to add small nonnegative numbers to the diagonal elements of the matrix in such a way that this stabilizes the inverse (i.e., the matrix to be inverted is regularized). This is sometimes done in other areas of applied statistics, notably in ridge analysis of response surfaces (see Khuri and Cornell, 1987) and in ridge regression (see, e.g., Myers, 1990). Hence we call the resulting controller a *ridge controller* by extension. Applied to the example discussed in equation (9.18), the ridge controller takes the form

$$\mathbf{X}_t = (B_0' B_0 + \mu I)^{-1} B_0' (I - C_1) \mathbf{Y}_t$$

where $\mu > 0$ is a small scalar quantity that makes $B_0' B_0 + \mu I$ positive definite and hence invertible. A pleasant property of the resulting controller is that by increasing the value of μ, we get increasingly more constrained solutions (i.e., the controllable factors X_i vary less the larger μ is). Thus μ is a tuning parameter that allows a process operator to *balance the variance of the outputs with the variance of the inputs.*[12] To find an adequate value of μ, it is suggested that a simulation of the controlled process be carried out first.

Example 9.12: Ridge Controller. Consider once again the process in Example 9.6. The corresponding ridge controller is

$$\mathbf{X}_t = -(B_0' B_0 + \mu I)^{-1} B_0' [B_1 \mathbf{X}_{t-1} + B_2 \mathbf{X}_{t-2} + (A_1 - C_1) \mathbf{Y}_t]$$

Table 9.7 illustrates numerically what the values of the controllable factors are for three iterations (the values of B_0, B_1, B_2, A_1, and C_1 are as in Example 9.6). The value $\mu = 10$ was used.

Performing a simple spreadsheet simulation, the standard deviation estimates in Table 9.8 were obtained from ten 500-time-unit-long simulations of the controller process. Figure 9.6 contrasts a single realization of the controlled process for $\mu = 1$ and $\mu = 10$. Evidently, as μ increases, the variability of the controllable factors decreases while the variability of the responses increases. This occurs up to a limit when we are in effect running the process uncontrolled. ☐

---

[12] In fact, μ can be thought as a Lagrange multiplier associated with the constraint $\mathbf{X}_t' \mathbf{X}_t < \rho$, where ρ is the radius of a sphere within which we want to keep the values of the controllable factors.

Table 9.7   **Numerical Computations in a Ridge Controller,**
$\mu = 10$, **Example 9.12**

Run, t	$\mathbf{Y}_t$	$\mathbf{X}_t$
0	$\begin{pmatrix} 0.0 \\ 0.0 \end{pmatrix}$	$\begin{pmatrix} 0.0 \\ 0.0 \\ 0.0 \end{pmatrix}$
1	$\begin{pmatrix} -0.105 \\ -0.094 \end{pmatrix}$	$\begin{pmatrix} -0.0018 \\ 0.0014 \\ -0.0008 \end{pmatrix}$
2	$\begin{pmatrix} -0.055 \\ -0.965 \end{pmatrix}$	$\begin{pmatrix} -0.0762 \\ 0.0588 \\ 0.0079 \end{pmatrix}$
3	$\begin{pmatrix} 1.096 \\ 0.416 \end{pmatrix}$	$\begin{pmatrix} -0.0279 \\ 0.0217 \\ 0.0155 \end{pmatrix}$
$\vdots$	$\vdots$	$\vdots$

Table 9.8   **Performance of the Ridge Controller Based on Ten 500-Time
Unit Simulations, Example 9.12.** [a]

μ	s_{y_1}	s_{y_2}	s_{x_1}	s_{x_2}	s_{x_3}
1000	1.11	2.15	0	0	0
100	1.14	2.07	0.02	0.01	0.00
10	1.13	1.95	0.15	0.12	0.02
1	1.22	1.40	0.68	0.52	0.30
0.1	1.06	1.08	2.26	1.72	2.40
0.02	1.04	1.05	10.26	7.3	10.6

[a] For values of $\mu < 0.02$, the inputs diverge.

9.5   MULTIVARIATE EWMA AND DOUBLE-EWMA FEEDBACK ADJUSTMENT SCHEMES

Multivariate generalizations of PID and EWMA feedback adjustment methods, discussed in a univariate setting in Chapters 6 and 7, can also be considered. In this section we briefly discuss how the EWMA and double-EWMA schemes of Chapter 7 generalize to a MIMO process.[13]

The process model assumed is similar to that of Butler and Stefani (1994) but extended to the multivariate case:

$$\mathbf{Y}_t = \boldsymbol{\alpha} + \boldsymbol{\beta}\mathbf{X}_{t-1} + \boldsymbol{\delta}t + \boldsymbol{\varepsilon}_t, \tag{9.19}$$

[13] Readers may wish to review Chapter 7 before continuing with this section.

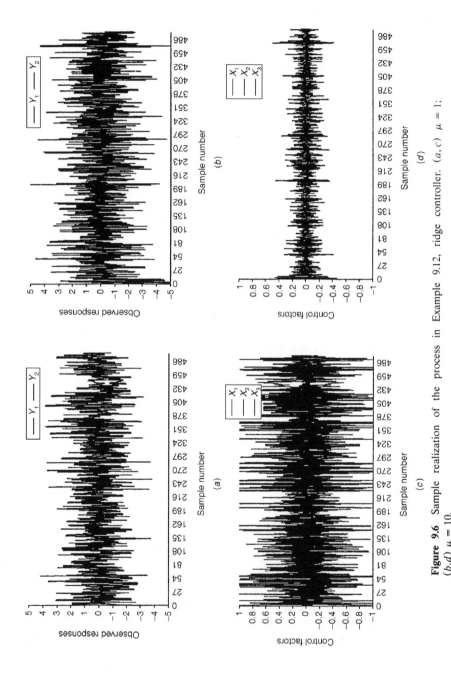

Figure 9.6 Sample realization of the process in Example 9.12, ridge controller. (*a*, *c*) $\mu = 1$; (*b*, *d*) $\mu = 10$.

where $\mathbf{Y}_t$ is a $r_2 \times 1$ vector of the measured quality characteristics or responses of run t, $\mathbf{X}_{t-1}$ is a $r_1 \times 1$ vector of levels of the controllable factors set at the end of run $(t - 1)$, $\boldsymbol{\delta}$ is a diagonal matrix containing the average drift per run for each of the responses, and $\{\boldsymbol{\varepsilon}_t\}_{t=1}^{\infty}$ is a multivariate white noise sequence. The parameter $\boldsymbol{\alpha}$ models the process offset and $\boldsymbol{\beta}$ models the input–output gain. In this section we distinguish, as we did in Chapter 7, between the true gains β and their estimates B.

The model above assumes a deterministic trend (DT) disturbance, $\mathbf{N}_t = \boldsymbol{\delta}t + \boldsymbol{\varepsilon}_t$. More general drift disturbances can be handled, just as in the univariate case. If $r_2 = r_1$, the square MIMO double-EWMA controller is given by

$$\mathbf{x}_t = \mathbf{B}^{-1}(\mathbf{T} - \mathbf{A}_t - \mathbf{D}_t) \tag{9.20}$$

where $\mathbf{A}_t$ and $\mathbf{D}_t$ are online estimates obtained using two multivariate double-EWMA equations at the end of each run:

$$\mathbf{A}_t = \boldsymbol{\Lambda}_1(\mathbf{Y}_t - \mathbf{B}\mathbf{X}_{t-1}) + (\mathbf{I} - \boldsymbol{\Lambda}_1)\mathbf{A}_{t-1} \tag{9.21}$$

$$\mathbf{D}_t = \boldsymbol{\Lambda}_2(\mathbf{Y}_t - \mathbf{B}\mathbf{X}_{t-1} - \mathbf{A}_{t-1}) + (\mathbf{I} - \boldsymbol{\Lambda}_2)\mathbf{D}_{t-1} \tag{9.22}$$

Here $\boldsymbol{\Lambda}_1$ and $\boldsymbol{\Lambda}_2$ are the two EWMA weight matrices. Although these matrices can have any form, it is convenient to select them of the form $\boldsymbol{\Lambda}_i = \lambda_i I$. This will make the sum $(\mathbf{A}_t + \mathbf{D}_t)$ an unbiased estimate of $\boldsymbol{\alpha} + \boldsymbol{\delta}$ $(t + 1)$ (i.e., the estimates provide an asymptotically unbiased one-step-ahead prediction of where the quality characteristics would have drifted in the absence of any control action). This is true only for the assumed linear drift model.

Example 9.13: Double-EWMA Controller. To illustrate numerically the computations required by the double-EWMA controller, consider a process with

$$\mathbf{B} = \beta = \begin{pmatrix} 1.0 & 0.2 \\ 0.3 & 1.0 \end{pmatrix}$$

and weight matrices $\boldsymbol{\Lambda}_1 = 0.2I$ and $\boldsymbol{\Lambda}_2 = 0.3I$ were used. Table 9.9 shows the computations required for the first five iterations for a hypothetical set of data. First, $\mathbf{A}$ and $\mathbf{D}$ are computed, and then the vector of controllable variables $\mathbf{X}$ is computed. □

The stability conditions, valid for nonsquare processes, were derived in del Castillo and Rajagopal (2001) and are as follows:

$$\left| 1 - 0.5\xi_{jj}(\lambda_1 + \lambda_2) + 0.5z \right| < 1 \qquad \forall j \in \{1, 2, \ldots, r_2\} \tag{9.23}$$

$$\left| 1 - 0.5\xi_{jj}(\lambda_1 + \lambda_2) - 0.5z \right| < 1 \qquad \forall j \in \{1, 2, \ldots, r_2\} \tag{9.24}$$

Table 9.9 Numerical Computations in a MIMO Double-EWMA Controller, $\Lambda_1 = (0.2)I$, $\Lambda_2 = (0.3)I$, Example 9.13

Run, t	Y_t	A_t	D_t	X_t
0	$\begin{pmatrix} 0.0 \\ 0.0 \end{pmatrix}$	$\begin{pmatrix} 0.0 \\ 0.0 \end{pmatrix}$	$\begin{pmatrix} 0.0 \\ 0.0 \end{pmatrix}$	
1	$\begin{pmatrix} 1.813 \\ 3.216 \end{pmatrix}$	$\begin{pmatrix} 0.363 \\ 0.643 \end{pmatrix}$	$\begin{pmatrix} 0.544 \\ 0.965 \end{pmatrix}$	$\begin{pmatrix} -0.622 \\ -1.421 \end{pmatrix}$
2	$\begin{pmatrix} 1.047 \\ 1.453 \end{pmatrix}$	$\begin{pmatrix} 0.681 \\ 1.127 \end{pmatrix}$	$\begin{pmatrix} 0.858 \\ 1.401 \end{pmatrix}$	$\begin{pmatrix} -1.099 \\ -2.198 \end{pmatrix}$
3	$\begin{pmatrix} 0.714 \\ -1.381 \end{pmatrix}$	$\begin{pmatrix} 0.995 \\ -1.131 \end{pmatrix}$	$\begin{pmatrix} 1.072 \\ 0.987 \end{pmatrix}$	$\begin{pmatrix} -1.749 \\ -1.593 \end{pmatrix}$
4	$\begin{pmatrix} -0.816 \\ 1.753 \end{pmatrix}$	$\begin{pmatrix} 1.047 \\ 1.679 \end{pmatrix}$	$\begin{pmatrix} 0.828 \\ 1.513 \end{pmatrix}$	$\begin{pmatrix} -1.315 \\ -2.797 \end{pmatrix}$
$\vdots$	$\vdots$	$\vdots$	$\vdots$	$\vdots$

where $z = \sqrt{\xi_{jj}^2(\lambda_1 + \lambda_2)^2 - 4\lambda_1\lambda_2\,\xi_{jj}}$, and ξ_{jj} is the jth diagonal element of $\xi = \beta\mathbf{B}^{-1}$.

Similar to Section 9.4, a "ridge" double-EWMA controller or a right pseudoinverse double-EWMA controller may be used when the number of inputs differs from the number of outputs. If a ridge controller is used, for example, the stability conditions are the same as above except that we should define instead $\xi = \beta(\mathbf{B}'\mathbf{B} + \mu\mathbf{I})^{-1}\,\mathbf{B}'$.

If $\Lambda_2 = \mathbf{0}$, a single multivariate EWMA controller results, which turns out to be a multivariate integral controller.

9.6 MULTIVARIATE SETUP ADJUSTMENT PROBLEM

The setup adjustment problem discussed in Section 5.6 and Appendix 8A has a natural extension to a multivariate setting. Suppose that in a discrete-part manufacturing process a machine produces items by applying r_2 heads or spindles simultaneously on the same number of parts or *streams* of product.[14] The machine may be off-target due to a bad setup operation by an unknown vector $\mathbf{d}$ of offsets. Estimating the vector $\mathbf{d}$ can be done using a multivariate Kalman filter approach in a direct generalization of the univariate Kalman filter offset estimate of Example 8.6. As each part is produced, we can reaim the machine at a new set point.

[14] In the SPC literature, this situation is called a *multiple stream process*.

Based on the same assumptions of a stable process as in Section 5.6 and Problem 8.6, the Kalman filter estimation of a $r_2 \times 1$ offset vector $\mathbf{d}$ is based on the simple state-space formulation:

$$\mathbf{d}_t = \mathbf{d}_{t-1}$$

$$\mathbf{Z}_t = \mathbf{Y}_t - \mathbf{U}_{t-1} = \mathbf{d}_t + \mathbf{v}_t$$

Here, $\mathbf{Y}_t$ is a $r_2 \times 1$ vector of deviations from target and $\mathbf{X}_t$ is a $r_2 \times 1$ vector of deviations between the targets and the aimed-at values (the set points) for the quality characteristics. The random vector $\mathbf{v}_t$ has a covariance matrix $\mathbf{V}$. This can be estimated from the variance of $\mathbf{Z}_t$. As in Problem 8.6, we have that from the Kalman filter formulation (Appendix 8A)

$$\mathbf{F}_t = \mathbf{I} \qquad \mathbf{G}_t = \mathbf{I} \qquad \mathbf{W}_t = \mathbf{0} \qquad \mathbf{V}_t = \mathbf{V}$$

so $\mathbf{R}_t = \mathbf{P}_{t-1}$, where $\mathbf{P}_t$ is the covariance matrix of the parameter estimates $\hat{\mathbf{d}}_t$. Applying the Kalman filter as described in Appendix 8A, the matrix of Kalman weights is given by

$$\mathbf{K}_t = \left[\mathbf{V} \mathbf{P}_0^{-1} + t\mathbf{I} \right]^{-1}$$

and the estimates are given by

$$\hat{\mathbf{d}}_t = \hat{\mathbf{d}}_{t-1} + \mathbf{K}_t \mathbf{Y}_t \tag{9.25}$$

In this case, if each diagonal element of $\mathbf{P}_0$ tends to infinity, Grubbs's (1954) harmonic rule is obtained for each of the r_2 quality characteristics. Same as before, the Bayesian interpretation is that, a priori, $\mathbf{d} \sim (\hat{\mathbf{d}}_0, \mathbf{P}_0)$.

The control rule is simply $\nabla \mathbf{X}_t = -\nabla \hat{\mathbf{d}}_t$, or $\mathbf{X}_t = -\hat{\mathbf{d}}_t$. This assumes that each quality characteristic is affected by one controllable factor only.

Example 9.14: Multiple-Stream Process. Consider the application of the multivariate adjustment scheme [equation (9.25) in conjunction with $\mathbf{X}_t = -\hat{\mathbf{d}}_t$] for a filling process in which measurements of the fill weight or volume generated by each of three spindles are taken. The spindles may be off-target after setting up the machine, and each may be off by different amounts and directions. Suppose it is known from historical records that the following variance–covariance matrix of the deviations from target is available:

$$\mathbf{V} = \begin{bmatrix} 5 & 3.5 & 2.7 \\ 3.5 & 3 & 2.5 \\ 2.7 & 2.5 & 2.8 \end{bmatrix}$$

Suppose that, a priori, our offset estimates are $\hat{\mathbf{d}}_0' = (0, 0, 0)$ with associated variance matrix $\mathbf{P}_0 = 5\mathbf{I}$. Finally, let us assume that the true offsets are

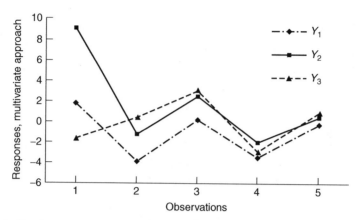

Figure 9.7 Simulated multiple-spindle process adjusted with multivariate Kalman filter scheme (Example 9.14), first five deviations from target.

$\mathbf{d}'(0, 10, 0)$. Then we have that the first Kalman matrix is given by

$$\mathbf{K}_1 = \left[\mathbf{VP}_0^{-1} + \mathbf{I}\right]^{-1} = \begin{bmatrix} 0.617195744 & -0.22588375 & -0.141246043 \\ -0.22588375 & 0.777237953 & -0.1709242 \\ -0.141246043 & -0.1709242 & 0.744701899 \end{bmatrix}$$

Using simulated values according to the data assumed in this problem, the first simulated observed vector of deviations from target is $\mathbf{Y}_1' = (1.758721335, 9.079666399, -1.724992545)$, where the error terms $\mathbf{v}_t$ where simulated from a multivariate normal distribution with mean zero and variance matrix $\mathbf{V}$. With this, we have

$$\hat{\mathbf{d}}_1 = \hat{\mathbf{d}}_0 + \mathbf{K}_1\mathbf{Y}_1 = (-0.721825397, 6.954637725, -3.084952368)'$$

and $\mathbf{X}_t = -\hat{\mathbf{d}}_t = (0.721825397, -6.954637725, 3.084952368)'$. Figure 9.7 illustrates the deviations from target of such simulated process for the first five observations. It is of interest to consider what would have happened if three univariate Kalman filters were used for each spindle separately. Figure 9.8 illustrates a simulation of the first five deviations from target of this alternative approach, using the same random numbers as in Figure 9.7. As can be seen, the convergence to target is faster if the multivariate information is utilized. As time passes, the differences between the two approaches are minimal, that is why the figures show only the first five deviations where differences are more notorious.

Suppose, more generally, that any of the $r_1 = r_2$ controllable factors can affect the r_2 responses via the equation

$$\mathbf{z}_t = \mathbf{Y}_t - \mathbf{BX}_{t-1} = \mathbf{d}_t + \mathbf{v}_t$$

where $\mathbf{B}$ is a $r_2 \times r_2$ matrix of input–output gains. In such a case, the

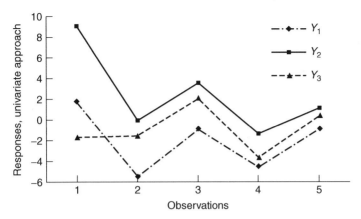

Figure 9.8 Simulated multiple-spindle process adjusted with three *independent* univariate Kalman filter schemes (Example 9.14), first five deviations from target.

adjustment rule is given by

$$\mathbf{X}_t = -\mathbf{B}^{-1}\hat{\mathbf{d}}_t$$

together with (9.25).

Suppose, finally, that the number of controllable factors is $r_1 > r_2$. Then a rule that can be used is a right inverse controller:

$$\mathbf{X}_t = -\mathbf{B}'(\mathbf{BB}')^{-1}\hat{\mathbf{d}}_t$$

used in conjunction with the estimate given by (9.25). Alternatively, a ridge controller as discussed in Section 9.5 can be applied instead. In either case, the matrix of gains **B** has to be estimated somehow. One way to do this is to use design of experiments and regression techniques prior to the startup of production, as done in the run-to-run control area in semiconductor manufacturing (del Castillo and Hurwitz, 1997; Moyne et al., 2000).

It should be pointed out that multivariate process monitoring methods can be integrated with multivariate process adjustment methods such as those in this chapter, providing an integrated SPC–EPC strategy. The integration takes place in the same sense as discussed in Chapter 1. The "Bibliography and Comments" section contains references about the multivariate counterparts of topics given in previous chapters, such as multivariate self-tuning and adaptive control.

PROBLEMS

9.1. Consider the process simulated in Example 9.6. Assume that no B polynomial and no controllable factors are present, so that we have a vector ARMA process. Determine whether or not this ARMA process is **(a)** stationary; **(b)** invertible.

9.2. Consider the fitted model in Example 9.7. Find and simulate both a right inverse and a ridge controller. Vary the parameter μ in the ridge controller and observe the behavior of the simulated processes.

9.3. Consider again the fitted model in Example 9.7. Simulate a multivariate double-EWMA controller applied to this process. Use $\Lambda_1 = \Lambda_2 = 0.3I$.

9.4. Show that the stability conditions for the multivariate double-EWMA controller of Section 9.5 also hold for a multivariate IMA(1, 1) disturbance.

9.5. Consider the process simulated in Example 9.14. Suppose that the initial offset vector is $\mathbf{d}' = (10, 10, 10)$ and we use $\hat{\mathbf{d}}_0' = (0, 0, 0)$. Use $P_0 = 5\,I$. Simulate this process under Grubbs's multivariate adjustment rule. Repeat if $\mathbf{d}' = (1, 0, 1)$ and $P_0 = 0.1I$ instead.

9.6. Suppose that we have a multivariate ARMAX process with $p = 4$, $l = 2$, and $q = 3$. Give the structure and dimensions of the matrices of the state-space model matrices we would get according to Akaike's (SAS PROC STATESPACE) formulation. In other words, from where will we get the estimates of the A_i, B_i, and C_i matrices?

9.7. Suppose that the multi-index of a state-space formulation (following Ljung and Matlab's Systems ID Toolbox is given by $\{d_i\} = \{3, 3, 3\}$. Find the elements of the state vector.

9.8. Repeat Problem 9.7 if the multiindex is $\{d_i\} = \{1, 3, 2\}$.

9.9. Use Matlab's Systems ID toolbox to fit a vector ARMAX model to the data in Problem 9.8 (file *MultiARMAXClosed.txt*).

9.10. Consider the process described in Example 9.8. Repeat Problem 9.2 for this process.

9.11. Repeat problem 9.3 for the process described in Example 9.8.

9.12. Simulate the right inverse controller of Example 9.11. What is the behavior of the controllable factors?

913. Use the SAS PROC STATESPACE automatic identification routine for the data sets described in Examples 9.1, 9.5, 9.6, and 9.8. Compare with the fitted models in this book.

9.14. Use Matlab's Systems ID Toolbox n4sid identification routine for the data sets described in Examples 9.1, 9.5, 9.6, and 9.8. Compare with the fitted models in this book.

BIBLIOGRAPHY AND COMMENTS

Dealing with multiple responses increases the complexity of the identification problem notably, so readers should not be mislead by the apparent simplicity with which we identified the true structure in the examples in this chapter. Contrary to the univariate case, there might be several multivariate ARMA formulations that lead to the same autocorrelation function. (see Reinsel, 1997; Hannan and Diestler, 1988). Identifiability problems in multivariate ARMAX modeling is also a difficult problem to handle, and this is probably why automated identification routines based on some statistic that measures the goodness of the fit (such as Akaike's AIC) are recommended by some authors. SAS PROC STATESPACE allows us precisely to do this. The difficulties have prompted intense research in the design of experiments for identifying multivariate ARMAX or state-space processes, a topic that we did not consider here. As mentioned in Chapter 8, experimental designs for identification of univariate processes are frequently based on binary pseudo-random sequences. This can provide very efficient estimates for univariate linear models and even for multiple-input, single-output processes [as demonstrated by Box et al. (1994), Chap. 11]. However, for multiple-input, multiple-output processes, they will not provide in general robust multivariate controllers (see Koung and MacGregor, 1993). Identifiability problems have also generated research toward finding a reduced space of variables that can be used for control purposes. Subspace methods (van Overschee and de Moor, 1996) is one family of methods that has received increasing attention in the last few years in the control literature. Methods based on the concept of canonical correlations and Akaike's work (Akaike, 1976) are being pursued by other authors, who have even developed commercial identification products (Larimore, 1999). Optimal control methods in the multivariate case can be very involved, and the type of controllers described in this chapter are among the simplest. The multivariate EWMA controller is simply a multivariate generalization of an integral controller. Double-EWMA controllers are recommended for processes with little or no process dynamics and drifting noise. The LQG (linear quadratic Gaussian) problem, mentioned in Chapter 5 but not discussed much in this book, can be solved readily for multivariable systems. Matlab's Control Systems Toolbox contains nice functions to help solving this type of problem, a solution that is based on the separation principle and on using the Kalman filter to estimate the state (Åström, 1970; Lewis, 1986). Many multivariate self-tuning and adaptive controllers have been discussed in the literature, based on both a polynomial representation and a state-space representation. This includes multivariate generalizations of Clarke and Gawthrop's (1975) controller and other controllers based on the idea of model predictive control. The resulting controllers are considerably more complex than those presented in this chapter (see Åström and Wittenmark, 1989; Yosuf and Omatu, 1992). Multivariate SPC methods are discussed by Fuchs and Kennet (1998).

Data Files and Spreadsheets Used in the Book

The following is a list of data files and spreadsheets that accompany this book. They can be found at Wiley's Web site.

File Name	Remarks	Used in:
BJ-F.txt	Box–Jenkins (1976) series F	Example1.7
Fadal.txt	Machining process	Example 1.8
Deming_funnel.xls	Spreadsheet simulation	Chapter 1
CNC.txt	Machining process	Example 1.12
Viscosity.txt	Montgomery et al. (1990) viscosity data	Example 3.13
BoylesData.txt	Boyles (2000) data set	Problem 3.3
ARMA.xls	AR and MA simulation	Problem 3.4
CMP.txt	CMP process: first column, removal rate; second column, platen speed	Example 4.3
Polymer.txt	Data based on Capilla et al. (1999); first column: melt index; second column, temperature	Example 4.6
BJ-J.txt	BJ (1976) series J; (X, Y) values, read by row	Problem 4.3
PIOptimization.xls	Optimization of PI controllers	Chapter 6
EWMAOptimization.xls	Optimization of single-EWMA controllers	Chapter 7
DEWMAOptimization.xls	Optimization of double-EWMA controllers	Chapter 7
SelfTune.xls	Simulation of ST controller	Problem 8.3
MultiAR.txt	Column 1, $Y_{1,t}$; column 2, $Y_{2,t}$	Example 9.1
MultiARMA.txt	Column 1, $Y_{1,t}$; column 2, $Y_{2,t}$	Example 9.5
MultiARMAX.txt	Columns 1–3, $X_{1,t}$ to $X_{3,t}$ column 4, $Y_{1,t}$; column 5, $Y_{2,t}$	Example 6
MultiARMAXClosed.txt	Columns 1–3, $X_{1,t}$ to $X_{3,t}$; column 4, $Y_{1,t}$; column 5, $Y_{2,t}$	Example 9.8

Bibliography

Abraham, B., and Ledolter, J. (1983). *Statistical Methods for Forecasting*. New York: Wiley.

Adams, B. M., and Tseng, I. T. (1998). "Robustness of forecast-based monitoring schemes," *Journal of Quality Technology*, 30(4), pp. 328–339.

Akaike, H. (1976). "Canonical correlation analysis of time series and the use of an information criterion," in *System Identification: Advances and Case Studies*, R. K. Mehra and D. G. Lainiotis, eds., San Diego, Calif.: Academic Press, pp. 27–96.

Akaike, H., and Nakagawa, T. (1988). *Statistical Analysis and Control of Dynamic Systems*. Boston: Kluwer Academic.

Alwan, L. C. (2000). *Statistical Process Analysis*. Boston: McGraw-Hill.

Alwan, L. C., and Roberts, H. V. (1988). "Time-series modeling for statistical process control," *Journal of Business and Economic Forecasting*, 6(1), pp. 87–95.

Åström, K. J. (1970). *Introduction to Stochastic Control Theory*. San Diego, Calif: Academic Press.

Åström, K. J., and Hägglund, T. (1995). *PID Controllers: Theory, Design, and Tuning*, 2nd ed. Research Triangle Park, N.C.: Instrument Society of America.

Åström, K. J., and Wittenmark, B. (1973). "On self tuning regulators," *Automatica*, 9, pp. 185–199.

Åström, K. J., and Wittenmark, B. (1989). *Adaptive Control*. Reading, Mass.: Addison-Wesley.

Åström, K. J. and Wittenmark, B. (1997). *Computer Controlled Systems: Theory and Design*, 3rd ed., Upper Saddle River, N.J.: Prentice Hall.

Barnard, G. A. (1959). "Control charts and stochastic processes," *Journal of the Royal Statistical Society*, 21(2), pp. 239–271.

Bartlett, M. S. (1946). "On the theoretical specification and sampling properties of autocorrelated time-series," *Journal of the Royal Statistics Society*, Ser. B, 8, pp. 27–41.

Bartlett, M. S. (1955). *An Introduction to Stochastic Processes, with Special Reference to Methods and Applications*. Cambridge: Cambridge University press.

Basseville, M., and Nikiforov, I. V. (1993). *Detection of Abrupt Changes: Theory and Application*. Upper Saddle River, N.J.: Prentice Hall.

Box, G. E. P. and Jenkins, G. M. (1963). "Further contributions to adaptive quality control: simultaneous estimation of dynamics: no-zero costs," *ISI Bulletin*, 34th session, Ottawa, Ontario, Canada, pp. 943–974.

Box, G. E. P. and Jenkins, G. M. (1970). *Time Series Analysis: Forecasting and Control*, Oakland, Calif.: Holden–Day.

Box, G. E. P. and Jenkins, G. M. (1976). *Time Series Analysis: Forecasting and Control*, new ed. Oakland, Calif.: Holden–Day.

Box, G. E. P., G. M. Jenkins, and Reinsel, G. (1994). *Time Series Analysis: Forecasting and Control*, 3rd ed. Upper Saddle River, N.J.: Prentice Hall.

Box, G. E. P., and Kramer, T. (1992). "Statistical process monitoring and feedback adjustment: a discussion," *Technometrics*, 34(3), pp. 251–267.

Box, G. E. P. and Luceño, A. (1997). *Statistical Control by Monitoring and Feedback Adjustment*. New York: Wiley.

Box, G. E. P., and MacGregor, J. F. (1974). "The analysis of closed-loop dynamic systems," *Technometrics*, 16, pp. 391–398.

Box, G. E. P., and MacGregor, J. F. (1976). "Parameter estimation with closed-loop operating data," *Technometrics*, 18, pp. 371–380.

Box, G. E. P., and Tiao, G. C. (1977). "A Canonical analysis of multiple time series," *Biometrika*, 64(2), pp. 355–365.

Box, G. E. P. and Wilson, K. G. (1951). "On the experimental attainment of optimum conditions," *Journal of the Royal Statistical Society*, Ser. B, 13, pp. 1–45.

Boyles, R. A. (2000). "Phase I analysis for autocorrelated processes," *Journal of Quality Technology*, 32(4), pp. 395–409.

Butler, S. W. and Stefani, J. A. (1994). "Supervisory run-to-run control of a polysilicon gate etch Using In situ ellipsometry," *IEEE Transactions on Semiconductor Manufacturing*, 7(2), 193–201.

Cameron, F., and Seborg, D. E. (1983). "A self-tuning controller with PID structure," *International Journal of Control*, 38(2), pp. 401–417.

Capilla, C., Ferrer, A., Romero, R., and Hualda, A. (1999). "Integration of statistical and engineering process control in a continuous polymerization process," *Technometrics*, 41(1), pp. 14–28.

Chatfield, C. (1989). *The Analysis of Time Series: An Introduction*, 4th ed. London: Chapman & Hall.

Clarke, D. W., and Gawthrop, P. J. (1975). "Self-tuning controller," *Proceedings of the Institution of Electrical Engineers*, 122(9), pp. 929–934.

Cleveland, W. S. (1972). "The inverse autocorrelations of a time series and their applications," *Technometrics*, 14, pp. 277–298.

Crowder, S. V. (1986). "Kalman filtering and statistical process control," Ph.D. dissertation, Department of Statistics, Iowa State University, Ames, Iowa.

Crowder, S. V. (1987). "A simple method for studying run-length distributions of exponentially weighted moving average charts," *Technometrics*, 29(4), 401–407.

Crowder, S. V. (1992). "An SPC model for short production runs: minimizing expected cost," *Technometrics*, 34, pp. 64–73.

Del Castillo, E. (1996). "Run length distributions and economic design of $\bar{X}$ charts with unknown process variance," *Metrika, International Journal of Theoretical and Applied Statistics*, 43(3), pp. 189–201.

Del Castillo, E. (1998). "A note on two process adjustment models" *Quality and Reliability Engineering International*, 14, pp. 23–28.

Del Castillo, E. (1999). "Long-run and transient analysis of a double EWMA feedback controller," *IIE Transactions*, 31(12), pp. 1157–1169.

Del Castillo, E. (2000). "A variance constrained PI controller that tunes itself," *IIE Transactions*, 32(6), pp. 479–491.

Del Castillo, E. (2001a). "Some properties of EWMA feedback quality adjustment schemes for drifting disturbances." *Journal of Quality Technology*, 33(2), 153–166.

Del Castillo, E. (2001b). "Closed-loop disturbance identification and controller tuning for discrete manufacturing processes," to appear in *Technometrics*.

Del Castillo, E., and Hurwitz, A. (1997). "Run to run process control: a review and some extensions," *Journal of Quality Technology*, 29(2), pp. 184–196.

Del Castillo, E. and Pan, R. (2001). "A unifying view of some process adjustment methods," Technical report. University Park, Pa.: Department of Industrial Engineering, Penn State University.

Del Castillo, E., and Rajagopal, R. (2001). "A multivariate double EWMA process adjustment scheme for drifting processes," Working paper, University Park, Pa.: Department of Industrial Engineering, Penn State University.

Del Castillo, E., Göb, R., and von Collani, E. (2000). "A methodological approach for the integration of SPC and EPC in discrete manufacturing processes," in *Statistical Process Monitoring and Optimization*, S. H. Park and G. G. Vining, eds. New York: Marcel Dekker.

Deming, W. E. (1986). *Out of the Crisis*. Cambridge, Mass: Center for Advanced Engineering Study MIT.

Duncan, D. B., and Horn, S. D. (1972). "Linear dynamic recursive estimation from the viewpoint of regression analysis," *Journal of the American Statistical Association*, 67(340), pp. 815–821.

Edgar, T. F., Butler, S. W., Campbell, W. J., Pfeiffer, C., Bode, C., Hwang, S. B., Balakrishnan, K. S., and Hahn, J. (2000). "Automatic control in microelectronics manufacturing: practices, challenges, and possibilities," *Automatica*, 36(11), pp. 1567–1603.

Faltin, F. W., Hahn, G. J., Tucker, W. T., and Vander Weil, S. A. (1993). "Algorithmic statistical process control: some practical observations," *International Statistical Institute*, 61, pp. 67–80.

Feld'baum, A.A. (1965). *Optimal Control Theory*. San Diego, Calif.: Academic Press.

Franklin, G. F., Powell, J. D., and Workman, M. L. (1998). *Digital Control of Dynamic Systems*, 3rd ed. Menlo Park, Calif.: Addison-Wesley.

Fuchs, C., and Kennet, R. S. (1998). *Multivariate Quality Control: Theory and Applications*. New York: Marcel Dekker.

Fuller, W. A. (1996). *Introduction to Statistical Time Series*, 2nd ed. New York: Wiley.

Garcia, C. E., and Morari, M., (1985). "Internal model control. 2. Design procedure for multivariable Systems," *Industrial and Engineering Chemistry, Process Design and Development*, 24, pp. 472–484.

Goldberg, S. (1986). *Introduction to Difference Equations*, New York: Dover Publications.

Goldsmith, P. L. amd Whitfield, H. (1961). "Average run lengths in cumulative chart quality control schemes," *Technometrics*, 3(1), pp. 11–20.

Goodwin, G. C., and Payne, R. L. (1977). *Dynamic System Identification, Experimental Design, and Data Analysis*. San Diego, Calif.: Academic Press.

Grubbs, F. E. (1954). "An optimum procedure for setting machines or adjusting processes," *Industrial Quality Control*, July; reprinted in *Journal of Quality Technology*, 1983, 15(4), pp. 186–189.

Hägglund, T., and Åström, K. J. (2000). "Supervision of adaptive control algorithms," *Automatica*, 36, pp. 1171–1180.

Hamilton, J. D. (1994). *Time Series Analysis*. Princeton, NJ: Princeton University Press.

Hannan, E. J., and Diestler, M. (1988). *The Statistical Theory of Linear Systems*. New York: Wiley.

Harris, T. J. (1989). "Assessment of control loop performance," *Canadian Journal of Chemical Engineering*, 67, pp. 856–861.

Hawkins, D. M., and Olwell, D. M. (1998). *Cumulative Sum Charts and Charting for Quality Improvement*. New York: Springer-Verlag.

Hunter, J. S. (1986). "The exponentially weighted moving average," *Journal of Quality Technology*, 18(4), pp. 203–210.

Hunter, J. S. (1994). "Beyond the Shewhart SPC paradigm," presentation given at the Department of Statistics, Southern Methodist University, Dallas, Texas.

Jenkins, G. M., and Watts, D. G. (1968). *Spectral Analysis and Its Applications*. San Francisco, Calif.: Holden-Day.

Jensen, K. L., and Vardeman, S. B. (1993). "Optimal adjustment in the presence of deterministic process drift and random adjustment error," *Technometrics*, 35, pp. 376–389.

Jiang, W., Tsui, K. L., and Woodall, W. H. (2001). "A new SPC monitoring method: the ARMA chart," to appear in *Technometrics*.

Jones, L. A., Champ C. W., and Rigdon, S. E. (2001). "The performance of exponentially weighted moving average charts with estimated parameters,," *Technometrics*, 43(2), pp. 156–167.

Kalman, R. E. (1958). "Design of a self-optimizing control system," *Transactions ASME*, 80, pp. 468–478.

Kalman, R. E. (1960). "A new approach to linear filtering and prediction problems," *Transactions ASME, Journal of Basic Engineering*, 82, pp. 35–45.

Katende, E., and Jutan, A. (1993). "A new constrained self-tuning PID controller," *Canadian Journal of Chemical Engineering*, 71, pp. 625–633.

Kendall, M., and Ord, J. K. (1990). *Time Series*. New York: Oxford University Press.

Khuri, A. I., and Cornell, J. A. (1987). *Response Surfaces: Designs and Analyses*. New York: Marcel Dekker.

Kuong, C. W., and MacGregor, J. F. (1993). "Design of identification experiments for robust control. a geometric approach for bivariate processes," *Industrial and Engineering Chemistry Research*, 32, pp. 1658–1666.

Larimore, W. E. (1999). "Automated multivariate system identification and industrial applications," *Proceedings of the* 1999 *American Control Conference*, San Diego, Calif., pp. 1148–1162.

Lewis, F. L. (1986). *Optimal Estimation with an Introduction to Stochastic Control Theory*. New York: Wiley.

Lewis, F. L. (1992). *Applied Optimal Control and Estimation*. Upper Saddle River, N.J.: Prentice Hall.

Ljung, L. (1977). "Analysis of recursive stochastic algorithms," *IEEE Transactions on Automatic Control*, AC-22, pp. 55–575.

Ljung, L. (1995). *System Identification Toolbox, for Use with Matlab*. Natick, Mass.: The Math Works, Inc.

Ljung, L. (1999). *System Identification: Theory for the User*, 2nd ed. Upper Saddle River, N.J.: Prentice Hall.

Ljung, L. (2001). *System Identification Toolbox, for Use with Matlab* (*version* 5). Natick, Mass.: The Math Works, Inc.

Ljung, L. and Söderström, T. (1983). *Theory and Practice of Recursive Identification*. Cambridge, Mass: MIT Press.

Lu, C. W., and Reynolds, M. R., Jr. (1999). "EWMA control charts for monitoring the mean of autocorrelated processes," *Journal of Quality Technology*, 31(2), pp. 166–188.

Lu, C. W. and Reynolds, M. R. Jr. (2001). "CUSUM charts for monitoring autocorrelated processes," *Journal of Quality Technology*, 33(3), pp. 316–334.

Lucas, J. M., and Crosier, R B. (1982). "Fast initial response for COSUM quality-control schemes: give your CUSUM a head start," *Technometrics*, 24, pp. 196–206.

Lucas, J. M., and Saccucci, M. S. (1990). "Exponentially weighted moving average control schemes properties and enhancements," *Technometrics*, 32(1), pp. 1–12.

Luceño, A. (1997). "Parameter estimation with closed-loop operating data under time varying discrete proportional–integral control," *Communications in Statistics, Simulation and Computation*, 26(1), pp. 215–232.

MacGregor, J. F. (1976). "Optimal choice of the sampling interval for discrete process control," *Technometrics*, 18(2), pp. 151–160.

MacGregor, J. F. (1988). "On-line statistical process control," *Chemical Engineering Progress*, October, pp. 21–31.

MacGregor, J. F., (1990). "A different view of the funnel experiment," *Journal of Quality Technology*, 22, pp. 255–259.

MacGregor, J. F., and Tidwell, P. W. (1977). "Discrete stochastic control with input constraint," *Proceedings of the Institution of Electrical Engineers*, 24(8), pp. 732–34.

Makridakis, S., Wheelwright, S. C., and McGee, V. E. (1983). *Forecasting, Methods and Applications*, 2nd ed. New York: Wiley.

Maragah, H. D. and Woodall, W. H. (1992). "The effect of autocorrelation on the retrospective X-chart," *Journal of Statistical Computing and Simulation*, 40, pp. 29–42.

Mastrangelo, C. and Montgomery, D. C. (1995). "Characterization of a moving centerline exponentially weighted moving average," *Quality and Reliability Engineering International*, 11, pp. 79–89.

Mehra, R. K. (1972). "Approaches to adaptive filtering," *IEEE Transactions on Automatic Control*, October, pp. 693–698.

Meinhold, R. J, and Singpurwalla, N. D. (1983). "Understanding the Kalman filter," *American Statistician*, 37(2), 123–127.

Moden, P. E. and Söderström, T. (1982). "Stationary performance of linear stochastic systems under single step optimal control," *IEEE Transactions on Automatic Control*, AC-27(1), pp. 214–216.

Montgomery, D. C. (2001). *Introduction to Statistical Quality Control*, 4th ed. New York: Wiley.

Montgomery, D. C., and Mastrangelo, C. M. (1991). "Some statistical process control methods for autocorrelated data" (with discussion) *Journal of Quality Technology*, 23, pp. 179–193.

Montgomery, D. C., Johnson, L. A., and Gardiner, J. S. (1990). *Forecasting and Time Series Analysis*, 2nd ed. New York: McGraw-Hill.

Montgomery, D. C., Keats, J. B., Runger, G. C., and Messina, W. S. (1994). "Integrating statistical process control and engineering process control," *Journal of Quality Technology*, 26(2), pp. 79–87.

Morari, M., and Zafiriou, E. (1989). *Robust Process Control*. Upper Saddle River, N.J.: Prentice Hall.

Moyne, J., del Castillo, E., and Hurwitz, A., eds. (2000). *Run to Run Control in Semiconductor Manufacturing*, Boca Raton, Fla.: CRC Press.

Myers, R. H. (1990). *Classical and Modern Regression with Applications*, 2nd ed. Belmont, Calif.: Duxbury Press.

Myers, R. H., and Milton, J. S. (1991). *A First Course in the Theory of Linear Statistical Models*. Boston: PWS-Kent.

Narendra, K. S., and Annaswamy, A. M. (1989). *Stable Adaptive Systems*. Upper Saddle River, NJ: Prentice Hall.

Ogata, K. (1995). *Discrete-Time Control Systems*, 2nd ed. Upper Saddle River, N.J.: Prentice Hall.

Ogunnaike, B. A. and Ray, W. H. (1994). *Process Dynamics, Modeling, and Control*. New York: Oxford University Press.

Palmor, Z. J., and Shinnar, R. (1979). "Design of sample data controllers," *Industrial and Engineering Chemistry Process Design and Development*, 18(1), pp. 8–30.

Pan, R. and del Castillo, E. (2001). "Identification and fine tuning of closed-loop processes under discrete EWMA and PI adjustments," to appear in *Quality and Reliability Engineering International*.

Pearson, R. K. (1999). *Discrete-Time Dynamical Models*. New York: Oxford University Press.

Plackett, R. L. (1950). "Some theorems in least squares," *Biometrika*, 37, pp. 149–157.

Quesenberry, C. P. (1997). *SPC Methods for Quality Improvement*. New York: Wiley.

Reinsel, G. C. (1997). *Elements of Multivariate Time Series Analysis*, 2nd ed. New York: Springer-Verlag.

Robbins, H., and Monro, S. (1951). "A stochastic approximation method," *Annals of Mathematical Statistics*, 22, pp. 400–407.

Sachs, E., Hu, A., and Ingolfsson, A. (1995). "Run by run process control: combining SPC and feedback control," *IEEE Transactions on Semiconductor Manufacturing*, 8(1), pp. 26–43.

Seborg, D. E., Edgar, T. F., and Mellichamp, D. A. (1989). *Process Dynamics and Control*. New York: Wiley.

Shah, S. L., and Cluett, W. R. (1991). "Recursive least squares based estimation schemes for self tuning control," *Canadian Journal of Chemical Engineering*, 69(1), p. 89.

Shewhart, W. A. (1931). *Economic Control of Quality of Manufacturing Product*. Princeton, N.J.: Van Nostrand Reinhold.

Shumway, R. H., and Stoffer, D. S. (2000). *Time Series Analysis and Its Applications*. New York: Springer-Verlag.

Siegmund, D. (1985). *Sequential Analysis*; *Tests and Confidence Intervals*. New York: Springer-Verlag.

Smith, T. H. and Boning, D. S. (1997). "Artificial neural network exponentially weighted moving average controller for semiconductor processes," *Journal of Vacuum Science and Technology*, 15(3).

Söderstrom, T. (1994). *Discrete-Time Stochastic Systems, Estimation, and Control*. London: Prentice Hall International.

Tiao, G. C., and Box, G. E. P. (1981), "Modeling multiple time series with applications," *J. Amer. Statist. Assoc.*, 78, pp. 802–816.

Trietsch, D., (1998). "The harmonic rule for process setup adjustment with quadratic loss," *Journal of Quality Technology*, 30(1), pp. 75–84.

Tsay, R. S., and Tiao, G. C. (1984), "Consistent estimates of autoregressive parameters and extended sample autocorrelation function for stationary and nonstationary ARMA models," *J. Amer. Statist. Assoc.*, 79, pp. 84–96.

Tsay, R. S., and Tiao, G. C., (1985). "Use of canonical analysis in time series model identification," *Biometrika*, 72, pp. 299–315.

Tsung, F., and Shi, J. J. (1999). "Integrated design of run to run PID controllers and SPC monitoring for process disturbance rejection," *IIE Transactions*, 31(6), pp. 517–527.

Tsung, F., Wu, H., and Nair, V. (1998). "On the efficiency and robustness of discrete proportional–integral control schemes," *Technometrics*, 40(3), pp. 214–222.

Tucker, W. T., Faltin, F. W., and Vander Wiel, S. A. (1993). "ASPC: an ellaboration," *Technometrics*, 35(4), pp. 363–375.

Vander Weil, S. A., (1996). "Monitoring processes that wander using integrated moving average models," *Technometrics*, 38(2), pp. 139–151.

Vander Wiel, S. A., Tucker, W. T., Faltin, F. W., and Doganaksoy, N. (1992). "Algorithmic statistical process control: concepts and an application," *Technometrics*, 34(3), pp. 286–297.

van Overschee, P., and de Moor, B. (1996). *Subspace Identification for Linear Systems*: *Theory, Implementation, Applications*, Boston: Kluwer Academic.

Wellstead, P. E., and Zarrop, M. B. (1991). *Self-Tuning Systems*. New York: Wiley.

West, M., and Harrison, J. (1997). *Bayesian Forecasting and Dynamic Models*. New York: Springer-Verlag.

Wiklund, S. J. (1994). "Control charts and process adjustments." Statistical studies 19. Umea, Sweden: Department of Statistics, University of Umea.

Wilson, G. T. (2001). "Univariate time series: autocorrelation, linear prediction, spectrum, and state-space model," in *A Course in Time Series Analysis*, D. Peña, G. C. Tiao, T. S. Tsay, eds. New York: Wiley.

Woodall, W. H. (1983). "The distribution of the run length of one-sided CUSUM procedures for continuous random variables," *Technometrics*, 25, pp. 295–301.

Woodall, W. H. (1986). "The design of CUSUM quality control charts," *Journal of Quality Technology*, 18, pp. 99–102.

Yaschin, E. (1993). "Performance of CUSUM control schemes for serially correlated observations," *Technometrics*, 35, pp. 37–52.

Young, P. (1984). *Recursive Estimation and Time-Series Analysis*. New York: Springer-Verlag.

Yusof, R., and Omatu, S. (1993). "A multivariate self-tuning PID controller," *International Journal of Control*, 57(6), pp. 1387–1403.

Index

WILEY SERIES IN PROBABILITY AND STATISTICS
ESTABLISHED BY WALTER A. SHEWHART AND SAMUEL S. WILKS

Editors
David J. Balding, Peter Bloomfield, Noel A. C. Cressie, Nicholas I. Fisher,
Iain M. Johnstone, J. B. Kadane, Louise M. Ryan, David W. Scott,
Adrian F. M. Smith, Jozef L. Teugels
Editors Emeriti: *Vic Barnett, J. Stuart Hunter, David G. Kendall*

The *Wiley Series in Probability and Statistics* is well established and authoritative. It covers many topics of current research interest in both pure and applied statistics and probability theory. Written by leading statisticians and institutions, the titles span both state-of-the-art developments in the field and classical methods.

Reflecting the wide range of current research in statistics, the series encompasses applied, methodological and theoretical statistics, ranging from applications and new techniques made possible by advances in computerized practice to rigorous treatment of theoretical approaches.

This series provides essential and invaluable reading for all statisticians, whether in academia, industry, government, or research.

*Now available in a lower priced paperback edition in the Wiley Classics Library.

*Now available in a lower priced paperback edition in the Wiley Classics Library.

DAVID · Order Statistics, *Second Edition*
*DEGROOT, FIENBERG, and KADANE · Statistics and the Law
DEL CASTILLO · Statistical Process Adjustment for Quality Control
DETTE and STUDDEN · The Theory of Canonical Moments with Applications in
    Statistics, Probability, and Analysis
DEY and MUKERJEE · Fractional Factorial Plans
DILLON and GOLDSTEIN · Multivariate Analysis: Methods and Applications
DODGE · Alternative Methods of Regression
*DODGE and ROMIG · Sampling Inspection Tables, *Second Edition*
*DOOB · Stochastic Processes
DOWDY and WEARDEN · Statistics for Research, *Second Edition*
DRAPER and SMITH · Applied Regression Analysis, *Third Edition*
DRYDEN and MARDIA · Statistical Shape Analysis
DUDEWICZ and MISHRA · Modern Mathematical Statistics
DUNN and CLARK · Applied Statistics: Analysis of Variance and Regression, *Second*
    Edition
DUNN and CLARK · Basic Statistics: A Primer for the Biomedical Sciences,
    Third Edition
DUPUIS and ELLIS · A Weak Convergence Approach to the Theory of Large Deviations
*ELANDT-JOHNSON and JOHNSON · Survival Models and Data Analysis
ETHIER and KURTZ · Markov Processes: Characterization and Convergence
EVANS, HASTINGS, and PEACOCK · Statistical Distributions, *Third Edition*
FELLER · An Introduction to Probability Theory and Its Applications, Volume I,
    Third Edition, Revised; Volume II, *Second Edition*
FISHER and VAN BELLE · Biostatistics: A Methodology for the Health Sciences
*FLEISS · The Design and Analysis of Clinical Experiments
FLEISS · Statistical Methods for Rates and Proportions, *Second Edition*
FLEMING and HARRINGTON · Counting Processes and Survival Analysis
FULLER · Introduction to Statistical Time Series, *Second Edition*
FULLER · Measurement Error Models
GALLANT · Nonlinear Statistical Models
GHOSH, MUKHOPADHYAY, and SEN · Sequential Estimation
GIFI · Nonlinear Multivariate Analysis
GLASSERMAN and YAO · Monotone Structure in Discrete-Event Systems
GNANADESIKAN · Methods for Statistical Data Analysis of Multivariate Observations,
    Second Edition
GOLDSTEIN and LEWIS · Assessment: Problems, Development, and Statistical Issues
GREENWOOD and NIKULIN · A Guide to Chi-Squared Testing
GROSS and HARRIS · Fundamentals of Queueing Theory, *Third Edition*
*HAHN · Statistical Models in Engineering
HAHN and MEEKER · Statistical Intervals: A Guide for Practitioners
HALD · A History of Probability and Statistics and their Applications Before 1750
HALD · A History of Mathematical Statistics from 1750 to 1930
HAMPEL · Robust Statistics: The Approach Based on Influence Functions
HANNAN and DEISTLER · The Statistical Theory of Linear Systems
HEIBERGER · Computation for the Analysis of Designed Experiments
HEDAYAT and SINHA · Design and Inference in Finite Population Sampling
HELLER · MACSYMA for Statisticians
HINKELMAN and KEMPTHORNE: · Design and Analysis of Experiments, Volume 1:
    Introduction to Experimental Design
HOAGLIN, MOSTELLER, and TUKEY · Exploratory Approach to Analysis
    of Variance
HOAGLIN, MOSTELLER, and TUKEY · Exploring Data Tables, Trends and Shapes

*Now available in a lower priced paperback edition in the Wiley Classics Library.

*Now available in a lower priced paperback edition in the Wiley Classics Library.

KOTZ, READ, and BANKS (editors) · Encyclopedia of Statistical Sciences: Update Volume 1

KOTZ, READ, and BANKS (editors) · Encyclopedia of Statistical Sciences: Update Volume 2

KOVALENKO, KUZNETZOV, and PEGG · Mathematical Theory of Reliability of Time-Dependent Systems with Practical Applications

LACHIN · Biostatistical Methods: The Assessment of Relative Risks

LAD · Operational Subjective Statistical Methods: A Mathematical, Philosophical, and Historical Introduction

LAMPERTI · Probability: A Survey of the Mathematical Theory, *Second Edition*

LANGE, RYAN, BILLARD, BRILLINGER, CONQUEST, and GREENHOUSE · Case Studies in Biometry

LARSON · Introduction to Probability Theory and Statistical Inference, *Third Edition*

LAWLESS · Statistical Models and Methods for Lifetime Data

LAWSON · Statistical Methods in Spatial Epidemiology

LE · Applied Categorical Data Analysis

LE · Applied Survival Analysis

LEE · Statistical Methods for Survival Data Analysis, *Second Edition*

LePAGE and BILLARD · Exploring the Limits of Bootstrap

LEYLAND and GOLDSTEIN (editors) · Multilevel Modelling of Health Statistics

LIAO · Statistical Group Comparison

LINDVALL · Lectures on the Coupling Method

LINHART and ZUCCHINI · Model Selection

LITTLE and RUBIN · Statistical Analysis with Missing Data

LLOYD · The Statistical Analysis of Categorical Data

MAGNUS and NEUDECKER · Matrix Differential Calculus with Applications in Statistics and Econometrics, *Revised Edition*

MALLER and ZHOU · Survival Analysis with Long Term Survivors

MALLOWS · Design, Data, and Analysis by Some Friends of Cuthbert Daniel

MANN, SCHAFER, and SINGPURWALLA · Methods for Statistical Analysis of Reliability and Life Data

MANTON, WOODBURY, and TOLLEY · Statistical Applications Using Fuzzy Sets

MARDIA and JUPP · Directional Statistics

MASON, GUNST, and HESS · Statistical Design and Analysis of Experiments with Applications to Engineering and Science

McCULLOCH and SEARLE · Generalized, Linear, and Mixed Models

McFADDEN · Management of Data in Clinical Trials

McLACHLAN · Discriminant Analysis and Statistical Pattern Recognition

McLACHLAN and KRISHNAN · The EM Algorithm and Extensions

McLACHLAN and PEEL · Finite Mixture Models

McNEIL · Epidemiological Research Methods

MEEKER and ESCOBAR · Statistical Methods for Reliability Data

MEERSCHAERT and SCHEFFLER · Limit Distributions for Sums of Independent Random Vectors: Heavy Tails in Theory and Practice

*MILLER · Survival Analysis, *Second Edition*

MONTGOMERY, PECK, and VINING · Introduction to Linear Regression Analysis, *Third Edition*

MORGENTHALER and TUKEY · Configural Polysampling: A Route to Practical Robustness

MUIRHEAD · Aspects of Multivariate Statistical Theory

MURRAY · X-STAT 2.0 Statistical Experimentation, Design Data Analysis, and Nonlinear Optimization

MYERS and MONTGOMERY · Response Surface Methodology: Process and Product Optimization Using Designed Experiments, *Second Edition*

*Now available in a lower priced paperback edition in the Wiley Classics Library.

*Now available in a lower priced paperback edition in the Wiley Classics Library.

*SERFLING · Approximation Theorems of Mathematical Statistics

SHAFER and VOVK · Probability and Finance: It's Only a Game!

SMALL and McLEISH · Hilbert Space Methods in Probability and Statistical Inference

STAPLETON · Linear Statistical Models

STAUDTE and SHEATHER · Robust Estimation and Testing

STOYAN, KENDALL, and MECKE · Stochastic Geometry and Its Applications, *Second Edition*

STOYAN and STOYAN · Fractals, Random Shapes and Point Fields: Methods of Geometrical Statistics

STYAN · The Collected Papers of T. W. Anderson: 1943–1985

SUTTON, ABRAMS, JONES, SHELDON, and SONG · Methods for Meta-Analysis in Medical Research

TANAKA · Time Series Analysis: Nonstationary and Noninvertible Distribution Theory

THOMPSON · Empirical Model Building

THOMPSON · Sampling, *Second Edition*

THOMPSON · Simulation: A Modeler's Approach

THOMPSON and SEBER · Adaptive Sampling

TIAO, BISGAARD, HILL, PEÑA, and STIGLER (editors) · Box on Quality and Discovery: with Design, Control, and Robustness

TIERNEY · LISP-STAT: An Object-Oriented Environment for Statistical Computing and Dynamic Graphics

TSAY · Analysis of Financial Time Series

UPTON and FINGLETON · Spatial Data Analysis by Example, Volume II: Categorical and Directional Data

VAN BELLE · Statistical Rules of Thumb

VIDAKOVIC · Statistical Modeling by Wavelets

WEISBERG · Applied Linear Regression, *Second Edition*

WELSH · Aspects of Statistical Inference

WESTFALL and YOUNG · Resampling-Based Multiple Testing: Examples and Methods for p-Value Adjustment

WHITTAKER · Graphical Models in Applied Multivariate Statistics

WINKER · Optimization Heuristics in Economics: Applications of Threshold Accepting

WONNACOTT and WONNACOTT · Econometrics, *Second Edition*

WOODING · Planning Pharmaceutical Clinical Trials: Basic Statistical Principles

WOOLSON and CLARKE · Statistical Methods for the Analysis of Biomedical Data, *Second Edition*

WU and HAMADA · Experiments: Planning, Analysis, and Parameter Design Optimization

YANG · The Construction Theory of Denumerable Markov Processes

*ZELLNER · An Introduction to Bayesian Inference in Econometrics

ZHOU, OBUCHOWSKI, and McCLISH · Statistical Methods in Diagnostic Medicine